THIRD EDITION

BASIC MEDICAL ENDOCRINOLOGY

THIRD EDITION
BASIC MEDICAL ENDOCRINOLOGY

H. Maurice Goodman

Department of Physiology
University of Massachusetts Medical School
Worcester, Massachusetts

ACADEMIC PRESS
An imprint of Elsevier Science

Amsterdam   Boston   London   New York   Oxford   Paris
San Diego   San Francisco   Singapore   Sydney   Tokyo

This book is printed on acid-free paper. ∞

Academic Press
An imprint of Elsevier Science.
525 B Street, Suite 1900, San Diego, California 92101-4495, USA
http://www.academicpress.com

Academic Press
84 Theobald's Road, London WClX 8RR, UK
http://www.academicpress.com

Library of Congress Catalog Card Number: 2002108537

International Standard Book Number: 0-12-290421-4

PRINTED IN HONG KONG
02 03 04 05 06 07   9 8 7 6 5 4 3 2 1

This volume is dedicated to the memory of Ernst Knobil,
whose teachings continue to inspire and
whose friendship will always be cherished.

CONTENTS

CHAPTER 1
Introduction

CHAPTER 2   *Ch 5, 6*
Pituitary Gland

CHAPTER 3   Ch 13
Thyroid Gland

CHAPTER 4   ch15
Adrenal Glands

CHAPTER 5   ch16
The Pancreatic Islets

CHAPTER 6
Principles of Hormonal Integration   Ch3,4

CHAPTER 7
Regulation of Sodium and Water Balance   Ch7

CHAPTER 8
Hormonal Regulation of Calcium Metabolism

CHAPTER 9
Hormonal Regulation of Fuel Metabolism

CHAPTER 10
Ch12
Hormonal Control of Growth

CHAPTER 11

Hormonal Control of Reproduction in the Male

Ch 17

CHAPTER 12   *Ch18*

Hormonal Control of Reproduction in the Female: The Menstrual Cycle

CHAPTER 13   *ch19*

Hormonal Control of Reproduction in the Female: Pregnancy and Lactation

PREFACE

Nearly a decade elapsed between publication of the second and third editions of *Basic Medical Endocrinology* due in large part to the turmoil in the publishing industry brought on by massive consolidation. Although this edition is new and the publisher is new, the aims of earlier editions of this work are unchanged. Its focus remains human endocrinology with an emphasis on cellular and molecular mechanisms presented in the context of integration of body functions. The intent is to provide a sufficient level of understanding of normal endocrine physiology to prepare students to study not only endocrine diseases, but also the cellular and molecular alterations that disrupt normal function. Such understanding is a prerequisite for institution of rational diagnostic procedures, therapeutic interventions, and research strategies. It is further hoped that this text provides the necessary background to facilitate continuing self-education in endocrinology.

A decade is a long time in this remarkable era of modern biology. Whole new vistas of inquiry have been opened since the previous edition of this text appeared, and new discoveries have mandated reinterpretation of many areas that were once thought to be solidly understood. Much of the progress of the past decade must be credited to ingenious application of rapidly evolving technology in molecular biology. Studies of gene expression and the charting of the genomes of several species, including our own, has provided a deluge of new information and new insights. The exquisite sensitivity and versatility of this technology has uncovered both hormone production and hormone receptors in unexpected places and revealed hitherto unappreciated roles for classical hormones. Classical techniques of organ ablation and extract injection have been reapplied using the once unthinkable technology of gene ablation or overexpression to explore the functions of individual proteins instead of individual glands. The decade has also witnessed the discovery of new hormones and expanded our appreciation of the physiological importance of extraglandular metabolism of hormones. The understanding of hormone actions has grown enormously and spawned the quest for "designer drugs" that target particular, critical, biochemical reactions in combating disease.

In light of these and many other developments, every chapter of this text has been extensively revised to present the well-established factual basis of endocrinology enriched by exciting, rapidly unfolding new information and insights. The challenge has been to digest and reduce the massive literature to illuminate the regulatory and integrative roles of the endocrine system without overloading the text with arcane detail. However, the text is designed to provide somewhat more than the minimum acceptable level of understanding and attempts to anticipate and answer some of the next level of questions that might occur to the thoughtful student.

Looking back over 40 years of teaching endocrine physiology, one cannot fail but to marvel at how far we have come and how resourceful is the human mind in probing the mysteries of life. As has always been true of scientific inquiry, obtaining answers to long-standing questions inevitably raises a host of new questions to challenge a new generation of endocrinologists. It is my hope that this text will provide a foundation for students to meet that challenge both in the clinic and in the laboratory.

H. Maurice Goodman
2002

PREFACE TO THE
FIRST EDITION

This volume is the product of more than 25 years of teaching endocrine physiology to first-year medical students. Its focus is human endocrinology with an emphasis on cellular and molecular mechanisms. In presenting this material, I have tried to capture some of the excitement of a dynamic, expanding discipline that is now in its golden age. It is hoped that this text provides sufficient understanding of normal endocrine physiology to prepare the student to study not only endocrine diseases but the cellular and molecular derangements that disrupt normal function and must therefore be reversed or circumvented by rational therapy. It is further hoped that this text provides the necessary background to facilitate continuing self-education in endocrinology.

Endocrinology encompasses a vast amount of information relating to at least some aspect of virtually every body function. Unfortunately, much of the information is descriptive and cannot be derived from first principles. Thorough, encyclopedic coverage is neither appropriate for a volume such as this one nor possible at the current explosive rate of expansion. On the other hand, limiting the text to the bare minimum of unadorned facts might facilitate memorization of what appear to be the essentials this year but would preclude acquisition of real understanding and offer little preparation for assimilating the essentials as they may appear a decade hence. I therefore sought the middle ground and present basic facts within enough of a physiological framework to foster understanding of both the current status of the field and those areas where new developments are likely to occur while hopefully avoiding the pitfall of burying key points in details and qualifications.

The text is organized into three sections. The first section provides basic information about organization of the endocrine system and the role of individual endocrine glands. Subsequent sections deal with complex hormonal interactions that govern maintenance of the internal environment (Part II) and growth and reproduction (Part III). Neuroendocrinology is integrated into discussions of specific glands or regulatory systems throughout the text rather than being treated as a separate subject. Although modern endocrinology has its roots in gastrointestinal (GI) physiology, the gut hormones are usually covered in texts of GI physiology rather than endocrinology; therefore, there is no chapter on intestinal hormones. In the interests of space and the reader's endurance, a good deal of fascinating material was omitted because it seemed either irrelevant to human biology or insufficiently understood at this time. For example, the pineal gland has intrigued generations of scientists and philosophers since Descartes, but it still has no clearly established role in human physiology and is therefore ignored in this text.

Human endocrinology has its foundation in clinical practice and research, both of which rely heavily on laboratory findings. Where possible, points are illustrated in the text with original data from the rich endocrine literature to give the reader a feeling for the kind of information on which theoretical and diagnostic conclusions are based. Original literature is not cited in the text, in part because such citations are distracting in an introductory text, and in part because proper citation might well double the length of this volume. For the reader who wishes to gain entrée to the endocrine literature or desires more comprehensive coverage of specific topics, review articles are listed at the end of each chapter.

H. Maurice Goodman
1988

PREFACE TO THE
SECOND EDITION

In the five years that have passed since the first edition of this text, the information explosion in endocrinology has continued unabated and may have even accelerated. Application of the powerful tools of molecular biology has made it possible to ask questions about hormone production and action that were only dreamed about a decade earlier. The receptor molecules that initiate responses to virtually all of the hormones have been characterized and significant progress has been made in unraveling the events that lead to the final cellular expression of hormonal stimulation. As more details of intracellular signaling emerge, the complexities of parallel and intersecting pathways of transduction have become more evident. We are beginning to understand how cells regulate the expression of genes and how hormones intervene in regulatory processes to adjust the expression of individual genes. Great strides have been made in understanding how individual cells talk to each other through locally released factors to coordinate growth, differentiation, secretion, and other responses within a tissue. In these regards, endocrinology and immunology share common themes and have contributed to each other's advancement.

In revising the text for this second edition of *Basic Medical Endocrinology*, I have tried to incorporate many of the exciting advances in our understanding of cellular and molecular processes into the discourse on integrated whole body function. I have tried to be selective, however, and include only those bits of information that deepen understanding of well-established principles or processes or that relate to emerging themes. Every chapter has been updated, but not surprisingly, progress has been uneven, and some have been revised more extensively than others. After reviewing the past five years of literature in as broad an area as encompassed by endocrinology, one cannot help but be humbled by the seemingly limitless capacity of the human mind to develop new knowledge, to assimilate new information into an already vast knowledge base, and to apply that knowledge to advancement of human welfare.

H. Maurice Goodman
1993

Introduction

OVERVIEW AND DEFINITIONS

As animals evolved from single cells to multicellular organisms, single cells took on specialized functions and became mutually dependent in order to satisfy individual cellular needs and the needs of the whole organism. Survival thus hinged on integration and coordination of individual specialized functions among all cells. Increased specialization of cellular functions was accompanied by decreased tolerance for variations in the cellular environment. Control systems evolved that allowed more and more precise regulation of the cellular environment, which in turn permitted the development of even more highly specialized cells, such as those of higher brain centers; the continued function of highly specialized cells requires that the internal environment be maintained constant within narrow limits—no matter what conditions prevail in the external environment. Survival of an individual requires a capacity to adjust and adapt to hostile external environmental conditions, and survival of a species requires coordination of reproductive function with those internal and external environmental factors that are most conducive to survival of offspring. Crucial to meeting these needs for survival as a multicellular organism is the capacity of specialized cells to coordinate cellular activities through some sort of communication.

Cells communicate with each other by means of chemical signals. These signals may be simple molecules such as modified amino or fatty acids, or they may be more complex peptides, proteins, or steroids. Communication takes place locally between cells within a tissue or organ, and at a distance in order to integrate the activities of cells or tissues in separate organs. For communication between cells whose surfaces are in direct contact, signals may be substances that form part of the cell surface, or they may be molecules that pass from the cytosol of one cell to another through gap junctions. For communication with nearby cells and also between contiguous cells, chemical signals are released into the extracellular fluid and reach their destinations by simple diffusion through extracellular fluid. Such communication is said to occur by *paracrine*, or local, secretion. Sometimes cells respond to their own secretions, and this is called *autocrine* secretion. For cells that are too far apart for the slow process of diffusion to permit meaningful communication, the chemical signal may enter the circulation and be transported in blood to all parts of the body. Release of chemical signals into the bloodstream is referred to as *endocrine*, or internal, secretion, and the signal secreted is called a *hormone*. We may define a hormone as a chemical substance that is released into the blood in small amounts and that, after delivery by the circulation, elicits a typical physiological response in other cells, which are often called *target* cells (Figure 1). Often these modalities are used in combination such that paracrine and autocrine secretions provide local fine tuning for events that are evoked by a hormonal signal that arrives from a distant source.

Because hormones are diluted in a huge volume of blood and extracellular fluid, achieving meaningful concentrations (10^{-10} to 10^{-7} M) usually requires coordinated secretion by a mass of cells, an *endocrine gland*. The secretory products of endocrine glands are released into the extracellular space and diffuse across the

A. Autocrine/Paracrine

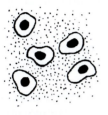

B. Endocrine

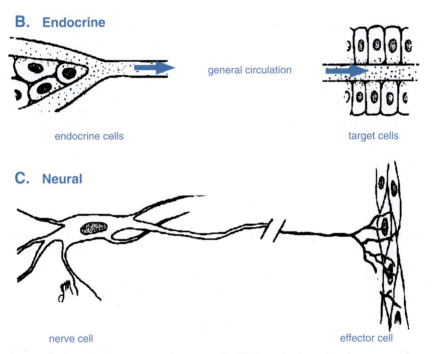

general circulation

endocrine cells   target cells

C. Neural

nerve cell   effector cell

Figure 1 Chemical communication between cells. (A) Autocrine/paracrine. Secretory product, shown as black dots, reaches nearby target cell by diffusion through extracellular fluid. (B) Endocrine. Secretory product reaches distant cells by transport via the circulatory system. (C) Neural secretory product released from terminals of long cell processes reaches target cells distant from the nerve cell body by diffusion across the synaptic cleft.

capillary endothelium into the bloodstream, thus giving rise to the terms "ductless glands" and "internal secretion." In contrast, *exocrine glands* deliver their products through ducts to the outside of the body or to the lumen of the gastrointestinal tract. Classical endocrine glands include the pituitary, thyroid, adrenals, parathyroids, gonads, and islets of Langerhans. It has become apparent, however, that this list is far too short. Virtually every organ, including brain, kidney, heart, and even fat, has an endocrine function in addition to its more commonly recognized role. Many aspects of gastrointestinal function are governed by hormones produced by cells of the gastric and intestinal mucosae. In fact, the word "hormone" was coined to describe a duodenal product, secretin, that regulates pancreatic fluid secretion. However, the gastrointestinal hormones are traditionally considered in textbooks of gastrointestinal physiology rather than endocrinology and hence will not be considered here.

It is only recently that endocrinologists have embraced the large number of locally produced hormonelike agents, called *growth factors* and *cytokines*, that regulate cell division, differentiation, function, and even programmed cell death, which is called *apoptosis*. These agents act locally in a paracrine or autocrine manner, but may also enter the circulation and affect the functions of distant cells, and hence function as hormones. Many of these secretions produce effects that impinge on actions of the classical hormones. Rapidly accumulating information about protein and gene structure has revealed relationships among these compounds, which can now be grouped into families or superfamilies. Some of the classical hormones, such as growth hormone and prolactin, belong to the same superfamily of proteins as some of the cytokines, whereas the insulin–like growth factors are closely related to insulin. At the molecular level, production, secretion, and actions of cytokines and growth factors are no different from those of the classical hormones.

Another mechanism has also evolved to breach the distance between cells and allow rapid communication. Some cells developed the ability to release their signals locally from the tips of long processes that extend great distances to nearly touch their target cells. This mechanism, of course, is how nerve cells communicate with each other or with effector cells. By releasing their signals (neurotransmitters) so close to receptive cells, nerve cells achieve both exquisite specificity and economy in the quantity of transmitter needed to provide a meaningful concentration within a highly localized space, the synapse. Although use of the action potential to trigger secretion is not unique for nerve cells, the electrical wave that travels along the axons enables these cells to transmit information rapidly over great distances between the perikarya and the nerve terminals. Despite these specialized features of nerve cells, it is important to note that the same cellular mechanisms are used for signal production and release as well as for reception and response during neural, endocrine, and paracrine communication.

Distinctions between the various modes of communication are limited only to the means of signal delivery to target cells, and even these distinctions are blurred in some cases. Neurotransmitters act in a paracrine fashion on postsynaptic cells and in some cases may diffuse beyond the synaptic cleft to affect other nearby cells or may even enter the blood and act as hormones, in which case they are called *neurohormones*. Moreover, the same chemical signals may be secreted by both endocrine and nerve cells and even in very small amounts by other cells that use them to communicate with neighboring cells in a paracrine or autocrine manner. Nature is parsimonious in this regard. Many peptides that have classically been regarded as hormones or neurohormones may also serve as paracrine regulators in a variety of tissues. Although adequate to cause localized responses, the minute quantities of these substances produced extraglandularly are usually too small to enter the blood and interfere with endocrine relationships.

Clearly, the boundaries between endocrinology and other fields of modern biology are both artificial and imprecisely drawn. Endocrinology has benefitted enormously from recent advances in other fields, particularly immunology, biochemistry, cell biology, and molecular biology. Early insights into endocrine function were gained from "experiments of nature," i.e., injury or inborn errors produced pathological conditions that were traced to defects in hormone secretion or action. Conversely, hormone-secreting tumors or deranged regulatory mechanisms produced early insights into the consequences of excess hormone production. Early endocrinologists were able to create similar experiments by excising a gland or administering glandular extracts and observing the consequences. Progress in biochemistry made it possible to study pure hormones, and application of immunological techniques allowed identification and measurement of various molecular species. The introduction of techniques of molecular biology brought breakthroughs in the understanding of hormone actions, and curiously brought us full circle back to the early approaches of studying the consequences of eliminating the source of a signaling molecule or administering an excess to gain insight into function. It is now possible to overexpress a hormone or other molecule by inserting its gene into developing mice to make them "transgenic." Conversely, it is possible to disrupt or "knock out" a particular gene and study the consequences of the lack of its protein product(s) in otherwise intact mice. It is even possible to limit expression of transgenes to particular organs and evoke their expression at desired stages of life. Similarly, it is now possible to knock out genes in particular organs and at particular times of life. In discussing hormone actions in subsequent chapters it will be necessary to refer to all of these experimental techniques and many others.

In this text we concentrate on the integrating function of the endocrine system and focus our discussion principally on that aspect of cellular communication that is carried out by the classical endocrine glands and their hormones (Table 1). Chapters 1 through 5 deal with basic information about various endocrine glands and their hormones. In the remaining chapters we consider the interaction of hormones and the integration of endocrine function to produce homeostatic regulation. Such regulation throughout the body is achieved by regulation of cellular functions, which, in turn, is achieved by actions of hormones on molecules within those cells. We therefore consider the actions of hormones on three levels (Figure 2). Throughout this text, emphasis is on normal function, and reference to disease is limited to those aspects that are logical extensions of normal physiology

Table 1

Chemical Nature of the Classic Hormones

Tyrosine derivatives	Steroids	Peptides (<20 amino acids)	Proteins (>20 amino acids)
Epinephrine	Testosterone	Oxytocin	Insulin
Norepinephrine	Estradiol	Vasopressin	Glucagon
Dopamine	Progesterone	Angiotensin	Adrenocorticotropic hormone
Triiodothyronine	Cortisol	Melanocyte-stimulating hormone	Thyroid-stimulating hormone
Thyroxine	Aldosterone		Thyrotropin-releasing hormone
	Vitamin D	Somatostatin	Follicle-stimulating hormone Luteinizing hormone Gonadotropin-releasing hormone Growth hormone Prolactin Corticotropin-releasing hormone Growth hormone-releasing hormone Parathyroid hormone Calcitonin Chorionic gonadotropin Choriosomatomammotropin

or that facilitate understanding of normal physiology. Endocrine disease is not simply a matter of too much or too little hormone; rather, disease occurs when there is an inappropriate amount of hormone for the prevailing physiological situation or when there is an inappropriate response by target tissues to a perfectly appropriate amount of hormone. Some aspects of endocrine disease are too poorly understood to be put in the context of normal physiology and are best left for a more detailed text of pathology or medicine.

Endocrinology is a subject that unfortunately involves a sometimes bewildering array of facts, not all of which can be derived from basic principles. To help organize and digest this necessarily large volume of material, the student might find the following outline of goals and objectives helpful.

GOALS AND OBJECTIVES

A. The student should be familiar with

1. Essential features of feedback regulation
2. Essentials of competitive binding assays

B. For each hormone, the student should know

1. Its cell of origin
2. Its chemical nature, including
   a. Distinctive features of its chemical composition
   b. Biosynthesis
   c. Whether it circulates free or bound to plasma proteins
   d. How it is degraded and removed from the body
3. Its principal physiological actions
   a. At the whole body level
   b. At the tissue level
   c. At the cellular level
   d. At the molecular level
   e. Consequences of inadequate or excess secretion
4. What signals or perturbations in the internal or external environment evoke or suppress its secretion
   a. How those signals are transmitted
   b. How that secretion is controlled
   c. What factors modulate the secretory response
   d. How rapidly the hormone acts
   e. How long it acts
   f. What factors modulate its action

BIOSYNTHESIS OF HORMONES

The classical hormones fall into three categories:

- Derivatives of the amino acid tyrosine
- Steroids, which are derivatives of cholesterol
- Peptides/proteins, which comprise the largest and most diverse class of hormones

Table 1 lists some examples of each category. A large number of other small molecules, including derivatives of amino acids and fatty acids, function as neurotransmitters or paracrine signals but fall outside the scope of the classical hormones. In most aspects of their synthesis, secretion, and molecular actions, these substances are indistinguishable from hormones. Relevant details of hormone

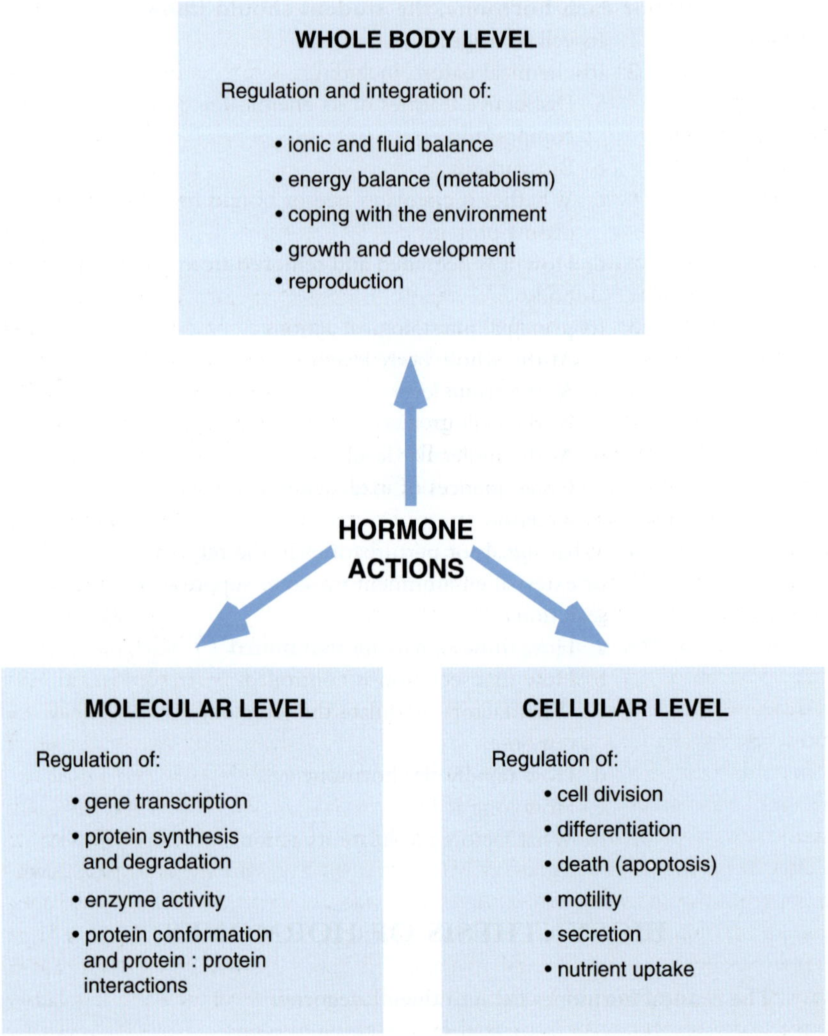

WHOLE BODY LEVEL

Regulation and integration of:

- ionic and fluid balance
- energy balance (metabolism)
- coping with the environment
- growth and development
- reproduction

**HORMONE
ACTIONS**

MOLECULAR LEVEL

Regulation of:

- gene transcription
- protein synthesis
  and degradation
- enzyme activity
- protein conformation
  and protein : protein
  interactions

CELLULAR LEVEL

Regulation of:

- cell division
- differentiation
- death (apoptosis)
- motility
- secretion
- nutrient uptake

Figure 2   Levels at which hormone actions are considered.

synthesis and storage, particularly for the amino acid and steroid hormones, are presented with the discussion of their glands of origin, but steps in biosynthesis, storage, and secretion common to all protein and peptide hormones are sufficiently general for this largest class of hormones to warrant some discussion here. A brief review of these steps also provides an opportunity for a general consideration of gene expression and protein synthesis and provides some background for understanding hormone actions. In-depth consideration of these complex processes is beyond the scope of this text, and is best left for the many excellent texts of cellular and molecular biology.

Protein and peptide hormones are encoded in genes, with each hormone usually represented only once in the genome. Information determining the amino acid sequence of proteins is encoded in the nucleotide sequence of deoxyribonucleic acid (DNA) (Figure 3). Nucleotides in DNA consist of a five-carbon sugar, deoxyribose, in ester linkage with a phosphate group and attached in N-glycosidic linkage to one of four organic bases: adenine (A), guanine (G), thymidine (T), or cytidine (C). The ability of the purine bases A and G to form complementary pairs with the pyrimidine bases T and C (Figure 4), respectively, on an adjacent strand of DNA is the fundamental property that permits accurate replication of DNA and transmission of stored information from generation to generation. A single strand of DNA consists of a chain of millions of nucleotides linked by phosphate groups that form ester bonds with hydroxyl groups at carbon 3 of one deoxyribose and carbon 5 of the next deoxyribose. The DNA in each chromosome is present as a pair of long strands oriented in opposite directions and is organized into *nucleosomes*, each of which consists of a stretch of about 180 nucleotides tightly wound around a complex of eight histone molecules. The nucleosomes are linked by stretches of about 30 nucleotides, and the whole double strand of nucleoproteins is tightly coiled in a higher order of organization to form the chromosomes.

Instructions for protein structure are transmitted from the DNA to cytoplasmic sites of protein synthesis, the ribosomes, in the messenger ribonucleic acid (mRNA) template. RNA differs in structure from DNA only in having ribose instead of deoxyribose as its sugar and uridine (U) instead of thymidine as one of its pyrimidine bases. The nucleotide sequence of the mRNA precursor is complementary to the nucleotide sequence of DNA. Messenger RNA synthesis proceeds linearly from an upstream "start site" designated by a particular sequence of nucleotides in DNA in a process called *transcription*. The start site is located downstream from the promoter region, which contains sequences to which regulatory proteins can bind, and a short sequence where RNA polymerase II and a large complex of proteins, the general transcription complex, bind. The DNA that is transcribed is composed of segments, *exons*, that encode structural and regulatory information; the exons are separated by intervening sequences of DNA, *introns*, which have no coding function (Figure 5). Transcription is regulated by nuclear proteins called *transcription factors* or *transactivating factors*, which bind to regulatory sites that are usually upstream from the promoter and stimulate or repress gene transcription. These proteins form complexes with multiple other transcription factors and proteins called *coactivators* or *corepressors*, which not only govern attachment and activity of the general transcription complex, but control the "tightness" of the DNA coil and hence the accessibility of genes to the transcription apparatus. Transcription proceeds from the start site through the introns and exons and a downstream flanking sequence, where a long polyadenine (polyA) tail is added.

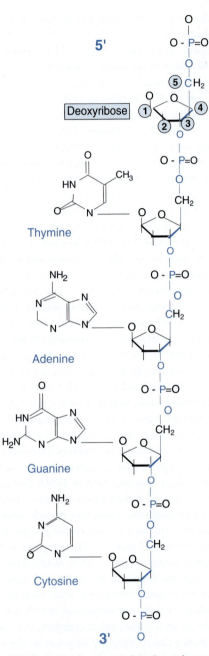

Figure 3 Composition of DNA. DNA is a polymer of the five-carbon sugar, deoxyribose, in diester linkage, with phosphate forming ester bonds with hydroxyl groups on carbons 3 and 5 on adjacent

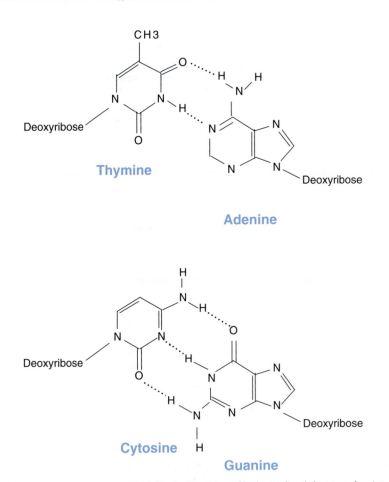

Figure 4 Complementary base pairing by the formation of hydrogen bonds between thymine and adenine and between cytosine and guanine. RNA contains uracil in place of the thymine found in DNA. Uracil and thymine differ in structure only by the presence of the methyl group (CH_3) found in thymine.

sugar molecules. The purine and pyrimidine bases are linked to carbon 1 of each sugar. The numbering system for the five carbons of deoxyribose is shown in blue at the top of the figure. The chemical bonds forming the backbone of the DNA chain are shown in blue. The 5′ and 3′ ends refer to the carbons in deoxyribose.

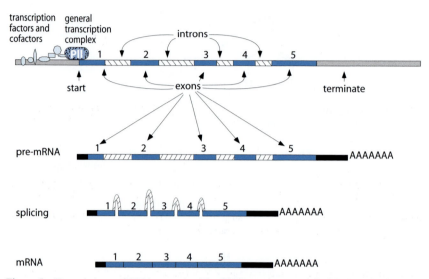

Figure 5   Transcription and RNA processing. The DNA strand contains all of the stored information for expression of the gene, including the promoter, distant regulatory elements (not shown), binding sites (response elements) for regulatory proteins, and the coding for the sequence of the protein (exons) interrupted by intervening sequences of DNA (introns). Exons are numbered 1–5. The primary RNA transcript contains the complementary sequence of bases coupled to a polyA tail at the 3′ end and a methyl guanosine cap at the 5′ end. Removal of the introns and splicing the remaining exons together produce the messenger RNA, which contains all of the information needed for translation, including the codons for the amino acid sequence of the protein and untranslated regulatory sequences at both ends. PII, RNA polymerase II.

A special "cap" structure containing methylated guanosine added to the opposite end of the RNA transcript permits its export from the nucleus after it is further modified by removal of the introns and attachment of the exons to each other in a process called *splicing*. Under some circumstances the splicing reactions may bypass some exons or parts of exons, which are then omitted from the final mRNA transcript. Because of such *alternate splicing*, a single gene can give rise to more than one mRNA transcript, and hence more than one protein product (Figure 6).

On export from the nucleus, the mRNA transcripts attach to ribosomes, where they are translated into protein (Figure 7). Ribosomes are large complexes of RNA and protein enzymes that "read" the mRNA code in triplets of nucleotides called *codons*. The translation initiation site begins with the codon for methionine. Each codon designates a specific amino acid. Triplets of complementary nucleotides (anticodons) are found in small RNA molecules called *transfer RNA* (tRNA), each of which binds a particular amino acid and delivers it to the ribosome. Alignment of amino acids in the proper sequence is achieved by the

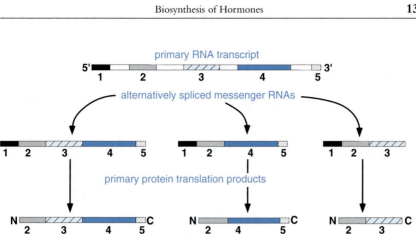

Figure 6 Alternative splicing of mRNA can give rise to different proteins. Numbers indicate exons. Exon 1 is untranslated. N, Amino terminus; C, carboxyl terminus.

complementary pairing of anticodons in the tRNA with codons in the mRNA. The tRNA thus delivers the correct amino acid to the carboxyl terminus of the growing peptide chain and holds it in position so that ribosomal enzymes can release it from the tRNA and link it to the peptide. Once the peptide bond is formed, the empty tRNA is released and the ribosome moves down the mRNA to the next codon, where the next tRNA molecule charged with its amino acid waits to bind to its complementary codon. Elongation of the chain continues until the ribosome reaches a "stop" codon, at which time it dissociates from the mRNA. As each ribosome moves down the mRNA, other ribosomes attach behind them to repeat the process. In this way, before it is degraded, a single mRNA molecule may be translated over and over again to yield many copies of a protein.

Protein and peptide hormones are initially synthesized as precursor molecules (prohormones and preprohormones) that are larger than the final secretory product. Proteins destined for secretion have a hydrophobic sequence of 12–25 amino acids at their amino termini (Figure 8). This *signal sequence* is recognized by a special structure that directs the growing peptide chain through a protein channel in the endoplasmic reticular membrane and into the cisternae of the endoplasmic reticulum. Postsynthetic processing begins in the endoplasmic reticulum as the hormone precursors are translocated to the Golgi apparatus for final processing and packaging for export. Processing includes cleavage to remove the signal peptide and interaction with other proteins that facilitate proper folding and formation of disulfide bonds linking cysteine residues. For some hormones, cleavage at appropriate loci removes those amino acid sequences that may have functioned to orient folding of the molecule so that disulfide bridges form in the right places. Clipping the protein by trypsinlike peptidases may yield more than one

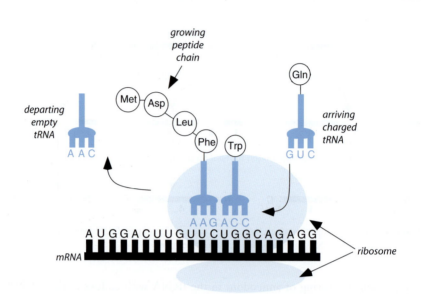

Figure 7   Translation. A molecule of transfer RNA (tRNA) charged with its specific amino acid, phenylalanine, and already linked to the growing peptide chain, is positioned on the mRNA by complementary pairing of its triplet of nucleotides with its codon of three nucleotides in the mRNA. A second molecule of tRNA charged with its specific amino acid, tryptophan, has docked at the adjacent triplet of nucleotides and awaits the action of ribosomal enzymes to form the peptide bond with phenylalanine. Linking the amino acid to the peptide chain releases it from its tRNA and allows the empty tRNA to dissociate from the mRNA. A third molecule of tRNA, which brought the preceding molecule of leucine, is departing from the left, while a fourth molecule of tRNA, carrying its cargo of glutamine, arrives from the right and waits to form the complementary bonds with the next codon in the mRNA that will bring the glycine in position to be joined to tryptophan at the carboxyl terminus of the peptide chain. The ribosome moves down the mRNA, adding one amino acid at a time until it reaches a stop codon. (Adapted from Alberts *et al.*, "Molecular Biology of the Cell." Garland Publishing, New York, copyright 1994. Reproduced by permission of Routledge, Inc., part of The Taylor & Francis Group.)

biologically active peptide molecule from a single precursor, as seen with the adrenocorticotropic and glucagon families of hormones (Chapters 2 and 5). For some secreted peptides, final clipping occurs in secretory granules with the result that one or more other molecules are released into the circulation along with the hormone. Other processing of peptide hormones may include glycosylation (addition of carbohydrate chains to asparagine residues) or coupling of subunits that are products of different genes, as seen with the pituitary glycoproteins (see Chapter 2).

Defects in processing of normal precursor molecules cause some rare inherited diseases. It is common to find precursor molecules (prohormones) in the circulation, sometimes in large amounts. This situation may be indicative of

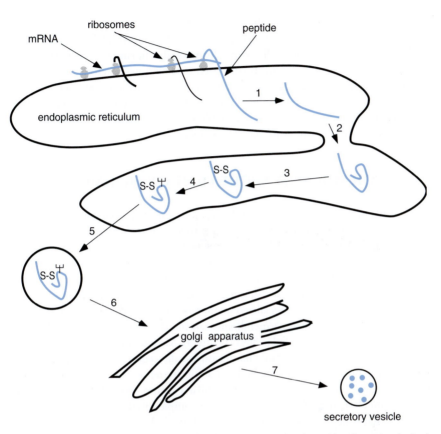

Figure 8 Posttranslational processing. The leader sequence, or signal peptide, of proteins destined for secretion enters the cisternae of the endoplasmic reticulum even as peptide elongation continues. In the endoplasmic reticulum (1) the leader sequence is removed; (2) the protein is folded with the assistance of protein chaperones, sulfhydryl bridges (3) may form, and carbohydrate (4) may be added (glycosylation). The partially processed protein (5) is then entrapped in vesicles that bud off the endoplasmic reticulum and (6) fuse with the Golgi apparatus, where glycosylation is completed, and (7) the protein is packaged for export in secretory vesicles in which the final stages of processing take place.

hyperactivity of endocrine cells or even aberrant production of hormone by nonendocrine tumor cells. Some prohormones have biological activity, and their effects may be the first manifestation of neoplasia.

Postsynthetic processing to the final biologically active form is not limited to the peptide hormones. Other hormones may be formed from their precursors after secretion. Postsecretory transformations to more active forms may occur in liver, kidney, fat, or blood, as well as in the target tissues. For example, thyroxine, the major secretory product of the thyroid gland, is converted extrathyroidally to triiodothyronine, which is the biologically active form of the hormone (see Chapter 3). Testosterone, the male hormone, is converted to dihydrotestosterone within some target tissues and may even be converted to the female hormone, estrogen, in other tissues (Chapter 11). These peripheral transformations, in addition to confounding the student of endocrinology, are vulnerable to derangement and hence must be considered as possible causes of endocrine disease.

STORAGE AND SECRETION OF HORMONES

With the notable exception of the steroids, most hormones are stored, often in large quantities, in their glands of origin, a factor that facilitated their original isolation and characterization. Protein and peptide hormones, and the tyrosine derivatives epinephrine and norepinephrine, are stored as dense granules in membrane-bound vesicles and are secreted in response to an external stimulus by the process of *exocytosis*. In this process, storage vesicles are translocated to the cell surface, where they dock with specialized membrane proteins. Membranes of the vesicles then fuse with the plasma membrane, causing the vesicle to open and empty its contents into the extracellular fluid (Figure 9). Movement of the secretory vesicle to the cell surface and membrane fusion usually require transient increases in cytosolic calcium concentrations, brought about by release of calcium from internal organelles and from influx of extracellular calcium through activated membrane channels. A detailed description of the complex molecular events that govern secretion is beyond the scope of this text but can be found in many fine texts of cell biology. It is obvious that synthesis of hormones must be coupled in some way with secretion, so that cells can replenish their supply of hormone. In general, the same cellular events that signal secretion also signal synthesis. In addition, some cells may be able to monitor how much hormone is stored and to adjust rates of synthesis or degradation accordingly.

Unlike the peptide hormones, which are encoded in genes, the steroid hormones are formed enzymatically through a series of modifications of their common precursor, cholesterol (see Chapter 4). In further contrast to the peptide hormones, there is little storage of steroid hormones in their cells of origin. Therefore, synthesis and secretion are aspects of the same process, and the lipid-soluble steroid hormones apparently diffuse across the plasma membrane as rapidly as they are formed. The synthetic process proceeds sufficiently rapidly that increased secretion can be observed as soon as a minute after the secretory stimulus has been applied, but the maximal rate of secretion is not reached for at least 10–15 minutes. In contrast, stored peptide and amine hormones may be released almost instantaneously.

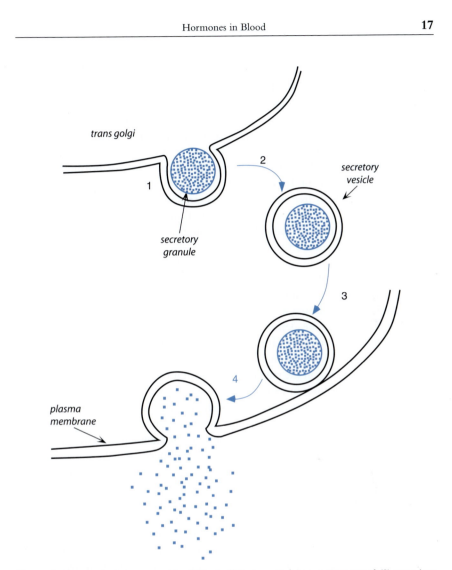

Figure 9 Exocytosis. Secretory vesicles (1) bud off the trans-Golgi compartment and (2) move into the cytosol, where they await a signal for secretion. (3) Secretion is usually accompanied by increased cellular calcium, which causes elements of the cytoskeleton to translocate secretory granules to the cell surface (4). The membrane surrounding the granule fuses with the plasma membrane, opening the secretory vesicle to the extracellular fluid and releasing the processed protein(s) along with enzymes and peptide fragments.

HORMONES IN BLOOD

Most hormones circulate in blood in free solution at nanomolar (10^{-9} M) or even picomolar (10^{-12} M) concentrations. Steroid hormones and thyroid hormones, which have a limited solubility in water, circulate bound specifically to large carrier proteins synthesized in the liver. Some protein and peptide hormones also circulate complexed with specific binding proteins (see Chapter 10). Bound hormones are in equilibrium with a small fraction, sometimes less than 1%, in free solution in plasma. Generally, only unbound hormones are thought to cross the capillary endothelium to reach their sites of biological action or degradation. Protein binding protects against loss of hormone by the kidney, slows the rate of hormone degradation by decreasing cellular uptake, and buffers changes in free hormone concentrations. In some instances binding proteins may affect hormonal responses by facilitating or impeding delivery of hormones to particular cells. Because biological responses are related to the concentration of hormone that reaches target cells, rather than the total amount present in blood, increases in abundance of binding proteins that occur during pregnancy, for example, or decreases seen with some forms of liver or kidney disease, may produce changes in total amounts of hormones circulating in blood, even though free, physiologically important concentrations may be normal.

Most hormones are destroyed rapidly after secretion and have a half-life in blood of less than 10 minutes. The half-life of a hormone in blood is defined as that period of time needed for its concentration to be reduced by half and depends on its rate of degradation and on the rapidity with which it can escape from the circulation and equilibrate with fluids in extravascular compartments. This process is sometimes called the metabolic clearance rate. Some hormones, e.g., epinephrine, have half-lives measured in seconds; others, e.g., thyroid hormones, have half-lives of the order of days. The half-life of a hormone in blood must be distinguished from the duration of its hormonal effect. Some hormonal effects are produced virtually instantaneously and may disappear as rapidly as the hormone is cleared from the blood. Other hormonal effects are seen only after a lag time that may last minutes or even hours, and the time of maximum effect may bear little relation to the time of maximum hormone concentration in the blood. Additionally, the time for decay of a hormonal effect is also highly variable; it may be only a few seconds, or it may require several days. Some responses persist well after hormonal concentrations have returned to basal levels. Understanding the time course of a hormone's survival in blood as well as the onset and duration of its action is obviously important for understanding normal physiology, endocrine disease, and the limitations of hormone therapy.

HORMONE DEGRADATION

Implicit in any regulatory system involving hormones or any other signal is the necessity for the signal to disappear once the appropriate information has been conveyed. Only a small amount of hormone is degraded as an aftermath to the process of signaling its biological effects. The remainder must therefore be inactivated and excreted. Degradation of hormones and their subsequent excretion are processes that are just as important as secretion. Inactivation of hormones occurs enzymatically in blood or intercellular spaces, in liver or kidney cells, as well as in the target cells. Degradation of peptide and protein hormones often involves uptake into cells by a mechanism of endocytosis that delivers them to the cellular sites of degradation, the lysosomes and proteosomes. Inactivation may involve complete metabolism of the hormone so that no recognizable product appears in urine, or it may be limited to some simple one- or two-step process such as addition of a methyl group or glucuronic acid. In the latter cases recognizable degradation products are found in urine and can be measured to obtain a crude index of the rate of hormone production.

MECHANISMS OF HORMONE ACTION

The ultimate mission of a hormone is to change the behavior of its target cells. Cellular behavior is determined by biochemical and molecular events that transpire within the cell, and these in turn are determined by the genes that are expressed, the biochemical reactions that carry out cellular functions, and the conformation and associations of the molecules that comprise the cell's physical structure. Hormonal messages must be converted to biochemical events that influence gene expression, biochemical reaction rates, and structural changes. The conversion of a hormonal message to cellular responses is called *signal transduction* and the series of biochemical changes that are set in motion are described as *signaling pathways*, although in reality *signaling network* might be a more accurate descriptor, because pathways branch and converge only to branch again. Signal transduction is a complex topic and the focus of intense investigation in many laboratories around the world. Detailed consideration is beyond the scope of this text. Instead, only general patterns of signal transduction are considered in the following section, but the topic will be revisited where appropriate in subsequent chapters in discussing individual hormones.

SPECIFICITY

Because all hormones travel in blood from their glands of origin to their target tissues, all cells must be exposed to all hormones. Yet under normal circumstances cells respond only to their appropriate hormones. Such specificity of hormone action resides primarily in the ability of receptors in the target cells to recognize only their own signal (Figure 10). We may define a hormone receptor as a molecule or complex of molecules, in or on a cell, that binds its hormone with great selectivity and in so doing is changed in such a manner that a characteristic response or group of responses is initiated. Hormone receptors are a subset of

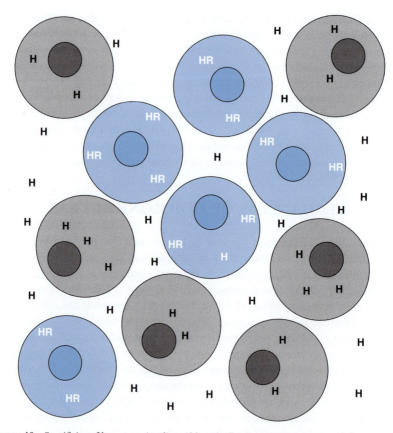

Figure 10   Specificity of hormone signaling. Although all cells come in contact with hormone (H), only the cells colored blue have receptors (R) and therefore can respond to the hormone.

the huge number of molecules that are utilized by all cells to receive specific information from other cells and the external environment. The mechanisms by which receptors operate and are regulated are not unique to endocrinology.

CHARACTERISTICS OF RECEPTORS

Hormone receptors are proteins or glycoproteins that are able to function as follows:

1. They distinguish their hormone from other molecules that may have very similar structures.

2. They bind to the hormone (sometimes called a *ligand*) even when its concentration is exceedingly low (10^{-8}–10^{-12} M).
3. They undergo a conformational change when bound to the hormone.
4. They catalyze biochemical events or transmit changes in molecular conformation to adjacent molecules, producing a biochemical change.

These aspects of receptor function may reside within a single molecule or in separate subunits of a receptor complex. The role of the hormone is simply to excite the receptor by binding to it. All of the biochemical changes initiated by the excited receptor derive from the properties of the receptor and not of the hormone. With modern technology it is now possible to create chimeric receptors in which the hormone recognition component of the receptor for one hormone can be fused to the signal-transducing component of the receptor for another hormone. The biochemical changes set in motion by hormone binding to such chimeric receptors are characteristic of the transduction component, and not the bound hormone (Figure 11). Under some pathophysiological circumstances an aberrant antibody may react with a receptor and produce a disease state that is indistinguishable from the disease that results from overproduction of the hormone, again indicating that the nature of the response is a property of the receptor.

Hormone receptors are found on the surface of the target cell, in the cytosol, or in the nucleus. Receptors that reside in the plasma membrane span its entire thickness, with the hormone recognition component facing outward. Components on the cytosolic face of the membrane communicate with other membrane or cytosolic proteins. Membrane receptors may be distributed over the entire surface of a cell or they may be confined to some discrete region, such as the basal surface of a renal tubular epithelial cell. Growing evidence suggests that some membrane receptors and the proteins they interact with may be confined to specialized "microdomains" within the plasma membrane, perhaps in microinvaginations called *caveolae*.

Only a few thousand receptor molecules are usually present in a target cell, but the number is not fixed. Cells can adjust the abundance of their hormone receptors, and hence their responsiveness to hormones according to changing physiological circumstances (see Chapter 6). Some receptors may be expressed only at certain stages of a cell's life cycle or as a consequence of stimulation by other hormones. Many cells adjust the number of receptors they express in accordance with the abundance of the signal that activates them. Frequent or intense stimulation may cause a cell to decrease, or *down-regulate*, the number of receptors expressed. Conversely, cells may *up-regulate* receptors in the face of rare or absent stimulation or in response to other signals.

Membrane receptors are internalized either alone or bound to their hormones (receptor-mediated endocytosis), and, like other cellular proteins, are broken down and replaced many times over during the lifetime of a cell.

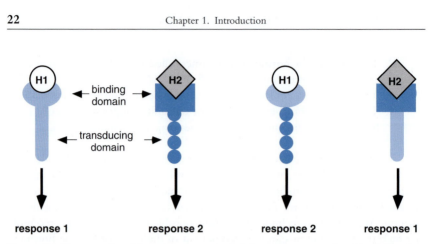

Figure 11   Hormones 1 and 2 produce separate and distinct responses transduced by their unique receptors. Chimeric receptors were produced by fusing the hormone binding domain of the receptor for hormone 1 to the signal transducing domain of the receptor for hormone 2 and vice versa. Hormone 1 now elicits a response formerly produced by hormone 2, and hormone 2 now produces the response formerly produced by hormone 1.

Adjustments in the relative rates of receptor synthesis or degradation may result in either up- or down-regulation of receptor abundance. Cells can also up- or down-regulate receptor function through reversible covalent modifications such as adding or removing phosphate groups. Membrane-associated receptors cycle between the plasma membrane and internal membranes, and their relative abundance on the cell surface can be adjusted by reversibly sequestering them in intracellular vesicles.

Although the mammalian organism expresses literally thousands of different receptor molecules that subserve a wide variety of functions in addition to endocrine signaling, our task in understanding receptor physiology is made somewhat simpler by the fact that there are relatively few general patterns of signaling. Based on the nucleotide sequence and organization of their genes and the structure of their proteins, receptors—like other proteins—can be organized into families or superfamilies that presumably arose from the same ancient progenitor gene. Even for distantly related receptors the general features of signal transduction follow common broad outlines that are seen with families of molecules that receive and transduce signals in eukaryotic cells of species ranging from yeast to humans.

HORMONAL ACTIONS MEDIATED BY
INTRACELLULAR RECEPTORS

The cholesterol derivatives (steroid hormones and vitamin D) are lipid soluble and are thought to enter cells by diffusion through the lipid bilayer of the plasma membrane. Similarly, the thyroid hormones, which are α-amino acids, have large nonpolar constituents and may penetrate cell membranes both by diffusion and by carrier-mediated transport. These hormones bind to receptors that are usually located in the cell nucleus and produce most, but not all, of their effects by altering rates of gene expression. Receptors bound to steroid hormones, in turn, bind to specific nucleotide sequences in DNA, called *hormone response elements* (HREs), located upstream of the transcription start sites of the genes they regulate. The end result of stimulation with these hormones is a change in genomic readout, which may be expressed in the formation of new proteins or modification of the rates of synthesis of proteins already in production. The sequence of events shown in Figure 12 is probably applicable to all steroid hormones.

Intracellular hormone receptors belong to a very large family of transcription factors found throughout the animal kingdom. Some members of this receptor family that have been isolated and characterized have no known ligands and thus are referred to as "orphan receptors." The most highly conserved region of these receptors is a stretch of about 65–70 amino acid residues that constitutes the DNA-binding domain. This region contains two molecules of zinc, each coordinated with four cysteine residues so that two loops of about 12 amino acids each are formed. These so-called zinc fingers can insert in a half-turn of the DNA helix and grasp the DNA at the site of the HRE. The hormone-binding domain, which is near the carboxyl terminus, also contains amino acid sequences that are necessary for activation of transcription. Between the DNA-binding domain and the amino terminus is the so-called hypervariable region, which, as its name implies, differs both in size and in amino acid sequence for each receptor.

The steroid hormone receptors constitute a closely related group within the family. In the unstimulated state steroid hormone receptors are noncovalently complexed with other proteins, including a dimer of the 90,000-Da heat shock protein (Hsp 90), which attaches adjacent to the hormone-binding domain (Figure 13). Heat shock proteins are abundant cellular proteins that are found in prokaryotes and all eukaryotic cells, and are so named because their synthesis abruptly increases when cells are exposed to high temperature or other stressful conditions. These proteins are thought to keep the receptor in a configuration that is favorable for binding the hormone and incapable of binding to DNA. Binding to its hormone causes the receptor to dissociate from Hsp 90 and the other proteins. The bound receptor then forms a homodimer with another liganded receptor molecule and undergoes a conformational change that increases its affinity for binding to DNA. After binding to the DNA, the receptor dimers recruit other nuclear regulatory proteins, including *coactivators,* which facilitate uncoiling of the DNA to make it accessible to the RNA polymerase complex. Receptors for at least four different steroid hormones bind to the identical HRE, and yet each governs expression of a unique complement of genes. Expression of genes that are specific for each hormone is determined by which receptor is present in a particular cell, by the cohort of nuclear transcription factors, coactivators,

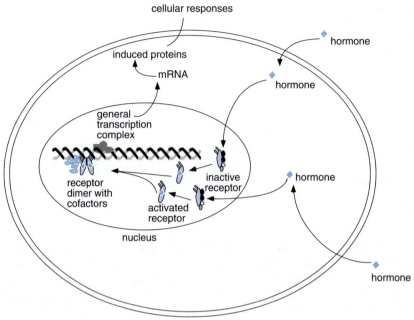

Figure 12   General scheme of steroid hormone action. Steroid hormones penetrate the plasma membrane and bind to intracellular receptors found largely in the nucleus (except adrenal steroid receptors).
Hormone binding activates the receptor, which forms complexes with other proteins and binds to
specific acceptor sites (hormone response elements) on DNA to initiate transcription and formation of
the proteins that express the hormonal response. The steroid hormone is then cleared from the cell.

and corepressors that are available to complex with the receptor in that cell, and by
the characteristics of the regulatory components in the DNA.

   Receptors for thyroid hormone and vitamin D and compounds related to
vitamin A (retinoic acid) belong to another closely related group within the same
family of proteins as the steroid receptors. Unlike the steroid hormone receptors,
these receptors are bound to their HREs in DNA even in the absence of hormone,
and do not form complexes with Hsp 90. In further distinction from the receptors
for steroid hormones, receptors for thyroid hormone and vitamin D may bind
to DNA either as homodimers or as heterodimers formed with a receptor for
9-*cis*-retinoic acid. In the absence of ligand, these DNA-bound receptors form
complexes with other nuclear proteins that may promote or inhibit transcription.
On binding its hormone, the receptor undergoes a conformational change that
displaces the associated proteins and allows others to bind, with the result that
transcription is either activated or suppressed.

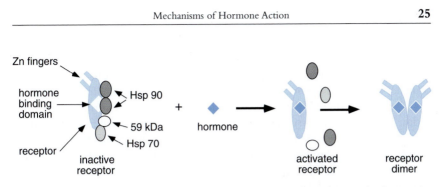

Figure 13   Activation of steroid hormone receptors. Inactive receptors associated with other proteins react with hormone, shed their associated proteins, change their configuration, and form dimers that can interact with DNA and a variety of nuclear peptide regulators of gene transcription. Zn, Zinc; Hsp, heat shock protein.

Many steps lie between activation of transcription and changes in cellular behavior. These include synthesis and processing of RNA, exporting it to cytosolic sites of protein synthesis, protein synthesis, protein processing, and delivery of the newly formed proteins to appropriate loci within the cells. These reactions necessarily occur sequentially and each takes time. Transcription proceeds at a rate of about 40 nucleotides per second, so that transcribing a gene that contains 10,000 nucleotides takes almost 5 minutes. Processing the preRNA to mature mRNA is even slower, so that nearly 20 minutes elapses from the time RNA synthesis is initiated to the time the mRNA exits the nucleus. Protein synthesis is much faster: about 15 amino acids per second are added to the growing peptide chain. All factors considered, changes in cellular behavior that result from steroid hormone action are usually not seen for at least 30 minutes after entry of the hormone into the cells. The final protein makeup of the cell at any time thereafter is also determined by rates of RNA and protein degradation. A complete catalog of which proteins are formed in any particular cell type as a result of hormone action should become available in the near future thanks to the successful completion of the Human Genome Project and the technology that permits screening of the entire library of mRNA expressed within a cell. Gaining an understanding of the physiological role of each of these proteins will take a bit longer.

As blood levels of hormones decline, intracellular concentrations also decline. Because binding is reversible, hormone dissociates from receptors and is cleared from the cell by diffusion into the extracellular fluid, usually after metabolic conversion to an inactive form. Unloaded steroid receptors dissociate from their DNA binding sites and regulatory proteins, and either recycle into new complexes with Hsp 90 and other proteins through some energy-dependent process or are degraded and replaced by new synthesis. RNA transcripts of hormone-sensitive genes are degraded usually within minutes to hours of their formation. Without continued hormonal stimulation of their synthesis, RNA templates for hormone-dependent proteins disappear, and the proteins they encode can no longer be formed. The proteins are degraded with half-lives that may range from seconds to days. Thus just as there is delay in onset, effects of the hormones that act through nuclear receptors may persist after the hormone has been cleared from the cell.

Accumulating evidence indicates that many of the hormones that were once thought to act only through nuclear receptors do in fact produce some rapid effects that are independent of changes in gene expression. For the most part the rapid responses that are produced are complementary to the delayed genomically mediated responses. It is likely that other, yet to be identified, receptors for these hormones are present on the cell surface or that some nuclear receptors are expressed on the cell surface as well as in the nucleus.

HORMONAL ACTIONS MEDIATED BY SURFACE RECEPTORS

The protein and peptide hormones and the amine derivatives of tyrosine cannot readily diffuse across the plasma membranes of their target cells. These hormones produce their effects by binding to receptors on the cell surface and rely on molecules on the cytosolic side of the membrane to convey the signal to the appropriate intracellular effector sites that bring about the hormonal response.

The G–Protein–Coupled Receptors

The most frequently encountered cell surface receptors belong to a very large superfamily of proteins that couple with guanosine nucleotide binding proteins (G-proteins) to communicate with intracellular effector molecules. This ancient superfamily of receptor molecules is widely expressed throughout eukaryotic phyla. G-Protein-coupled receptors are crucial for sensing external environmental signals such as light, taste, and odor as well as signals transmitted by hormones. G-Protein-coupled receptors receive signals carried by a wide range of neurotransmitters, immune modulators, and paracrine factors. More than 1000 varieties of G-protein-coupled receptors may be expressed in humans, and about 30% of all effective pharmaceutical agents are said to target actions mediated by receptors in this superfamily. All G-protein-coupled receptors are composed of single strands of protein and contain seven stretches of about 25 amino acids that are each thought to form membrane-spanning α-helices (Figure 14). The single long peptide chain that constitutes the receptor thus threads through the membrane seven times, creating three extracellular and three intracellular loops. For this reason, these receptors are sometimes called *heptahelical* receptors, or serpentine receptors. The amino-terminal tail is extracellular and along with the external loops may contain covalently bound carbohydrate. The carboxyl tail lies within the cytoplasm. The lengths of the loops and the carboxyl and amino-terminal tails vary in characteristic ways among the subgroups of these receptors. Outward-facing components of the receptor, including parts of the α-helices, contribute to the hormone recognition and binding site. The cytosolic loops and carboxyl tail bind to specific G-proteins near the interface of the membrane and the cytosol.

G-Proteins are heterotrimers composed of alpha, beta, and gamma subunits. Lipid moieties covalently attached to the alpha and gamma subunits insert into the inner leaflet of the plasma membrane bilayer and tether the G-proteins to the membrane (Figure 14). The alpha subunits are enzymes that catalyze the conversion of guanosine triphosphate (GTP) to guanosine diphosphate (GDP). In the unactivated or resting state, the catalytic site in the alpha subunit is occupied by GDP. When the receptor binds to its hormone, a conformational change transmitted across the membrane allows its cytosolic domain to interact with the alpha subunit of the G-protein in a way that causes the alpha subunit to release the GDP in exchange for a molecule of GTP and to dissociate from the beta/gamma subunits, which remain tightly bound to each other (Figure 15). Though tethered to the membrane, the dissociated subunits apparently can move laterally along the inner surface of the membrane. In its GTP-bound state, the alpha subunit interacts with and modifies the activity of membrane-associated enzymes that initiate the hormonal response. The liberated beta/gamma complex can also bind to cellular proteins and modify their activities, and both the free alpha subunits and the beta/gamma subunits can bind to ion channel proteins and cause the channels to open or close.

Hydrolysis of GTP to GDP restores the resting state of the alpha subunit, allowing it to reassociate with the beta/gamma subunits to reconstitute the

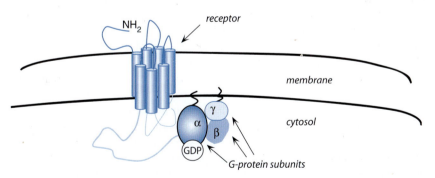

Figure 14   G-Protein-coupled receptor. The seven transmembrane α-helices are connected by three extracellular and three intracellular loops of varying length. The extracellular loops may be glycosylated, and the intracellular loops and C-terminal tail may be phosphorylated. The receptor is coupled to a G-protein consisting of a GDP-binding alpha subunit closely bound to a beta/gamma component. The alpha and gamma subunits are tethered to the membrane by lipid groups.

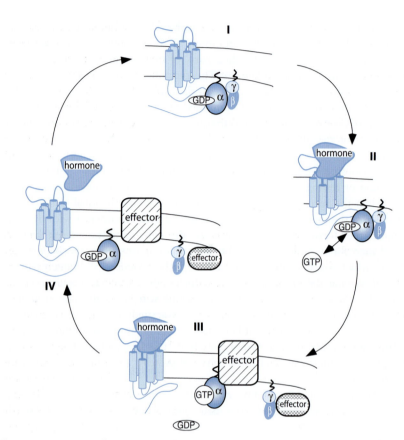

Figure 15   Activation of G-protein-coupled receptor. (I) Resting state. (II) Hormone binding produces a conformational change in the receptor that causes (III) the alpha subunit to exchange ADP for GTP, dissociate from the beta/gamma subunit, and interact with its effector molecule. The beta/gamma subunit also interacts with its effector molecule. (IV) The alpha subunit converts GTP to GDP, which allows it to reassociate with the beta/gamma subunit, and the hormone dissociates from the receptor, restoring the resting state (I).

heterotrimer. GTPase activity of the alpha subunit is relatively slow. Consequently the alpha subunit may interact multiple times with effector enzymes before it returns to its resting state. In addition, because some G-proteins may be as much as 100 times as abundant as the receptors with which they associate, a single hormone-bound receptor may interact sequentially with multiple G-proteins before the hormone dissociates from the receptor. These characteristics provide mechanisms for amplification of the signal. That is, interaction of a single hormone molecule with a single receptor molecule may result in multiple signal-generating events within a cell.

At least three different classes of G-protein alpha subunits are involved in transduction of hormonal signals. Each class includes the products of several closely related genes. Alpha subunits of the "s" (stimulatory) class ($G\alpha_s$) stimulate the transmembrane enzyme, adenylyl cyclase, to catalyze the synthesis of cyclic 3',5'-adenosine monophosphate (cyclic AMP, or cAMP) from ATP (Figure 16). Alpha subunits of the "i" (inhibitory) class ($G\alpha_i$) inhibit the activity of adenylyl cyclase. Alpha subunits belonging to the "q" class stimulate the activity of the membrane-bound enzyme phospholipase Cβ (PLCβ), which catalyzes hydrolysis of the membrane phospholipid phosphatidylinositol 4,5-bisphosphate to liberate inositol 1,4,5-trisphosphate (IP$_3$) and diacylglycerol (DAG) (Figure 17).

The Second Messenger Concept

For a hormonal signal that is received at the cell surface to be effective, it must be transmitted to the intracellular organelles and enzymes that produce the cellular response. To reach intracellular effectors, the G-protein-coupled receptors rely on intermediate molecules called *second messengers*, which are formed and/or released into the cytosol in response to hormonal stimulation (the first message). Second messengers activate intracellular enzymes and also amplify signals. A single hormone molecule interacting with a single receptor may result in the formation of tens or hundreds of second messenger molecules, each of which might activate an enzyme that in turn catalyzes formation of hundreds or thousands of molecules of product. Most of the responses that are mediated by second messengers are achieved by regulating the activity of enzymes in target cells, usually by adding a phosphate group. The resulting conformational change increases or decreases enzymatic activity. Enzymes that catalyze the transfer of the terminal phosphate

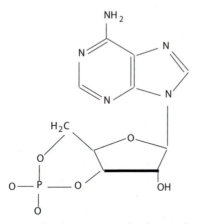

Figure 16   Cyclic adenosine monophosphate (cyclic AMP).

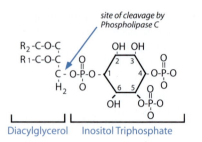

Figure 17 Phosphatidylinositol bisphosphate. When cleaved by phospholipase C, inositol 1,4,5-triphosphate and diacylglycerol are formed. R_1 and R_2 are long chain fatty acids.

from adenosine triphosphate (ATP) to a hydroxyl group in serine or threonine residues in proteins are called *protein kinases.* Hydroxyl groups of tyrosine residues may also be phosphorylated in this way, and the enzymes that catalyze this phosphorylation are called *tyrosine kinases.* The human genome is thought to contain more than a thousand genes that encode protein kinases, but only a few are activated by second messengers. Many protein kinases are activated by phosphorylation catalyzed by other protein kinases. *Protein phosphatases* remove phosphate groups from these residues and thus restore them to their unstimulated state. For the most part, protein phosphatases are constitutively active, but some specific protein phosphatases are directly or indirectly activated in response to hormonal stimulation.

Unlike responses that require synthesis of new cellular proteins, responses that result from phosphorylation–dephosphorylation reactions occur very quickly, and therefore most second messenger-mediated responses are turned on and off without appreciable latency. However, second messengers can also promote the phosphorylation of transcription factors and thus regulate transcription of specific genes in much the same way as discussed for the nuclear receptors. These responses require the same time-consuming processes as are needed for nuclear receptor-mediated changes and are seen only after a delay.

Although a very large number of hormones and other first messages act through surface receptors, to date only a few substances have been identified as second messengers. This is because receptors for many different extracellular signals utilize the same second messenger. When originally proposed, the hypothesis that the same second messenger might mediate different actions of many different hormones, each of which produces a unique pattern of cellular responses, was met with skepticism. The idea did not gain widespread acceptance until it was recognized that the special nature of a cellular response is determined by the particular enzymatic machinery with which a cell is endowed, rather than by the signal that turns on that machinery. Thus, when activated, a hepatic cell makes glucose, and a smooth muscle cell contracts or relaxes.

The Cyclic AMP System

The first of the second messengers to be recognized was cyclic AMP. The broad outlines of cyclic AMP-mediated cellular responses to hormones are shown in Figure 18. Cyclic AMP transmits the hormonal signal by activating the enzyme protein kinase A (PKA). When cellular concentrations of cyclic AMP are low, two catalytic subunits of protein kinase A are firmly bound to a dimer of regulatory subunits, which keep the tetrameric holoenzyme in the inactive state. The catalytic and regulatory subunits of protein kinase A are products of separate genes. Reversible binding of two molecules of cyclic AMP to each regulatory subunit liberates the catalytic subunits. Cyclic AMP is degraded to 5′-AMP by the enzyme cyclic AMP phosphodiesterase. As cyclic AMP concentrations fall, bound cyclic AMP separates from the regulatory subunits, which then reassociate with the catalytic subunits, restoring basal activity.

The regulatory regions of many genes contain a cyclic AMP response element (CRE) analogous to the HREs that bind nuclear hormone receptors, as discussed above. One or more forms of CRE binding proteins (CREBs) are found in

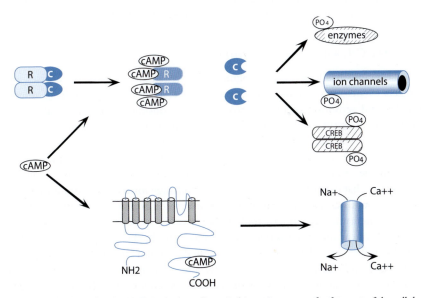

Figure 18 Effects of cyclic AMP. Activation of protein kinase A accounts for for most of the cellular actions of cyclic AMP (upper portion of the figure). Inactive protein kinase consists of two catalytic units (C), each of which is bound to a dimer of regulatory units (R). When two molecules of cyclic AMP bind to each regulatory unit, active catalytic subunits are released. Phosphorylation of enzymes, ion channels, and transcription factors of the CREB family activates or inactivates these proteins. Cyclic AMP also binds to the alpha subunits of cyclic nucleotide-gated ions channels (lower portion of the figure) causing them to open and allow influx of sodium and calcium.

the nuclei of most cells and are substrates for PKA. Dimers of phosphorylated CREBs bind to the CREs of regulated genes and recruit other nuclear proteins to form complexes that regulate gene transcription in the same manner as described for the nuclear hormone receptors.

Desensitization and Down-Regulation

In addition to simple dissociation of hormone from its receptor, signaling is often terminated and receptors desensitized to further stimulation by active cellular processes. G-Protein-coupled receptors may be inactivated by phosphorylation of one of their intracellular loops, catalyzed by a special G-protein receptor kinase, which uncouples the receptor from the alpha subunit and promotes binding to a cytoplasmic protein of the β-arrestin family. Binding to β-arrestin may lead to receptor internalization and down-regulation by sequestration in intracellular vesicles. Sequestered receptors may recycle to the cell surface, or, when cellular stimulation is prolonged, they may be degraded in lysosomes. β-Arrestins may have the additional function of serving as a scaffold for binding to a variety of other proteins, including the mitogen-activated protein kinases (see below), and thereby provide another pathway for signaling between G-protein-coupled receptors and the nucleus.

The Calcium:Calmodulin System

Calcium has long been recognized as a regulator of cellular processes and triggers events such as muscular contraction, secretion, polymerization of microtubules, and activation of various enzymes. The concentration of free calcium in cytoplasm of resting cells is very low, about one ten-thousandth of its concentration in extracellular fluid. This steep concentration gradient is maintained primarily by the actions of calcium ATPases that transfer calcium out of the cell or into storage sites within the endoplasmic reticulum and by sodium–calcium exchangers that extrude one calcium ion in exchange for three sodium ions. When cells are stimulated by some hormones, their cytosolic calcium concentration rises abruptly, increasing perhaps 10-fold or more within seconds. This is accomplished by release of calcium from intracellular storage sites in the endoplasmic reticulum and by influx of calcium through activated calcium channels in the plasma membrane. Although calcium can directly affect the activity of some proteins, it generally does not act alone. Virtually all cells are endowed with a protein called *calmodulin,* which reversibly binds four calcium ions. When complexed with calcium, the configuration of calmodulin is modified in a way that enables it to bind to protein kinases and other enzymes and to activate them. The behavior of calmodulin–dependent protein kinases is quite similar to that of protein kinase A. Calmodulin kinase II may catalyze the phosphorylation of many of the same substrates as PKA, including CREB and other nuclear transcription factors.

On cessation of hormonal stimulation, calcium channels in the endoplasmic reticular and plasma membranes close, and constitutively active calcium pumps in these membranes restore cytoplasmic concentrations to low resting levels. A low cytosolic concentration favors release of calcium from calmodulin, which then dissociates from the various enzymes it has activated.

The DAG and IP₃ System

Both products of PLC-catalyzed hydrolysis of phosphatidylinositol 4,5-bisphosphate, DAG, and IP₃ behave as second messengers (Figure 19). IP₃ diffuses through the cytosol to reach its receptors in the membranes of the endoplasmic reticulum. Activated IP₃ receptors function as calcium ion channels through which calcium passes into the cytoplasm. Because of its lipid solubility DAG remains associated with the plasma membrane and promotes the translocation of another protein kinase, protein kinase C, from the cytosol to the plasma membrane by

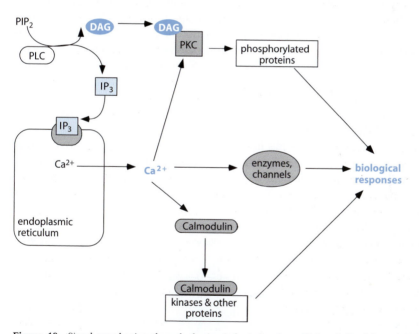

Figure 19 Signal transduction through the inositol trisphosphate (IP₃) and diacylglycerol (DAG) second messenger system. Phosphatidylinositol 4,5-bisphosphate (PIP₂) is cleaved into IP₃ and DAG by the action of a phospholipase C (PLC). DAG activates protein kinase C (PKC), which then phosphorylates a variety of proteins to produce various cell-specific effects. IP₃ binds to its receptor in the membrane of the endoplasmic reticulum, causing release of Ca²⁺, which further activates PKC, directly activates or inhibits enzymes or ion channels, or binds to calmodulin, which then binds to and activates protein kinases and other proteins.

increasing its affinity for phosphatidylserine in the membrane, and also activates it. Protein kinase C has also been called the calcium/phospholipid-dependent protein kinase because the initially discovered members of this enzyme family require both phosphatidylserine and calcium to be fully activated. The simultaneous increase in cytosolic calcium concentration resulting from the action of IP_3 complements DAG in stimulating the catalytic activity of some members of the protein kinase C family. Some members of the protein kinases C family are stimulated by DAG even when cytosolic calcium remains at resting levels.

IP_3 is cleared from cells by stepwise dephosphorylation to inositol. DAG is cleared by addition of a phosphate group to form phosphatidic acid, which may then be converted to a triglyceride or resynthesized into a phospholipid. Phosphatidylinositides of the plasma membrane are regenerated by combining inositol with phosphatidic acid, which may then undergo stepwise phosphorylation of the inositol.

The phosphatidylinositol precursor of IP_3 and DAG also contains a 20-carbon polyunsaturated fatty acid called arachidonic acid (Figure 20). This fatty acid is typically found in ester linkage with carbon 2 of the glycerol backbone of phospholipids and may be liberated by the action of a diacylglyceride lipase from the DAG formed in the breakdown of phosphatidylinositol. Liberation of arachidonic acid is the rate-determining step in the formation of the thromboxanes, the prostaglandins, and the leukotrienes (see Chapter 4). These compounds, which are produced in virtually all cells, diffuse across the plasma membrane and behave as local regulators of nearby cells. Thus the same hormone:receptor interaction that produces DAG and IP_3 as second messages to communicate with cellular organelles frequently also results in the formation of arachidonate derivatives that inform neighboring cells that a response has been initiated. It is important to recognize that phosphatidylinositol is only one of several membrane phospholipids that contain arachidonate. Arachidonic acid is also released from more abundant

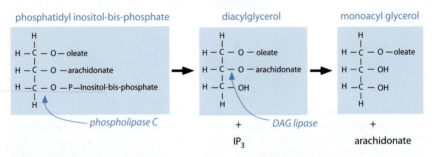

Figure 20 Diacylglycerol (DAG) is formed from phosphatidylinositol 4,5-bisphosphate by the action of phospholipase C, may be cleaved by DAG lipase to release arachidonate, the precursor of the prostaglandins and leukotrienes.

membrane phospholipids by the actions of the phospholipase A_2 class of enzymes that can be activated by calcium, by phosphorylation by protein kinase C, and by beta/gamma subunits of G-proteins.

Cyclic GMP

Though considerably less versatile than cyclic AMP, cyclic guanosine 3',5'-monophosphate (cyclic GMP) plays an analogous role in many cells. The formation of cyclic GMP from guanosine triphosphate is catalyzed by the enzyme guanylyl cyclase. Guanylyl cyclase and cyclic GMP-dependent protein kinase activities are present in many cells, but the activation of guanylyl cyclase is quite different from that of adenylyl cyclase. Guanylyl cyclase activity is an intrinsic property of the transmembrane receptor for atrial natriuretic hormone and is activated without the intercession of a G-protein. Guanylyl cyclase is also present in a soluble form within the cytoplasm of many cells and is activated by nitric oxide (NO). Increased formation of cyclic GMP in vascular smooth muscle is associated with relaxation and may account for vasodilator responses to the atrial natriuretic hormone (see Chapter 8).

Receptors That Signal through Tyrosine Kinase

Some hormones transmit their messages from the cell surface to intracellular effectors without the agency of second messengers. Receptors for these hormones rely on physical association between proteins (protein:protein interactions) to activate enzymes that phosphorylate transcription factors and other cytosolic proteins in much the same way as already discussed. The tyrosine kinase-dependent receptors have a single membrane-spanning region and either contain intrinsic protein tyrosine kinase enzymatic activity in their intracellular domains or are associated with cytosolic protein tyrosine kinases. Generally, the tyrosine kinase-dependent receptors are synthesized as dimers or form dimers when activated by their ligands. Hormone binding to the extracellular domain of the receptor activates protein tyrosine kinases that catalyze the phosphorylation of hydroxyl groups of tyrosine residues in the cytosolic portion of the receptor, in the associated kinase (autophosphorylation), or in other cytosolic proteins that complex with the phosphorylated receptor.

The protein substrates for receptor-activated tyrosine kinases may have catalytic activity or may act as scaffolds to which other proteins are recruited and positioned so that enzymatic modifications are facilitated. As a result, large signaling complexes are formed. Phosphorylated tyrosines act as docking sites for proteins that contain so-called src homology 2 (SH2) domains. SH2 domains are named for the particular configuration of the tyrosine phosphate binding region originally discovered in v-src, the cancer-inducing protein tyrosine kinase of the Rous sarcoma virus. SH2 domains represent one type of a growing list of modules within a protein that recognize and bind to specific complementary motifs in other proteins. A typical SH2 domain consists of about 100 amino acid residues and recognizes a phosphorylated tyrosine in the context of the three or four amino acid residues that are downstream from the tyrosine. There are multiple SH2 groups that recognize phosphorylated tyrosines in different contexts. Typically, multiple tyrosines are phosphorylated so that several different SH2-containing proteins are recruited and initiate multiple signaling pathways.

Although responses to activation of tyrosine kinase include modifications of cellular metabolism without nuclear participation, they often involve a change in genomic readout that promotes cell division (mitogenesis) or differentiation. One way that these receptors communicate with the genome is through activation of the mitogen-activated protein (MAP) kinase pathway (Figure 21). MAP kinase is

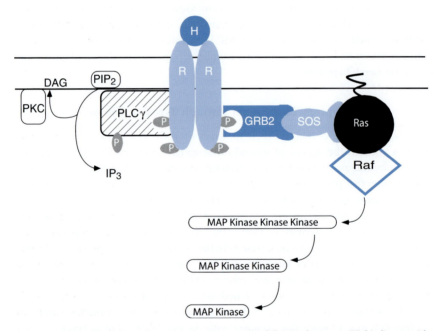

Figure 21 Phosphorylation of tyrosines on a receptor (R) following hormone (H) binding provides docking sites for the attachment of proteins that transduce the hormonal signal. The growth factor receptor binding protein 2 (GRB2) binds to a phosphorylated tyrosine in the receptor, and binds at its other end to the nucleotide exchange factor SOS, which stimulates the small G-protein Ras to exchange its GDP for GTP. Thus activated, Ras in turn activates the protein kinase Raf to phosphorylate mitogen-activated protein (MAP) kinase kinase kinase and initiate the MAP kinase cascade that ultimately phosphorylates nuclear transcription factors. The gamma isoform of phospholipase C (PLCγ) docks on the phosphorylated receptor and is then tyrosine phosphorylated and activated to cleave phosphatidylinositol 4,5-bisphosphate (PIP_2), releasing diacylglycerol (DAG) and inositol trisphosphate (IP_3) and activating protein kinase C (PKC), as shown in Fig. 19.

a cytosolic enzyme that is activated by phosphorylation of both serine and tyrosine residues and then enters the nucleus, where it phosphorylates and activates certain transcription factors. Activation of MAP kinase follows an indirect route that involves a small G-protein called Ras, which was originally discovered as a constitutively activated protein present in many tumors.

Ras proteins belong to a family of small G-proteins that function as biochemical switches to regulate processes such as entry of proteins into the nucleus, sorting and trafficking of intracellular vesicles, and cytoskeletal rearrangements. The small G-proteins are members of the same superfamily as the alpha subunit of the heterotrimeric G-proteins discussed above, but do not form complexes with beta/gamma subunits and do not interact directly with hormone receptors. The small G-proteins are GTPases that are in their active state when bound to GTP and in their inactive state when bound to GDP. Instead of liganded receptors, the small G-proteins are activated by proteins called *nucleotide exchange factors*, which cause G-proteins to dissociate from GDP and bind GTP. They remain activated as they slowly convert GTP to GDP, but inactivation can be accelerated by interaction with *GTPase-activating proteins* (GAPs).

One of the proteins that docks with phosphorylated tyrosine residues is the growth factor receptor binding protein 2 (GRB2). GRB2 is an adaptor protein that has an SH2 group at one end and other binding motifs at its opposite end, which enable it to bind other proteins, including a nucleotide exchange factor called SOS. By means of these protein:protein interactions the activated receptor can thus communicate with and activate SOS, which causes Ras to exchange GTP for GDP. The effector for the Ras that is thus activated is the enzyme Raf kinase, which phosphorylates and activates the first of a cascade of MAP kinases, ultimately resulting in phosphorylation of nuclear transcription factors.

The gamma form of phospholipase C is another effector protein that is recruited to tyrosine-phosphorylated receptors by way of its SH2 group. It is also a substrate for tyrosine kinases and is activated by tyrosine phosphorylation. Activation of this member of the phospholipase C family of proteins results in hydrolysis of phosphatidylinositol bisphosphate to produce DAG and IP_3 in the same manner as already discussed for the beta forms of the enzyme associated with G-protein-coupled receptors. In this manner tyrosine kinase-dependent receptors can stimulate cellular changes that are mediated by PKC and the calcium:calmodulin second messenger system, including phosphorylation of nuclear transcription factors by calmodulin kinase (Figure 21).

Another mechanism for modifying gene expression involves activation of a family of proteins called Stat proteins. Stat is an acronym for signal transducer and activator of transcription. The Stat proteins are transcription factors that reside in the cytosol in their inactive state. They have an SH2 group that enables them to bind to tyrosine-phosphorylated proteins. When bound to the receptor/kinase complex, Stat proteins become tyrosine phosphorylated, whereupon they dissociate from their docking sites, form homodimers, and enter the nucleus, where they activate transcription of specific genes (Figure 22).

Yet another important mediator of tyrosine kinase signaling is phosphatidylinositol-3 (PI-3) kinase, which catalyzes the phosphorylation of carbon 3 of the inositol of phosphatidylinositol bisphosphate in cell membranes to form phosphatidylinositol 3,4,5-trisphosphate (PIP_3). PI-3 kinase consists of a regulatory subunit that contains an SH2 domain and a catalytic subunit. Binding of the regulatory subunit to phosphorylated tyrosines of the receptor-associated complex activates the catalytic subunit. Activation of PI-3 kinase plays a key role in transducing signals from tyrosine kinase-dependent receptors to downstream events.

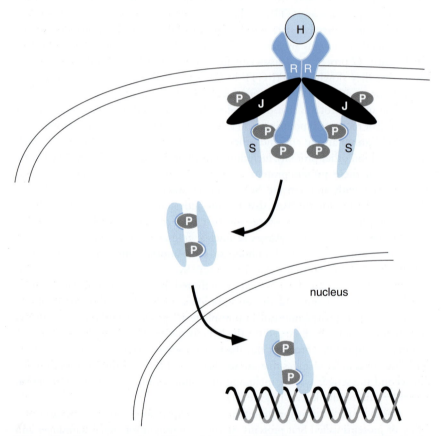

Figure 22   Hormone receptors (R) associate with the JAK family of cytosolic protein tyrosine kinases (J), and following binding of hormone and dimerization become phosphorylated on tyrosines. Proteins of the Stat family (S) of transcription factors that reside in the cytosol in the unstimulated state bind to the phosphorylated receptor, and on being tyrosine phosphorylated by JAK, dissociate from the receptor, form homodimers, and migrate to the nucleus, where they activate gene transcription.

The enzymes activated by PIP_3 are protein kinases that regulate cellular metabolism, vesicle trafficking, cytoskeletal changes, and nucleotide exchange factors that control the activity of the small G-proteins.

Integration of Hormonal Signals

As must already be quite obvious, the simple interaction of a hormone with its receptor sets in motion intracellular signaling pathways that are both intricate and complex. Most cells express receptors for multiple hormones and other signaling molecules and are simultaneously bombarded with excitatory, inhibitory, or a conflicting mixture of excitatory and inhibitory inputs from different hormones, whose signaling pathways may run in parallel, intersect, coincide, diverge, and perhaps intersect again before influencing the final effector molecules. Some signaling pathways must compete for common substrates as well as for the final effector molecules that express the final alterations in cellular behavior. Target cells must integrate all inputs by summing them algebraically, and then respond accordingly. For example, in the hepatocyte, both glucagon and epinephrine stimulate adenylyl cyclase, each by way of its own G-protein-coupled receptor. The effects of these signals combine to produce a more intense activation of adenylyl cyclase than would result from either one alone. At the same time, these cells may also be receiving some input from insulin, one of whose actions is activation of cyclic AMP phosphodiesterase, which breaks down cyclic AMP. In the pancreatic beta cell, which secretes insulin, epinephrine binds to yet another receptor, the α_2 receptor, which is coupled to adenylyl cyclase through a $G\alpha_i$, and to β receptors, which are coupled to adenylyl cyclase through a $G\alpha_s$. The two receptors thus transmit conflicting information, and, in this case, the inhibitory influence of the α_2 receptor is "stronger" and prevails. Figure 23 shows an example of how several signaling pathways might operate simultaneously.

Integration occurs at various levels along signal transduction pathways, with cross-talk among the various G-protein receptor-mediated signaling pathways, or among the various tyrosine phosphorylation-dependent pathways, and among G-protein and tyrosine kinase-mediated pathways and nuclear receptor-mediated pathways. Integration thus is not limited to the rapidly expressed responses that result from phosphorylation/dephosphorylation reactions, but can also occur at the level of gene transcription or may involve a mixture of the two. In responding simultaneously to multiple inputs, cells are able to preserve the signaling fidelity of individual hormones even when their transduction pathways appear to share common effector molecules. This is possible in part because many of the protein molecules involved in complex intracellular signal transmission do not float around freely in a cytoplasmic "soup," but are anchored at specific cellular loci by interactions with the cytoskeleton or the membranes of intracellular organelles. Protein kinase A, for example, may be localized to specific regions in the cell by specialized

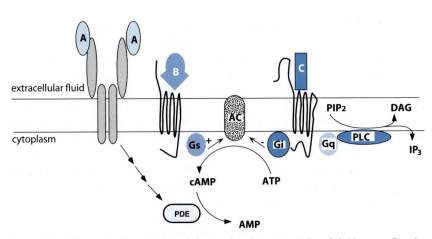

Figure 23   Cells may simultaneously receive inputs from hormones A, B, and C. Hormone B, acting through a G-protein-coupled receptor, activates adenylyl cyclase (AC) through the α stimulatory subunit (α_s, shown here as Gs). Hormone C binds to its G-protein-coupled receptor, which, through the inhibitory subunit (α_i, shown here as Gi), inhibits adenylyl cyclase, and through the α_q (Gq) subunit activates phospholipase C (PLC), resulting in the cleavage of phosphatidylinositol 4,5-bisphosphate (PIP$_2$) and the release of diacylglycerol (DAG) and inositol trisphosphate (IP$_3$). Hormone A, acting through a tyrosine kinase receptor, activates cyclic AMP phosphodiesterase (PDE), which degrades cAMP. The cell must then integrate all of these signals into an integrated response.

proteins, the A kinase anchoring proteins (AKAPs), and various forms of protein kinase C are localized by interacting with receptors for activated C kinase (RACKs).

Although there are only a few basic patterns of signal transduction, there is great versatility in their operation. Virtually all of the intracellular signaling molecules discussed are present in the human body in multiple forms, often referred to as isoforms. Isoforms may be products of separate genes, or they may arise by alternate splicing of mRNA from the same gene or by posttranslational processing. Each isoform has unique regulatory properties. Most cells express more than one of the nine isoforms of adenylyl cyclase, and although all isoforms of adenylyl cyclase are activated by any of the four different isoforms of Gα_s, some are also activated or inhibited by different isoforms of protein kinase C, or by calcium or by some of the more than 50 possible combinations of G-protein beta/gamma subunit isoforms. There are also multiple uniquely regulated isoforms of cyclic AMP phosphodiesterase, which shares responsibility with adenylyl cyclase for regulating cellular cyclic AMP concentrations. Similar statements can be made for the phospholipases C and the proteins that participate in tyrosine kinase and nuclear receptor signaling systems. Seven different proteins arise from alternate splicing of the CREB gene. In all, it has been estimated that more than 20% of the genome is devoted to signal transduction, and it is evident that there is ample complexity to account for all of the regulated behaviors governed by hormones.

REGULATION OF HORMONE SECRETION

For hormones to function as carriers of critical information, their secretion must be turned on and off at precisely the right times. The organism must have some way of knowing when there is a need for a hormone to be secreted, how much is needed, and when that need has passed. The necessary components of endocrine regulatory systems are illustrated in Figure 24. As we discuss hormonal control it is important to identify and understand the components of the regulation of each hormonal secretion because (1) derangements in any of the components are the bases of endocrine disease and (2) manipulation of any component provides an opportunity for therapeutic intervention.

NEGATIVE FEEDBACK

Secretion of most hormones is regulated by negative feedback. Negative feedback means that some consequence of hormone secretion acts directly or indirectly on the secretory cell in a negative way to inhibit further secretion. A simple example from everyday experience is the thermostat. When the temperature in a room falls below some preset level, the thermostat signals the furnace to produce

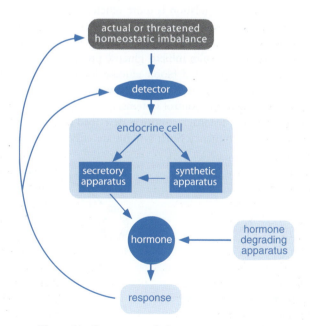

Figure 24 Components of a hormone response system.

heat. When room temperature rises to the preset level, the signal from the thermostat to the furnace is shut off, and heat production ceases until the temperature again falls. This is a simple closed-loop feedback system and is analogous to the regulation of glucagon secretion. A fall in blood glucose detected by the alpha cells of the islets of Langerhans causes them to release glucagon, which stimulates the liver to release glucose and thereby increase blood glucose concentrations (Figure 25). With restoration of blood glucose to some predetermined level, or *set point*, further secretion of glucagon is inhibited. This simple example involves only secreting cells and responding cells. Other systems may be considerably more complex and involve one or more intermediary events, but the essence of negative feedback regulation remains the same: hormones produce biological effects that directly or indirectly inhibit their further secretion.

A problem that emerges with this system of control is that the thermostat maintains room temperature constant only if the natural tendency of the temperature is to fall. If the temperature were to rise, it could not be controlled by simply turning off the furnace. This problem is at least partially resolved in hormonal systems, because at physiological set points the basal rate of secretion usually is not zero. In the example above, when there is a rise in blood glucose concentration, glucagon secretion can be diminished and therefore diminish the impetus on the liver to release glucose. Some regulation above and below the set point can therefore be accomplished with just one feedback loop; this mechanism is seen in some endocrine control systems. Regulation is more efficient, however, with a second, opposing loop, which is activated when the controlled variable deviates in the opposite direction. For the example with regulation of blood glucose, that second loop is provided by insulin. Insulin inhibits glucose production by the liver and is secreted in response to an elevated blood glucose level (Figure 26). Protection against deviation in either direction is often achieved in biological systems by the opposing actions of antagonistic control systems.

Closed-loop negative feedback control as described above can maintain conditions only in a state of constancy. Such systems are effective in guarding against

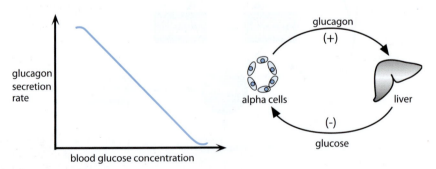

Figure 25   Negative feedback of glucose production by glucagon. (−), Inhibition; (+), stimulation.

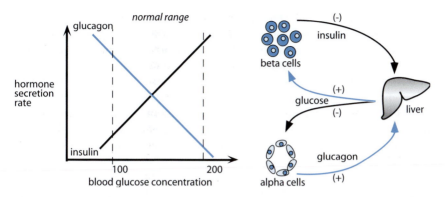

Figure 26 Negative feedback regulation of blood glucose concentration by insulin and glucagon. (−), Inhibition; (+) stimulation.

upward or downward deviations from some predetermined set point, but changing environmental demands often require temporary deviation from constancy. This can be accomplished in some cases by adjusting the set point and in other cases by a signal that overrides the set point. For example, epinephrine secreted by the adrenal medulla in response to some emergency inhibits insulin secretion and increases glucagon secretion even though the concentration of glucose in the blood may already be high. Whether the set point is changed or overridden, deviation from constancy is achieved by the intervention of some additional signal from outside the negative feedback system. In most cases that additional signal originates with the nervous system.

Hormones also initiate or regulate processes that are not limited to steady or constant conditions. Virtually all of these processes are self-limiting, and their control resembles negative feedback, but of the open-loop type. For example, oxytocin is a hormone that is secreted by hypothalamic nerve cells, the axons of which terminate in the posterior pituitary gland. Its secretion is necessary for the extrusion of milk from the lumen of the mammary alveolus into secretory ducts so that the infant suckling at the nipple can receive milk. In this case, sensory nerve endings in the nipple detect the signal and convey afferent information to the central nervous system, which in turn signals release of oxytocin from axon terminals in the pituitary gland. Oxytocin causes contraction of myoepithelial cells, resulting in delivery of milk to the infant. When the infant is satisfied, the suckling stimulus at the nipple ceases.

POSITIVE FEEDBACK

Positive feedback means that some consequence of hormonal secretion acts on the secretory cells to provide augmented drive for secretion. Rather than being self-limiting, as with negative feedback, the drive for secretion becomes progressively more intense. Positive feedback systems are unusual in biology because they terminate with some cataclysmic, explosive event. A good example of a positive feedback system involves oxytocin and its other effect: causing contraction of uterine muscle during childbirth (Figure 27). In this case the stimulus for oxytocin secretion is dilation of the uterine cervix. On receipt of this information through sensory nerves, the brain signals the release of oxytocin from nerve endings in the posterior pituitary gland. Enhanced uterine contraction in response to oxytocin results in greater dilation of the cervix, which strengthens the signal for oxytocin release, and so on until the infant is expelled from the uterine cavity.

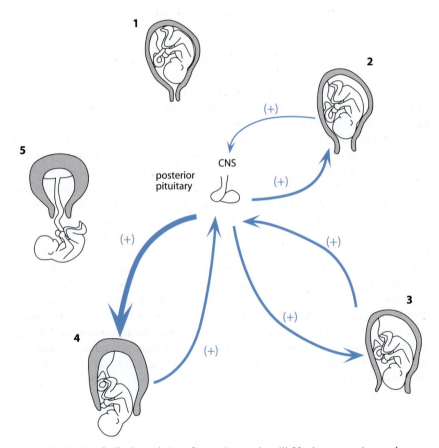

Figure 27 Positive feedback regulation of oxytocin secretion. (1) Uterine contractions at the onset of parturition apply mild stretch to the cervix. (2) In response to sensory input from the cervix, oxytocin is secreted from the posterior pituitary gland, and stimulates (+) further contraction of the uterus, which in turn stimulates secretion of more oxytocin (3), leading to further stretching of the cervix, and even more oxytocin secretion (4), until the fetus is expelled (5).

FEED FORWARD

Feed forward controls can be considered as anticipatory or preemptive and prepare the body for an impending change or demand. For example, following a meal rich in glucose, secretory cells in the mucosa of the gastrointestinal tract secrete a hormone that signals the pancreas to secrete insulin (see Chapter 5). Having increased insulin present in the blood by the time the glucose is absorbed thus moderates the change in blood glucose that might otherwise occur if insulin were secreted after the blood glucose concentrations started to increase. Unlike feedback systems, feed forward systems are unaffected by the consequences of the changes they evoke, and simply are shut off when the stimulus disappears.

MEASUREMENT OF HORMONES

Whether it is for the purpose of diagnosing a patient's disease or for research to gain understanding of normal physiology, it is often necessary to measure how much hormone is present in some biological fluid. Chemical detection of hormones in blood is difficult. With the exception of the thyroid hormones, which contain large amounts of iodine, there is no unique chemistry that sets hormones apart from other bodily constituents. Furthermore, hormones circulate in blood at minute concentrations, which further complicates the problem of their detection. Consequently, the earliest methods developed for measuring hormones are bioassays and depend on the ability of a hormone to produce a characteristic biological response. For example, induction of ovulation in the rabbit in response to an injection of urine from a pregnant woman is an indication of the presence of the placental hormone chorionic gonadotropin and is the basis for the rabbit test, which was used for many years as an indicator of early pregnancy. Before hormones were identified chemically they were quantitated in units of the biological responses they produced. For example, a unit of insulin is defined as one-third of the amount needed to lower blood sugar in a 2-kg rabbit to convulsive levels within 3 hours. Although bioassays are now seldom used, some hormones, including insulin, are still standardized in terms of biological units. Terms such as milliunits and microunits are still in use.

IMMUNOASSAYS

As knowledge of hormone structure increased, it became evident that peptide hormones are not identical in all species. Small differences in amino acid sequence, which may not affect the biological activity of a hormone, were found to produce antibody reactions with prolonged administration.

Hormones isolated from one species were recognized as foreign substances in recipient animals of another species, which often produced antibodies to the foreign hormone. Antibodies are exquisitely sensitive and can recognize and react with tiny amounts of the foreign material (antigens) that evoked their production, even in the presence of large amounts of other substances that may be similar or different. Techniques have been devised to exploit this characteristic of antibodies for the measurement of hormones, and to detect antibody–antigen reactions even when minute quantities of antigen (hormone) are involved.

Radioimmunoassay

Reaction of a hormone with an antibody results in a complex with altered properties such that it is precipitated out of solution or behaves differently when subjected to electrophoresis or adsorption to charcoal or other substances. A typical radioimmunoassay takes advantage of the fact that iodine of high specific radioactivity can be incorporated readily into tyrosine residues of peptides and proteins and thereby permits detection and quantitation of tiny amounts of hormone. Hormones present in biological fluids are not radioactive, but can compete with radioactive hormone for a limited number of antibody binding sites. To perform a radioimmunoassay, a sample of plasma containing an unknown amount of hormone is mixed in a test tube with a known amount of antibody and a known amount of radioactive iodinated hormone. The unlabeled hormone present in the plasma competes with the iodine-labeled hormone for binding to the antibody. The more hormone present in the plasma sample, the less iodinated hormone can bind to the antibody. Antibody-bound radioactive iodine is then separated from unbound iodinated hormone by any of a variety of physicochemical means, and the ratio of bound to unbound radioactivity is determined. The amount of hormone present in plasma can be estimated by comparison with a standard curve constructed using known amounts of unlabeled hormone instead of the biological fluid samples (Figure 28).

Although this procedure was originally devised for protein hormones, radioimmunoassays are now available for all of the known hormones. Production of specific antibodies to nonprotein hormones can be induced by first attaching these compounds to some protein, e.g., serum albumin. For hormones that lack a site capable of incorporating iodine, such as the steroids, another radioactive label can be used or a chemical tail containing tyrosine can be added. Methods are even available to replace the radioactive iodine with fluorescent tags or other labels that can be detected with great sensitivity.

The major limitation of radioimmunoassays is that immunological rather than biological activity is measured by these tests, because the portion of the hormone molecule recognized by the antibody probably is not the same as the portion

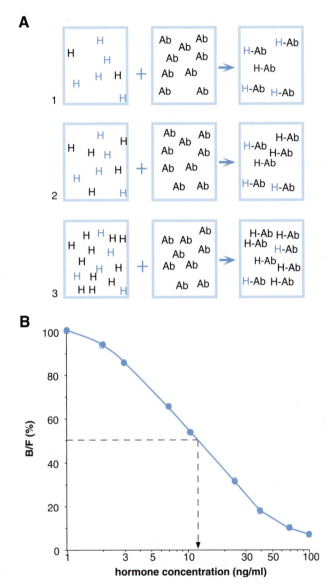

Figure 28 (A) Competing reactions that form the basis of the radioimmunoassay. Labeled hormone (H), shown in blue, competes with hormone in the biological sample (shown in black) for a limiting amount of antibodies (Ab). As the concentration of hormone in the biological sample rises (rows 1, 2, and 3) decreasing amounts of the labeled hormone appear in the hormone–antibody (H–Ab) complex and the ratio of bound/free labeled hormone (B/F) decreases. (B) A typical standard curve used to estimate the amount of hormone in the biological sample. A B/F ratio of 70% corresponds to 5 ng/ml in this example.

recognized by the hormone receptor. Thus a protein hormone that may be biologically inactive may retain all of its immunological activity. For example, the biologically active portion of parathyroid hormone resides in the amino-terminal one-third of the molecule, but the carboxyl-terminal portion formed by partial degradation of the hormone has a long half-life in blood and accounts for nearly 80% of the immunoreactive parathyroid hormone in human plasma. Until this problem was understood and appropriate adjustments were made, radioimmunoassays grossly overestimated the content of parathyroid hormone in plasma (see Chapter 8). Similarly, biologically inactive prohormones may be detected. By and large, discrepancies between biological activity and immunoactivity have not presented insurmountable difficulties and in several cases even led to increased understanding.

Immunometric Assays

Even greater sensitivity in hormone detection has been attained with the development of assays that can take advantage of exquisitely sensitive detectors that

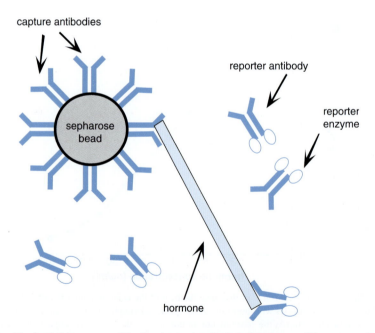

Figure 29 Sandwich-type assay. The first (capture) antibody is linked to a solid support such as an agarose bead. The hormone to be measured is shown below the bead. The second (reporter) antibody is linked to a reporter enzyme, which, on reacting with a test substrate, gives a colored product. In this model, the amount of reported antibody captured is directly proportional to the amount of hormone in the sample being tested.

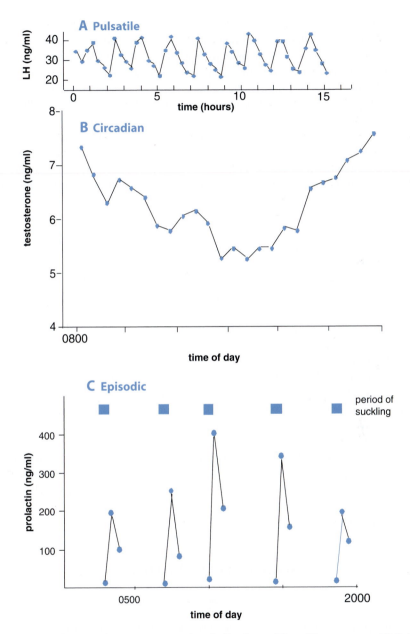

Figure 30 Changes in hormone concentrations in blood may follow different patterns. (A) Daily rhythm in luteinizing hormone (LH) secretion. (From Bremer *et al.*, *J. Clin. Endocrinol. Metab.* **56**, 1278, 1983, by permission of The Endocrine Society.) (B) Hourly rhythm of testosterone secretion. (From Yamaji *et al.*, *Endocrinology* **90**, 771, 1972, by permission of the author.) (C) Episodic secretion of prolactin. (From Hwang *et al.*, *Proc. Natl. Acad. Sci.* U.S.A. **68**, 1902, 1971, by permission of The Endocrine Society.)

can be coupled to antibodies. Such assays require the use of two different antibodies that recognize different immunological determinants in the hormone. One antibody is coupled to a solid support such as an agarose bead or is adsorbed onto the plastic of a multiwell culture dish. The biological sample containing the unknown amount of hormone is then added under conditions in which there is a large excess of antibody, so that essentially all of the hormone can be bound by the antibody. The second antibody, linked to a fluorescent probe or an enzyme that can generate a colored product, is than added and allowed to bind to the hormone that is held in place by the first antibody, so that the hormone is sandwiched between the two antibodies and acts to link them together. In this way the amount of antibody-linked detection system that is held to the solid support is directly proportional to the amount of hormone present in the test sample (Figure 29). These assays are sometimes called sandwich assays, or enzyme-linked immunosorbent assays (ELISAs), when the second antibody is coupled to an enzyme that converts a substrate to a colored product.

HORMONE LEVELS IN BLOOD

It is evident now that hormone concentrations in plasma fluctuate from minute to minute and may vary widely in the normal individual over the course of a day. Hormone secretion may be episodic, pulsatile, or follow a daily rhythm (Figure 30). In most cases it is necessary to make multiple serial measurements of hormones before a diagnosis of a hyper- or hypofunctional state can be confirmed. Endocrine disease occurs when the concentration of hormone in blood is inappropriate for the physiological situation rather than because the absolute amounts of hormone in blood are high or low. It is also becoming increasingly evident that the pattern of hormone secretion, rather than the amount secreted, may be of great importance in determining hormone responses. This subject is discussed further in Chapter 11. It is noteworthy that for the endocrine system as well as the nervous system additional information can be transmitted by the frequency of signal production as well as by the signal.

SUGGESTED READING

Conn, P. M. (ed.) (1999). "Handbook of Physiology, Section 7: Endocrinology, Volume 1: Cellular Endocrinology." American Physiological Society and Oxford University Press, New York. (This volume provides in-depth coverage of many of the topics considered in this chapter.)

Dannies, P. S. (1999). Protein hormone storage in secretory granules: Mechanisms for concentration and sorting. *Endocr. Rev.* **20**, 3–21.

Gerber, S. H., and Sidhof, T. C. (2002). Molecular determinants of regulated exocytosis. *Diabetes* **51** (Suppl. 1), S3–S11.

Gether, U. (2000). Uncovering molecular mechanisms involved in activation of G protein-coupled receptors. *Endocr. Rev.* **21**, 90–113.

McKenna, N., Rainer, J., Lanz, B., and O'Malley, B. W. (1999). Nuclear receptor coregulators: Cellular and molecular biology. *Endocr. Rev.* **20**, 321–344.

Pearson, G., Robinson, F., Beers Gibson, T., Xu, B., Karandikar, M., Berman, K., and Cobb, M. H. (2001). Mitogen-activated protein (MAP) kinase pathways: Regulation and physiological functions. *Endocr. Rev.* **22**, 153–183.

Pekary, A. E., and Hershman, J. M. (1995). Hormone assays. *In* "Endocrinology and Metabolism" (P. Felig, J. D. Baxter, and L. A. Frohman, eds.), 3rd Ed., pp. 201–218. McGraw-Hill, New York.

Pratt, W. B., and Toft, D. O. (1997). Steroid receptor interactions with heat shock protein and immunophilin chaperones. *Endocr. Rev.* **18**, 306–360.

Spiegel, A. M. (2000). G protein defects in signal transduction. *Horm. Res.* **53** (Suppl. 3), 17–22.

CHAPTER 2

Pituitary Gland

OVERVIEW

The pituitary gland has often been characterized as the "master gland" because its hormone secretions control the growth and activity of three other endocrine glands: the thyroid, adrenals, and gonads. However, because the secretory activity of the pituitary gland is also controlled by hormones which originate in either the brain or the target glands, it is perhaps better to think of the pituitary gland as the relay between the control centers in the central nervous system and the peripheral endocrine organs. The pituitary hormones are not limited in their activity to regulation of endocrine target glands; they also act directly on nonendocrine target tissues. Secretion of all of these hormones is under the control of signals arising in both the brain and the periphery.

MORPHOLOGY

The pituitary gland is located in a small depression in the sphenoid bone, the *sella turcica*, just beneath the hypothalamus, and is connected to the hypothalamus by a thin stalk called the *infundibulum*. The pituitary is a compound organ consisting of a neural or *posterior lobe* derived embryologically from the brain stem, and a larger anterior portion, the *adenohypophysis*, which derives embryologically from the primitive foregut. The cells located at the junction of the two lobes comprise the *intermediate lobe*, which is not readily identifiable as an anatomical entity in humans (Figure 1).

Histologically, the anterior lobe consists of large polygonal cells arranged in cords and surrounded by a sinusoidal capillary system. Most of the cells contain secretory granules, although some are only sparsely granulated. Based on their characteristic staining with standard histochemical dyes and immunofluorescent stains, it is possible to identify the cells that secrete each of the pituitary hormones. It was once thought that there was a unique cell type for each of the pituitary hormones, but it is now recognized that some cells may produce more than one hormone. Although particular cell types tend to cluster in central or peripheral regions of the gland, the functional significance, if any, of their arrangement within the anterior lobe is not known.

The posterior lobe consists of two major portions: the infundibulum, or stalk, and the infundibular process, or neural lobe. The posterior lobe is richly endowed with nonmyelinated nerve fibers that contain electron-dense secretory granules. The cell bodies from which these fibers arise are located in the bilaterally paired supraoptic and paraventricular nuclei of the hypothalamus. These cells

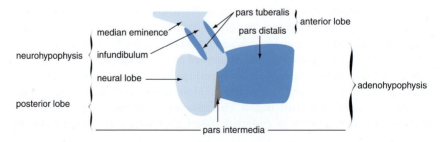

Figure 1 Midsagittal section of the human pituitary gland, indicating the nomenclature of the various parts.

are characteristically large compared to other hypothalamic neurons and hence are called *magnocellular*. Secretory material synthesized in cell bodies in the hypothalamus is transported down the axons and stored in bulbous nerve endings within the posterior lobe. Dilated terminals of these fibers lie in close proximity to the rich capillary network, which has a fenestrated endothelium that allows secretory products to enter the circulation readily.

The vascular supply and innervation of the two lobes reflect their different embryological origins and provide important clues in understanding their physiological regulation. The anterior lobe is sparsely innervated and lacks any secretomotor nerves. This fact might argue against a role for the pituitary as a relay between the central nervous system and peripheral endocrine organs, except that communication between the anterior pituitary and the brain is through vascular, rather than neural, channels.

The anterior lobe is linked to the brain stem by the hypothalamo–hypophyseal portal system, through which it receives most of its blood supply (Figure 2). The superior hypophyseal arteries deliver blood to an intricate network of capillaries, the primary plexus, in the median eminence of the hypothalamus. Capillaries of the primary plexus converge to form long hypophyseal portal vessels, which course down the infundibular stalk to deliver their blood to capillary sinusoids interspersed among the secretory cells of the anterior lobe. The inferior hypophyseal arteries supply a similar capillary plexus in the lower portion of the infundibular stem. These capillaries drain into short portal vessels, which supply a second sinusoidal capillary network within the anterior lobe. Nearly all of the blood that reaches the anterior lobe is carried in the long and short portal vessels. The anterior lobe receives only a small portion of its blood supply directly from the paired trabecular arteries, which branch off the superior hypophyseal arteries. In contrast, the circulation in the posterior pituitary is unremarkable. It is supplied with blood by the inferior hypophyseal arteries. Venous blood drains from both lobes through a number of short veins into the nearby cavernous sinuses.

The portal arrangement of blood flow is important because blood that supplies the secretory cells of the anterior lobe first drains the hypothalamus. Portal blood can thus pick up released central nervous system neuronal chemical signals and deliver them to secretory cells of the anterior pituitary. As might be anticipated, because hypophyseal portal blood flow represents only a tiny fraction of the cardiac output, when delivered in this way only minute amounts of neural secretions are needed to achieve biologically effective concentrations in pituitary sinusoidal blood. More than 1000 times more secretory material would be needed if it were dissolved in the entire blood volume and delivered through the arterial circulation. This arrangement also provides a measure of specificity to hypothalamic secretion, because pituitary cells are the only ones exposed to concentrations that are high enough to be physiologically effective.

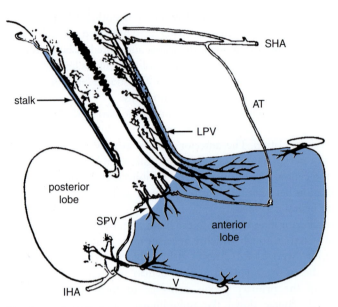

Figure 2   Vascular supply of the human pituitary gland. Note the origin of long portal vessels (LPV) from the primary capillary bed and the origin of short portal vessels (SPV) from the capillary bed in the lower part of the stalk. Both sets of portal vessels break up into sinusoidal capillaries in the anterior lobe. SHA and IHA, Superior and inferior hypophyseal arteries, respectively; AT, trabecular artery, which forms an anastomotic pathway between SHA and IHA; V, venous sinuses. (Redrawn from Daniel, P. M., and Prichard, M. M. L., *Am. Heart J.* **72**, 147, 1966, with permission.)

HORMONES OF THE ANTERIOR
PITUITARY GLAND

There are six anterior pituitary hormones for which physiological importance is clearly established. These hormones govern the functions of the thyroid and adrenal glands, the gonads, the mammary glands, and bodily growth. They have been called "trophic" or "tropic," from the Greek *trophos*, to nourish, or *tropic*, to turn toward. Both terms are generally accepted. We thus have, for example, thyrotrophin, or thyrotropin, which is also more accurately called thyroid-stimulating hormone (TSH). Because its effects are exerted throughout the body, or *soma* in Greek, growth hormone (GH) has also been called the somatotropic hormone (STH), or somatotropin. Table 1 lists the anterior pituitary hormones and their various synonyms. The various anterior pituitary cells are named for the hormones they contain. Thus we have thyrotropes, corticotropes, somatotropes, and lactotropes. Because a substantial number of growth hormone-producing cells also secrete

Table 1

Hormones of the Anterior Pituitary Gland

Hormone	Target	Major actions in humans
Glycoprotein family		
Thyroid-stimulating hormone (TSH), also called thyrotropin	Thyroid gland	Stimulates synthesis and secretion of thyroid hormones
Follicle-stimulating hormone (FSH)	Ovary	Stimulates growth of follicles and estrogen secretion
	Testis	Acts on Sertoli cells to promote maturation of sperm
Luteinizing hormone (LH)	Ovary	Stimulates ovulation of ripe follicle and formation of corpus luteum; stimulates estrogen and progesterone synthesis by corpus luteum
	Testis	Stimulates interstitial cells of Leydig to synthesize and secrete testosterone
Growth hormone/prolactin family		
Growth hormone (GH), also called somatotropic hormone (STH)	Most tissues	Promotes growth in stature and mass; stimulates production of insulin-like growth factor (IGF-I); stimulates protein synthesis; usually inhibits glucose utilization and promotes fat utilization
Prolactin	Mammary glands	Promotes milk secretion and mammary growth
Proopiomelanocortin family		
Adrenocorticotropic hormone (ACTH), also known as adrenocorticotropin or corticotropin	Adrenal cortex	Promotes synthesis and secretion of adrenal cortical hormones
β-Lipotropin	Fat	Physiological role not established

prolactin, they are called somatomammotropes. Some evidence suggests that somatomammotropes are an intermediate stage in the interconversion of somato-tropes and lactotropes. The two gonadotropins are found in a single cell type, called the gonadotrope.

All of the anterior pituitary hormones are proteins or glycoproteins. They are synthesized on ribosomes and translocated through various cellular compart-ments, where they undergo posttranslational processing. They are packaged in membrane-bound secretory granules and are secreted by exocytosis. The pituitary gland stores relatively large amounts of hormone, sufficient to meet physiological demands for many days. Over the course of many decades these hormones have been extracted, purified, and characterized for research purposes. Now even the structure of their genes is known, and we can group the anterior pituitary hormones by families.

GLYCOPROTEIN HORMONES

The glycoprotein hormone family includes TSH, whose only known physiological role is to stimulate secretion of thyroid hormone, and the two gonadotropins, follicle-stimulating hormone (FSH) and luteinizing hormone (LH). Although named for their function in women, both gonadotropic hormones are crucial for the function of testes as well as ovaries. In women FSH promotes growth of ovarian follicles and in men it promotes formation of spermatozoa by the germinal epithelium of the testis. In women LH induces ovulation of the ripe follicle and formation of the corpus luteum from remaining glomerulosa cells in the collapsed, ruptured follicle. It also stimulates synthesis and secretion of the ovarian hormones estrogen and progesterone. In men LH stimulates secretion of the male hormone, testosterone, by interstitial cells of the testis. Consequently, it has also been called interstitial cell-stimulating hormone (ICSH), but this name has largely disappeared from the literature. The actions of these hormones are discussed in detail in Chapters 11 and 12.

The three glycoprotein hormones are synthesized and stored in pituitary basophils and, as their name implies, each contains sugar moieties covalently linked to asparagine residues in the polypeptide chains. All three are composed of two peptide subunits, designated alpha and beta, which, though tightly coupled, are not covalently linked. The alpha subunit of all three hormones is identical in its amino acid sequence, and is the product of a single gene located on chromosome 6. The beta subunits of each are somewhat larger than the alpha subunit and confer physiological specificity. Both alpha and beta subunits contribute to receptor binding and both must be present in the receptor binding pocket to produce a biological response. Beta subunits are encoded in separate genes located on different chromosomes: TSH β on chromosome 1, FSH β on chromosome 11, and LH β on chromosome 19, but there is a great deal of homology in their amino acid sequences. Both subunits contain carbohydrate moieties that are considerably less constant in their composition than are their peptide chains. Alpha subunits are synthesized in excess over beta subunits, and hence it is synthesis of beta subunits that appears to be rate limiting for production of glycoprotein hormones. Pairing of the two subunits begins in the rough endoplasmic reticulum and continues in the Golgi apparatus, where processing of carbohydrate components

of the subunits is completed. The loosely paired complex then undergoes spontaneous refolding in secretory granules into a stable, active hormone.
Control of expression of the alpha and beta subunit genes is not perfectly coordinated, and free alpha and beta subunits of all three hormones may be found in
blood plasma.

The placental hormone human chorionic gonadotropin (hCG) is closely
related chemically and functionally to the pituitary gonadotropic hormones. It,
too, is a glycoprotein and consists of an alpha and a beta chain. The alpha chain is
a product of the same gene as the alpha chain of pituitary glycoprotein hormones.
The peptide sequence of the beta chain is identical to that of LH except that it is
longer by 32 amino acids at its carboxyl terminus. Curiously, although there is only
a single gene for each beta subunit of the pituitary glycoprotein hormones,
the human genome contains seven copies of the hCG beta gene, all located on
chromosome 19 in close proximity to the LH beta gene. Not surprisingly, hCG has
biological actions that are similar to those of LH, as well as a unique action on the
corpus luteum (Chapter 13).

GROWTH HORMONE AND PROLACTIN

Growth hormone is required for attainment of normal adult stature (see
Chapter 10) and produces metabolic effects that may not be directly related to its
growth-promoting actions. Metabolic effects include mobilization of free fatty
acids from adipose tissue and inhibition of glucose metabolism in muscle and
adipose tissue. (The role of GH in energy balance is discussed in Chapter 9.)
Somatotropes, which secrete GH, are by far the most abundant anterior pituitary
cells, and account for about half of the cells of the adenohypophysis. Growth
hormone is secreted throughout life and is the most abundant of the pituitary
hormones. The human pituitary gland stores between 5 and 10 mg of GH, an
amount that is 20 to 100 times greater than amounts of other anterior pituitary
hormones.

Structurally, GH is closely related to another pituitary hormone, prolactin (PRL), which is required for milk production in postpartum women (see
Chapter 13). The functions of PRL in men or nonlactating women are not firmly
established, but a growing body of evidence suggests that it may stimulate cells of
the immune system. These pituitary hormones are closely related to the placental
hormone human chorionic somatomammotropin (hCS), which has both growth-
promoting and milk-producing activity in some experimental systems. Because
of this property, hCS is also called human placental lactogen (hPL). Although the
physiological function of this placental hormone has not been established with
certainty, it may regulate maternal metabolism during pregnancy and prepare the
mammary glands for lactation (see Chapter 13).

Growth hormone, PRL, and hCS appear to have evolved from a single ancestral gene that duplicated several times—the GH and PRL genes before the emergence of the vertebrates, and the hCS and GH genes after the divergence of the primates from other mammalian groups. The human haploid genome contains two GH and three hCS genes, all located on the long arm of chromosome 17, and a single PRL gene located on chromosome 6. These genes are similar in the arrangement of their transcribed and nontranscribed portions as well as in their nucleotide sequences. All are composed of five exons separated by four introns located at homologous positions. All three hormones are large, single-stranded peptides containing two internal disulfide bridges at corresponding parts of the molecule. PRL also has a third internal disulfide bridge. GH and hCS have about 80% of their amino acids in common, and a region 146 amino acids long is similar in hGH and PRL. Only one of the GH genes (hGH N) is expressed in the pituitary, but because an alternative mode of splicing of the RNA transcript is possible, two GH isoforms are produced. The larger form is the 22-kDa molecule (22K GH), which is about 10 times more abundant than the smaller, 20-kDa molecule (20K GH), which lacks amino acids 32 to 46. The other GH gene (hGH V) appears to be expressed only in the placenta and is the predominant form of GH in the blood of pregnant women. It encodes a protein that appears to have the same biological actions as the pituitary hormone, although it differs from the pituitary hormone in 13 amino acids and also in that it may be glycosylated.

Considering the similarities in their structures, it is not surprising that GH shares some of the lactogenic activity of PRL and hCS. However, human GH also has about two-thirds of its amino acids in common with GH molecules of cattle and rats, but humans are completely insensitive to cattle or rat GH and respond only to the GH produced by humans or monkeys. This requirement of primates for primate GH is an example of species specificity and largely results from the change of a single amino acid in GH and a corresponding change of a single amino acid in the binding site in the GH receptor. Because of species specificity, human GH for research and therapy was in short supply until the advent of recombinant DNA technology, which made possible an almost limitless supply.

ADRENOCORTICOTROPIN FAMILY

Portions of the cortex of the adrenal glands are controlled physiologically by adrenal corticotropic hormone (ACTH), which is also called corticotropin or adrenocorticotropin. ACTH belongs to a family of pituitary peptides that also includes α- and β-melanocyte-stimulating hormone (MSH), β- and α-lipotropin (LPH), and β-endorphin. Of these, ACTH is the only peptide for which a physiological role in humans is established. The MSHs, which disperse melanin pigment in melanocytes in the skin of lower vertebrates, have little importance in this regard in humans and are not secreted in significant amounts. β–LPH is named for its stimulatory effect on mobilization of lipids from adipose tissue in rabbits, but the physiological importance of this action is uncertain. The 91–amino–acid chain of β–LPH contains at its carboxyl end the complete amino acid sequence of β–endorphin (from *end*ogenous m*orphin*e), which reacts with the same receptors as morphine.

The ACTH–related peptides constitute a family because (1) they contain regions of homologous amino acid sequences, which may have arisen through exon duplication, and (2) because they all arise from the transcription and translation of the same gene (Figure 3). The gene product is proopiomelanocortin (POMC), which consists of 239 amino acids after removal of the signal peptide. The molecule contains 10 doublets of basic amino acids (arginine and lysine in various combinations), which are potential sites for cleavage by trypsin–like

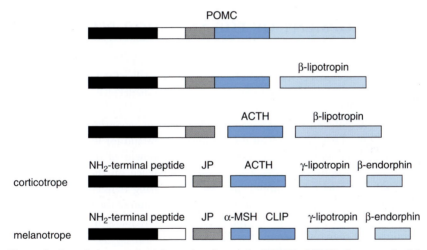

Figure 3   Proteolytic processing of proopiomelanocortin (POMC). POMC after removal of the signal peptide is shown on the first line. The first cleavage by prohormone convertase 1 releases β–lipotropin. The second cleavage releases ACTH. A third cleavage releases the joining peptide (JP) to produce the principal secretory products of the corticotropes of the anterior pituitary gland. Third and fourth cleavages take place in the melanotropes of the intermediate lobe and split ACTH into α–melanocyte–stimulating hormone (α–MSH) and the corticotropin–like intermediate lobe peptide (CLIP), and divide β–lipotropin into γ–lipotropin and β–endorphin. Some cleavage of β–lipotropin also takes place in the corticotrope. Additional posttranslational processing (not shown) includes removal of the carboxyl–terminal amino acid from each of the peptides, and glycosylation and phosphorylation of some of the peptide fragments. In neural tissue the NH$_2$–terminal peptide, depicted by the clear area, is also released to produce γ$_3$–MSH.

endopeptidases, called *prohormone convertases*. POMC is expressed by cells in the anterior lobe of the pituitary and in the intermediate lobe, and by various cells in the central nervous system, but tissue-specific differences in the way the molecule is processed after translation give rise to differences in the final secretory products. More than seven different enzymes carry out these posttranslational modifications. The predominant products of human corticotropes are ACTH and β-LPH. Because final processing of POMC occurs in the secretory granule, β-LPH is secreted along with ACTH. Cleavage of β-LPH also occurs to some extent in human corticotropes, so that some β-endorphin may also be released, particularly when ACTH secretion is brisk. The intermediate lobe in some animals gives rise principally to α- and β-MSH. Because the intermediate lobe of the pituitary gland of humans is thought to be nonfunctional except perhaps in fetal life, it is not discussed further here. Some of the POMC peptides produced in hypothalamic neurons may play an important role in regulating food intake (see Chapter 9) and in coordinating overall responses to stress.

DEVELOPMENT OF THE ANTERIOR PITUITARY

The various cell types of the anterior pituitary arise from a common primordium whose initial development begins when the cells of the oral ectoderm of Rathke's pouch come in contact with the cells of the diencephalon. Expression of several regionally specific transcription factors in different combinations appears to determine the different cellular lineages. Deficiencies in expression of two of these factors account for several mutant dwarf mouse strains and for human syndromes of combined pituitary hormone deficiency. Development of thyrotropes, lactotropes, and somatotropes shares a common dependence on the homeodomain transcription factors called prop-1 and pit-1. Appearing transiently early in the development process, prop-1 appears to foretell expression of the pituitary-specific pit-1, and its name derives from "prophet of pit-1." The transcription factor pit-1 is required not only for differentiation of these cell lineages, but also for continued expression of GH, PRL, and the beta subunit of TSH throughout life; pit-1 also regulates expression of the receptor for the hypothalamic hormone that controls GH synthesis and secretion. Genetic absence of pit-1 results in failure of the somatotropes, lactotropes, and thyrotropes to develop and hence absence of GH, PRL, and TSH. Absence of prop-1 results in deficiencies of these three hormones as well as deficiencies in gonadotropin production. Cells destined to become corticotropes and gonadotropes depend on expression of combinations of other transcription factors, as is also true for the divergence of the pit-1-dependent cell types into their mature phenotypes. A detailed consideration of pituitary organogenesis is beyond the scope of this text, but can be found in the article by Anderson and Rosenfeld cited at the end of this chapter.

REGULATION OF ANTERIOR PITUITARY FUNCTION

Secretion of the anterior pituitary hormones is regulated by the central nervous system and by hormones produced in peripheral target glands. Input from the central nervous system provides the primary drive for secretion and peripheral input plays a secondary, though vital, role in modulating secretory rates. Secretion of all of the anterior pituitary hormones except PRL declines severely in the absence of stimulation from the hypothalamus, as can be produced, for example, when the pituitary gland is removed surgically from its natural location and reimplanted at a site remote from the hypothalamus. In contrast, PRL secretion is dramatically increased. The persistent high rate of secretion of PRL under these circumstances indicates not only that the pituitary glands can revascularize and survive in a new location but also that PRL secretion is normally under tonic inhibitory control by the hypothalamus.

Secretion of each of the anterior pituitary hormones follows a diurnal pattern entrained by activity, sleep, or light–dark cycles. Secretion of each of these hormones also occurs in a pulsatile manner, probably reflecting synchronized pulses of hypothalamic neurohormone release into hypophyseal portal capillaries. Pulse frequency varies widely, from about two pulses per hour for ACTH to one pulse every 3 or 4 hours for TSH, GH, and PRL. Modulation of secretion in response to changes in the internal or external environment may be reflected as changes in the amplitude or frequency of secretory pulses, or by episodic bursts of secretion. In this chapter we discuss only general aspects of the regulation of anterior pituitary function. A detailed description of the control of the secretory activity of each hormone is given in subsequent chapters in conjunction with a discussion of its role in regulating physiological processes.

HYPOPHYSIOTROPIC HORMONES

As already mentioned, the central nervous system communicates with the anterior pituitary gland by means of neurosecretions released into the hypothalamo–hypophyseal portal system. These neurosecretions, called *hypophysiotropic hormones*, are listed in Table 2. The fact that only small amounts of the hypophysiotropic hormones are synthesized, stored, and secreted frustrated efforts to isolate and identify them for nearly 25 years. Their abundance in the hypothalamus is less than 0.1% of that of even the scarcest pituitary hormone in the anterior lobe.

THYROTROPIN RELEASING HORMONE

Thyrotropin-releasing hormone (TRH), the first of the hypothalamic neurohormones to be characterized, was found to be a tripeptide. It was isolated,

Table 2

Hypophysiotropic Hormones

Hormone	Amino acids	Hypothalamic source	Physiological actions on the pituitary
Corticotropin-releasing hormone (CRH)	41	Parvoneurons of the paraventricular nuclei	Stimulates secretion of ACTH and β-lipotropin
Gonadotropin-releasing hormone (GnRH), originally called luteinizing hormone-releasing hormone (LHRH)	10	Arcuate nuclei	Stimulates secretion of FSH and LH
Growth hormone-releasing hormone (GHRH)	44	Arcuate nuclei	Stimulates GH secretion
Growth hormone-releasing peptide (ghrelin)	28	?	Increases response to GHRH and may directly stimulate GH secretion
Somatotropin release-inhibiting factor (SRIF); somatostatin	14 or 28	Anterior hypothalamic periventricular system	Inhibits secretion of GH
Prolactin-stimulating factor (?)	?	?	Stimulates prolactin secretion (?)
Prolactin-inhibiting factor (PIF)	?	Dopamine secretion; tuberohypophyseal neurons	Inhibits prolactin secretion
Thyrotropin-releasing hormone (TRH)	3	Parvoneurons of the paraventricular nuclei	Stimulates secretion of TSH and prolactin
Arginine vasopressin (AVP)	9	Parvoneurons of the paraventricular nuclei	Acts in concert with CRH to stimulate secretion of ACTH

identified, and synthesized almost simultaneously in the laboratories of Roger Guillemin and Andrew Schally, who were subsequently recognized for this monumental achievement with the award of a Nobel Prize. Guillemin's laboratory processed 25 kg of sheep hypothalami to obtain 1 mg of TRH. Schally's laboratory extracted 245,000 pig hypothalami to yield only 8.2 mg of this tripeptide. The human TRH gene encodes a 242-residue preprohormone molecule that contains six copies of TRH. TRH is synthesized primarily in parvocellular (small) neurons in the paraventricular nuclei of the hypothalamus, and is stored in nerve terminals in the median eminence. TRH is also expressed in neurons widely dispersed throughout the central nervous system and probably acts as a neurotransmitter that mediates a variety of other responses. Actions of TRH that regulate TSH secretion and thyroid function are discussed further in Chapter 3.

GONADOTROPIN RELEASING HORMONE

Gonadotropin-releasing hormone (GnRH) was the next hypophysiotropic hormone to be isolated and characterized. Hypothalamic control over secretion of both FSH and LH is exerted by this single hypothalamic decapeptide. Endocrinologists originally had some difficulty accepting the idea that both gonadotropins are under the control of a single hypothalamic releasing hormone, because FSH and LH appear to be secreted independently under certain circumstances. Most endocrinologists have now abandoned the idea that there must be separate FSH- and LH-releasing hormones, because other factors can account for partial independence of LH and FSH secretion. The frequency of pulses of GnRH release determines the ratio of FSH and LH secreted. In addition, target glands secrete hormones that selectively inhibit secretion of either FSH or LH. These complex events are discussed in detail in Chapters 11 and 12.

The GnRH gene encodes a 92-amino-acid preprohormone that contains the 10-amino-acid GnRH peptide and an adjacent 56-amino-acid GnRH-associated peptide (GAP), which may also have some biological activity. GAP is found with GnRH in nerve terminals and may be secreted along with GnRH. Cell bodies of the neurons that release GnRH into the hypophysial portal circulation reside primarily in the arcuate nucleus in the anterior hypothalamus, but GnRH-containing neurons are also found in the preoptic area and project to extrahypothalamic regions, where GnRH release may be related to various aspects of reproductive behavior. GnRH is also expressed in the placenta. Curiously, humans and some other species have a second GnRH gene, but it is expressed elsewhere in the brain and appears to have no role in gonadotropin release.

GROWTH HORMONE RELEASING HORMONE, SOMATOSTATIN, AND GHRELIN

Growth hormone secretion is controlled by the growth hormone-releasing hormone (GHRH) and a GH release-inhibiting hormone, somatostatin, which is also called somatotropin release-inhibiting factor (SRIF). In addition, a newly discovered peptide called *ghrelin* may act both on the somatotropes, to increase GH secretion by augmenting the actions of GHRH, and on the hypothalamus, to increase secretion of GHRH. The physiological role of this novel peptide, which is synthesized both in the hypothalamus and in the stomach, has not been established. GHRH is a member of a family of gastrointestinal and neurohormones that includes vasoactive intestinal peptide (VIP), glucagon (see Chapter 5), and the probable ancestral peptide in this family, pituitary adenylate cyclase-activating peptide (PACAP). The GHRH-containing neurons are found predominantly in the arcuate nuclei, and to a lesser extent in the ventromedial nuclei of the hypothalamus. Curiously, GHRH was originally isolated from a pancreatic tumor, and is normally expressed in the pancreas, the intestinal tract, and other tissues, but the physiological role of extrahypothalamically produced GHRH is unknown.

Somatostatin was originally isolated from hypothalamic extracts based on its ability to inhibit GH secretion. The somatostatin gene codes for a 118-amino-acid preprohormone from which either a 14-amino-acid or a 28-amino-acid form of somatostatin is released by proteolytic cleavage. Both forms are similarly active. The remarkable conservation of the amino acid sequence of the somatostatin precursor and the presence of processed fragments that accompany somatostatin in hypothalamic nerve terminals have suggested to some investigators that additional physiologically active peptides may be derived from the somatostatin gene. The somatostatin gene is widely expressed in neuronal tissue as well as in the pancreas (see Chapter 5) and in the gastrointestinal tract. The somatostatin that regulates GH secretion originates in neurons present in the preoptic, periventricular, and paraventricular nuclei. It appears that somatostatin is secreted nearly continuously, and restrains GH secretion except during periodic brief episodes that coincide with increases in GHRH secretion. Coordinated episodes of decreased somatostatin release and increased GHRH secretion produce a pulsatile pattern of GH secretion.

CORTICOTROPIN RELEASING HORMONE, ARGININE VASOPRESSIN, AND DOPAMINE

Corticotropin-releasing hormone (CRH) is a 41-amino-acid polypeptide derived from a preprohormone of 192 amino acids. CRH is present in greatest abundance in the parvocellular neurons in the paraventricular nuclei, the axons of which project to the median eminence. About half of these cells also express arginine vasopressin (AVP), which also acts as a corticotropin-releasing hormone. AVP has other important physiological functions and is a hormone of the posterior pituitary gland (see below). The wide distribution of CRH-containing neurons in the central nervous system suggests that CRH has other actions besides regulation of ACTH secretion.

The simple monoamine neurotransmitter dopamine appears to satisfy most of the criteria for a PRL inhibitory factor, the existence of which was suggested by the persistent high rate of PRL secretion by pituitary glands transplanted outside the sella turcica. It is likely that there is also a PRL-releasing hormone, but although several candidates have been proposed, general agreement on its nature or even its existence is still lacking.

SECRETION OF HYPOPHYSIOTROPIC HORMONES

Although, in general, the hypophysiotropic hormones affect the secretion of one or another pituitary hormone specifically, TRH can increase the secretion of PRL at least as well as it increases the secretion of TSH. The physiological meaning of this experimental finding is not understood. Under normal physiological conditions, PRL and TSH appear to be secreted independently, and increased PRL secretion is not necessarily seen in circumstances that call for increased TSH secretion. However, in laboratory rats and possibly in human beings as well, suckling at the breast increases both PRL and TSH secretion in a manner suggestive of increased TRH secretion. In the normal individual, somatostatin may inhibit secretion of other pituitary hormones in addition to GH, but again the physiological significance of this action is not understood. With disease states, specificity of responses of various pituitary cells for their own hypophysiotropic hormones may break down, or cells might even begin to secrete their hormones autonomously.

The neurons that secrete the hypophysiotropic hormones are not autonomous. They receive input from many structures within the brain as well as from circulating hormones. Neurons that are directly or indirectly excited by actual or impending changes in the internal or external environment, from emotional changes, and from generators of rhythmic activity, signal to hypophysiotropic neurons by means of classical neurotransmitters as well as by neuropeptides. In addition, neuronal activity is modulated by hormonal changes in the general circulation. Integration of responses to all of these signals may take place in the hypophysiotropic neurons or information may be processed elsewhere in the brain and relayed to the hypophysiotropic neurons. Conversely, hypophysiotropic neurons or neurons that release hypophysiotropic peptides as their neurotransmitters communicate with other neurons dispersed throughout the central nervous system to produce responses that presumably are relevant to the physiological circumstances that call forth pituitary hormone secretion.

Hypophysiotropic hormones increase both secretion and synthesis of pituitary hormones. All appear to act through stimulation of G-protein-coupled receptors (see Chapter 1) on the surfaces of anterior pituitary cells to increase the formation of cyclic AMP or inositol trisphosphate/diacylglyceride second messenger systems. Release of hormone almost certainly is the result of an influx of calcium, which triggers and sustains the process of exocytosis. The actions of hypophysiotropic hormones on their target cells in the pituitary are considered further in later chapters.

FEEDBACK CONTROL OF ANTERIOR PITUITARY FUNCTION

We have already indicated that the primary drive for secretion of all of the anterior pituitary hormones except PRL is stimulation by the hypothalamic releasing hormones. In the absence of the hormones of their target glands, secretion of TSH, ACTH, and the gonadotropins gradually increases manyfold. Secretion of these pituitary hormones is subject to negative feedback inhibition by secretions of their target glands. Regulation of secretion of anterior pituitary hormones in the normal individual is achieved through the interplay of stimulatory effects of releasing hormones and inhibitory effects of target gland hormones (Figure 4). Regulation of the secretion of pituitary hormones by hormones of target glands could be achieved equally well if negative feedback signals acted at the level of (1) the hypothalamus, to inhibit secretion of hypophysiotropic hormones, or (2) the pituitary gland, to blunt the response to hypophysiotropic stimulation. Actually, some combination of the two mechanisms applies to all of the anterior pituitary hormones except PRL.

In experimental animals it appears that secretion of GnRH is variable and highly sensitive to environmental influences, e.g., day length, or even the act of mating. In humans and other primates secretion of GnRH after puberty appears to be somewhat less influenced by changes in the internal and external environment, but there is ample evidence that GnRH secretion is modulated by factors in both the internal and the external environments. It has been shown experimentally in rhesus monkeys and human subjects that all the complex changes in the rates of FSH and LH secretion characteristic of the normal menstrual cycle can occur when the pituitary gland is stimulated by pulses of GnRH delivered at an invariant frequency and amplitude. For such changes in pituitary secretion to occur, changes in secretion of target gland hormones that accompany ripening of the follicle, ovulation, and luteinization must modulate the responses of gonadotropes to GnRH (Chapter 12). However, in normal humans it is evident

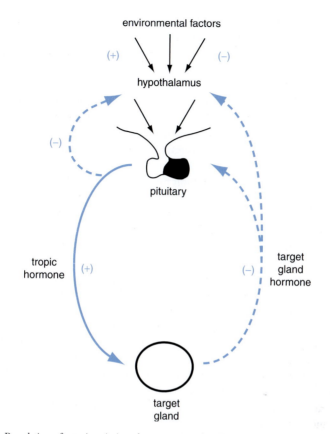

Figure 4   Regulation of anterior pituitary hormone secretion. Environmental factors may increase or decrease pituitary activity by increasing or decreasing hypophysiotropic hormone secretion. Pituitary secretions increase the secretion of target gland hormones, which may inhibit further secretion by acting at either the hypothalamus or the pituitary. Pituitary hormones may also inhibit their own secretion by a short feedback loop. (From Goodman, H. M., "Medical Physiology," 14th Ed. Mosby, St. Louis, 1980, with permission.)

that target gland hormones also act at the level of the hypothalamus to regulate both the amplitude and frequency of GnRH secretory bursts.

Adrenal cortical hormones exert a negative feedback effect both on pituitary corticotropes, where they decrease transcription of POMC, and on hypothalamic neurons, where they decrease CRH synthesis and secretion (see Chapter 4). The rate of CRH secretion is also profoundly affected by changes in the internal and external environments. Physiologically, CRH is secreted in increased amounts in response to nonspecific stress. This effect is seen even in the absence of the adrenal glands, and hence the inhibitory effects of its hormones, indicating that CRH secretion must be controlled by positive inputs as well as by the negative effects of adrenal hormones.

Control of GH secretion is more complex because it is under the influence of a releasing hormone, GHRH, probably a release-enhancing hormone, ghrelin, and a release-inhibiting hormone. In addition, GH secretion is under negative feedback control by products of its actions in peripheral tissues. As is discussed in detail in Chapter 10, GH evokes production of a peptide, called insulin-like growth factor (IGF), which mediates the growth-promoting actions of GH. IGF exerts powerful inhibitory effects on GH secretion by decreasing the sensitivity of soma-totropes to GHRH. It also acts on the somatostatin-secreting neurons to increase the release of somatostatin and to inhibit the release of GHRH. GH is also a meta-bolic regulator, and products of its metabolic activity, such as increased glucose or free fatty acid concentrations in blood, may also inhibit its secretion.

Modulating effects of target gland hormones on the pituitary gland are not limited to inhibiting secretion of their own provocative hormones. Target gland hormones may modulate pituitary function by increasing the sensitivity of other pituitary cells to their releasing factors or by increasing the synthesis of other pituitary hormones. Hormones of the thyroid and adrenal glands are required for normal responses of the somatotropes to GHRH. Similarly, estrogen secreted by the ovary in response to FSH and LH increases PRL synthesis and secretion.

In addition to feedback inhibition exerted by target gland hormones, there is evidence that pituitary hormones may inhibit their own secretion. In this so-called short-loop feedback system, pituitary cells respond to increased concen-trations of their own hormones by decreasing further secretion. The physiological importance of short-loop feedback systems has not been established, nor has that of the postulated ultrashort-loop feedback, in which high concentrations of hypophysiotropic hormones may inhibit their own release.

PHYSIOLOGY OF THE POSTERIOR PITUITARY

OXYTOCIN AND VASOPRESSIN

The posterior pituitary gland secretes two hormones. They are oxytocin, which means "rapid birth," in reference to its action to increase uterine contrac-tions during parturition, and vasopressin, in reference to its ability to contract vascular smooth muscle and thus increase blood pressure. Because the human hormone has an arginine in position 8 instead of the lysine found in the corres-ponding hormone originally isolated from pigs, it is called arginine vasopressin. Both hormones are nonapeptides that have a disulfide bond linking cysteines at positions 1 and 6 to form a ring with six amino acids and a three-amino-acid side chain. They differ by only two amino acids. Similarities in the structure and organization of their genes and in their posttranslational processing make it virtually certain that oxytocin and vasopressin evolved from a single ancestral gene. The genes that encode them occupy adjacent loci on chromosome 20, but in opposite transcriptional orientation.

Each of the posterior pituitary hormones has other actions in addition to the action for which it was named. Oxytocin also causes contraction of the myoepithelial cells that envelop the secretory alveoli of the mammary glands and thus enables the suckling infant to receive milk. AVP is also called antidiuretic hormone (ADH) for its action to promote reabsorption of "free water" by renal tubules (see Chapter 7). These effects of oxytocin and AVP are mediated by different heptihelical receptors that are coupled to different G-protein-dependent second messenger systems. V_1 receptors signal vascular muscle contraction by means of the inositol trisphosphate/diacylglycerol pathway (Chapter 1), whereas V_2 receptors utilize the cyclic AMP system to produce the antidiuretic effect in renal tubules. Oxytocin acts through a single class of G-protein receptors that signal through the inositol trisphosphate/diacylglycerol pathway. Physiological actions of these hormones are considered further in Chapters 7 and 13.

Oxytocin and AVP are stored in and secreted by the neurohypophysis, but are synthesized in magnocellular neurons, the cell bodies of which are present in both the supraoptic and paraventricular nuclei of the hypothalamus. Cells in the supraoptic nuclei appear to be the major source of neurohypophyseal vasopressin, whereas cells in the paraventricular nuclei may be the principal source of oxytocin. After transfer to the Golgi apparatus the oxytocin and AVP prohormones are packaged in secretory vesicles along with the enzymes that cleave them into the final secreted products. The secretory vesicles are then transported down the axons to the nerve terminals in the posterior gland (Figure 5), where they are stored in relatively large amounts. It has been estimated that sufficient AVP is stored in the neurohypophysis to provide for 30–50 days of secretion at basal rates or 5–7 days at maximal rates of secretion. Oxytocin and AVP are stored as 1:1 complexes with 93- to 95-residue peptides called *neurophysins*, which actually are adjacent segments of their prohormone molecules. The neurophysins are cosecreted with AVP or oxytocin, but have no known hormonal actions. The neurophysins, however, play an essential role in the posttranslational processing of the neurohypophyseal hormones. The amino acid sequence of the central portion of the neurophysins is highly conserved across many vertebrate species, and mutations in this region of the preprohormone are responsible for hereditary deficiencies in AVP that produce the disease diabetes insipidus (see Chapter 7), even though expression of the AVP portion of the preprohormone is normal.

As already discussed, AVP is also synthesized in small cells of the paraventricular nuclei and is delivered by the hypophyseal portal capillaries to the anterior pituitary gland, playing a role in regulating ACTH secretion. AVP is produced in considerably larger amounts in the magnocellular neurons and is carried directly

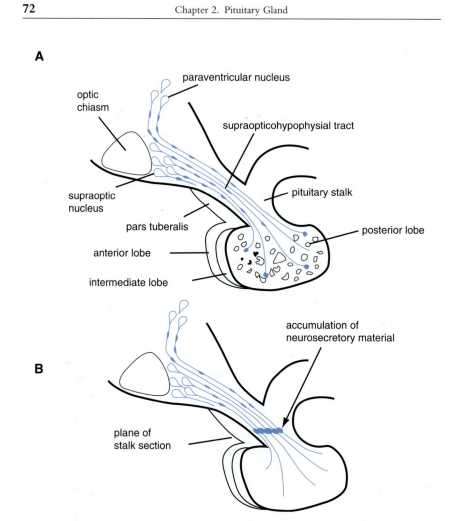

Figure 5 Hypothalamic–neurohypophyseal relations, showing the distribution of secretory granules before (A) and after (B) stalk section. Disappearance of neurosecretory granules from the posterior lobe and their accumulation proximal to the site of stalk section provide definitive evidence that the secretory material stored in the posterior lobe originates in hypothalamic neurons. (From Scharrer, E., and Scharrer, B., *Recent Prog. Horm. Res.* **10**, 193–240, 1954, with permission.)

into the general circulation by the veins that drain the posterior lobe. It is unlikely that AVP that originates in magnocellular neurons acts as a hypophysiotropic hormone, but it can reach the corticotropes and stimulate ACTH secretion when its concentration in the general circulation increases sufficiently. Oxytocin, like AVP, may also be synthesized in parvocellular neurons at other sites in the nervous system and may be released from axon terminals that project to a wide range of sites within the central nervous system. Oxytocin may also be produced in some reproductive tissues, where it acts as a paracrine factor.

REGULATION OF POSTERIOR PITUITARY FUNCTION

Because the hormones of the posterior pituitary gland are synthesized and stored in nerve cells, it should not be surprising that their secretion is controlled in

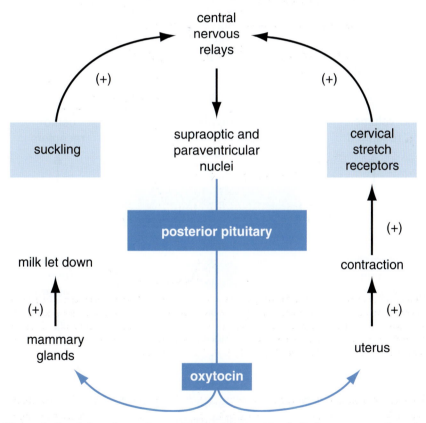

Figure 6 Regulation of oxytocin secretion, showing a positive feedback arrangement. Oxytocin stimulates the uterus to contract and causes the cervix to stretch. Increased cervical stretch is sensed by neurons in the cervix and they signal the hypothalamus, which signals more oxytocin secretion. Oxytocin secreted in response to suckling forms an open-loop feedback system in which positive input is interrupted when the infant is satisfied and stops suckling. Further details are given in Chapter 13.

the same way as that of more conventional neurotransmitters. Action potentials that arise from synaptic input to the cell bodies within the hypothalamus course down the axons in the pituitary stalk, trigger an influx of calcium into nerve terminals, and release the contents of neurosecretory granules. Vasopressin and oxytocin are released along with their respective neurophysins, other segments of the precursor molecule, and presumably the enzymes responsible for cleavage of the precursor. Tight binding of AVP and oxytocin to their respective neurophysins is favored by the acidic pH of the secretory granule, but on secretion, the higher pH of

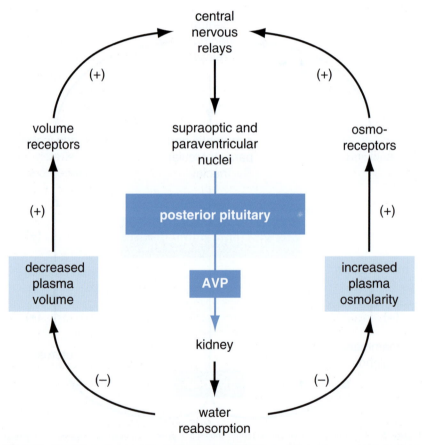

Figure 7 Regulation of vasopressin secretion. Increased blood osmolality or decreased blood volume is sensed in the brain or thorax, respectively, increasing vasopressin secretion. Vasopressin, acting principally on the kidney, produces changes that restore osmolality and volume, thereby shutting down further secretion in a negative feedback arrangement. Further details are given in Chapter 7.

the extracellular environment allows the hormones to dissociate from their neurophysins and to circulate in an unbound form. Oxytocin and vasopressin are rapidly cleared from the blood with a half-life of about 2 minutes.

As discussed in Chapter 1, signals for the secretion of oxytocin originate in the periphery and are transmitted to the brain by sensory neurons. After appropriate processing in higher centers, cells in the supraoptic and paraventricular nuclei are signaled to release their hormone from nerve terminals in the posterior pituitary gland (Figure 6). The importance of neural input to oxytocin secretion is underscored by the observation that it may also be secreted in a conditioned reflex. A nursing mother sometimes releases oxytocin in response to cries of her baby even before the infant begins to suckle. Oxytocin is also secreted at the same low basal rate in both men and women, but no physiological role for oxytocin in men and nonlactating women has been established.

Signals for vasopressin secretion in response to increased osmolality of the blood are thought to originate in hypothalamic neurons, and possibly in the AVP secretory cells. Osmoreceptor cells are exquisitely sensitive to small increases in osmolality and signal the cells in the paraventricular and supraoptic nuclei to secrete AVP. AVP is also secreted in response to decreased blood volume. Although the specific cells responsible for monitoring blood volume have not been identified, volume monitors appear to be located within the thorax and relay their information to the central nervous system in afferent neurons in the vagus nerves. The control of AVP secretion is shown in Figure 7, and is discussed more fully in Chapter 7. Under basal conditions blood levels of AVP fluctuate in a diurnal rhythmic pattern that closely resembles that of ACTH.

SUGGESTED READING

Andersen, B., and Rosenfeld, M. G. (2001). POU domain factors in the neuroendocrine system: Lessons from developmental biology provide insights into human disease. *Endocr. Rev.* **22**, 2–35.

Eipper, B. A., and Mains, R. E. (1980). Structure and biosynthesis of Pro-ACTH/endorphin and related peptides. *Endocr. Rev.* **1**, 1–27.

Fiddes, J. C., and Talmadge, K. (1984). Structure, expression, and evolution of the genes for the human glycoprotein hormones. *Recent Prog. Horm. Res.* **40**, 43–74.

Gharib, S. D., Wierman, M. E., Shupnik, M. A., and Chin, W. W. (1990). Molecular biology of the pituitary gonadotropins. *Endocr. Rev.* **11**, 177–199.

Ling, N., Zeytin, F., Böhlen, P., Esch, F., Brazeau, P., Wehrenberg, W. B., Baird, A., and Guillemin, R. (1985). Growth hormone releasing factors. *Annu. Rev. Biochem.* **54**, 404–424.

Miller, W. L., and Eberhardt, N. L. (1983). Structure and evolution of the growth hormone gene family. *Endocr. Rev.* **4**, 97–130.

Osorio, M., Kopp, P., Marui, S., Latronico, A. C., Mendonca, B. B., and Arnhold, I. J. (2000). Combined pituitary hormone deficiency caused by a novel mutation of a highly conserved residue (F88S) in the homeodomain of PROP-1. *J. Clin. Endocrinol. Metab.* **85**, 2779–2785.

Palkovits, M. (1999). Interconnections between the neuroendocrine hypothalamus and the central autonomic system. *Front. Neuroendocrinol.* **20**, 270–295.

Shupnik, M. A., Ridgway, E. C., and Chin, W. W. (1989). Molecular biology of thyrotropin. *Endocr. Rev.* **10**, 459–475.

Vale, W., Rivier, C., Brown, M. R., Spiess, J., Koob, G., Swanson, L., Bilezikjian, L., Bloom, F., and Rivier, J. (1984). Chemical and biological characterization of corticotropin releasing factor. *Recent Prog. Horm. Res.* **40**, 245–270.

Vitt, U. A., Hsu, S. Y., and Hsueh, A. J. (2001). Evolution and classification of cystine knot-containing hormones and related extracellular signaling molecules. *Mol. Endocrinol.* **15**, 681–694.

CHAPTER 3

Thyroid Gland

OVERVIEW

In the adult human, normal operation of a wide variety of physiological processes affecting virtually every organ system requires appropriate amounts of

thyroid hormone. Governing all of these processes, thyroid hormone acts as a mod-
ulator, or gain control, rather than an all-or-none signal that turns the process on
or off. In the immature individual, thyroid hormone plays an indispensable role in
growth and development. Its presence in optimal amounts at a critical time is an
absolute requirement for normal development of the nervous system. In its role in
growth and development too, its presence seems to be required for the normal
unfolding of processes whose course it modulates but does not initiate. Because
thyroid hormone affects virtually every system in the body in this way, it is difficult to give a simple, concise answer to the naive but profound question: What does
thyroid hormone do? The response of most endocrinologists would be couched in
terms of consequences of hormone excess or deficiency. Indeed, deranged function of the thyroid gland is among the most prevalent of endocrine diseases and
may affect as many as 4–5% of the population in the United States. In regions of
the world where the trace element iodine is scarce, the incidence of deranged
thyroid function may be even higher.

MORPHOLOGY

The human thyroid gland is located at the base of the neck and wraps around
the trachea just below the cricoid cartilage (Figure 1). The two large lateral lobes that
comprise the bulk of the gland lie on either side of the trachea and are connected
by a thin isthmus. A third structure, the pyramidal lobe, which may be a remnant of
the embryonic thyroglossal duct, is sometimes also seen as a fingerlike projection
extending headward from the isthmus. The thyroid gland in the normal human
being weighs about 20 g but is capable of enormous growth, sometimes achieving
a weight of several hundred grams when stimulated intensely over a long period of
time. Such enlargement of the thyroid gland, which may be grossly obvious, is called
a goiter, and is one of the most common manifestations of thyroid disease.

The thyroid gland receives its blood supply through the inferior and superior thyroid arteries, which arise from the external carotid and subclavian arteries.
Relative to its weight, the thyroid gland receives a greater flow of blood than do
most other tissues of the body. Venous drainage is through the paired superior,
middle, and inferior thyroid veins into the internal jugular and innominate veins.
The gland is also endowed with a rich lymphatic system that may play an important role in delivery of hormone to the general circulation. The thyroid gland
also has an abundant supply of sympathetic and parasympathetic nerves. Some
studies suggest that sympathetic stimulation or infusion of epinephrine or norepinephrine may increase secretion of thyroid hormone, but it is probably only of
minor importance in the overall regulation of thyroid function.

The functional unit of the thyroid gland is the follicle, which is composed
of epithelial cells arranged as hollow vesicles of various shapes, ranging in size

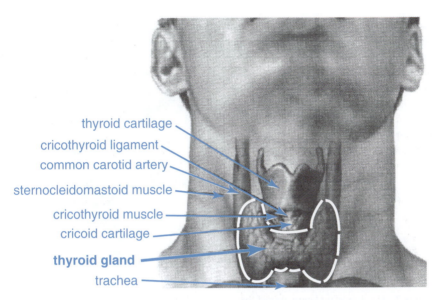

thyroid cartilage

cricothyroid ligament

common carotid artery

sternocleidomastoid muscle

cricothyroid muscle

cricoid cartilage

thyroid gland

trachea

Figure 1 Gross anatomy of the thyroid gland. (From Netter, F. H., "Atlas of Human Anatomy," 2nd Ed. Novartis, Summit, New Jersey, copyright 1989. Icon Learning Systems, LLC, a subsidiary of MediMedia USA Inc. Reprinted with permission from Icon Learning Systems, LLC, illustrated by Frank H. Netter, MD. All rights reserved.)

from 0.02 to 0.3 mm in diameter; each follicle is filled with a glycoprotein colloid called thyroglobulin (Figure 2). There are about 3 million follicles in the adult human thyroid gland. Epithelial cells lining each follicle may be cuboidal or columnar, depending on their functional state, with the height of the epithelium being greatest when its activity is highest. Each follicle is surrounded by a dense capillary network separated from epithelial cells by a well-defined basement membrane. Groups of densely packed follicles are bound together by connective tissue septa to form lobules that receive their blood supply from a single small artery. The functional state of one lobule may differ widely from that of an adjacent lobule. Secretory cells of the thyroid gland are derived embryologically and phylogenetically from two sources. Follicular cells, which produce the classical thyroid hormones thyroxine and triiodothyronine, arise from endoderm of the primitive pharynx. Parafollicular cells located between the follicles arise from neuroectoderm and produce the polypeptide hormone calcitonin, which is discussed in Chapter 8.

THYROID HORMONES

The thyroid hormones are α-amino acid derivatives of tyrosine (Figure 3). The thyronine nucleus consists of two benzene rings in ether linkage, with an

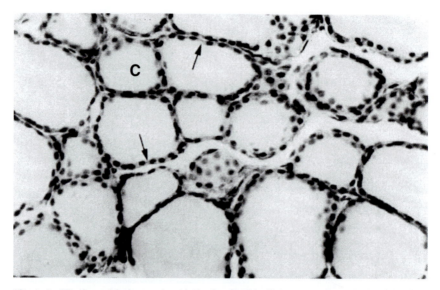

Figure 2 Histology of the human thyroid. Simple cuboidal cells (arrows) make up the follicles. Thyroid colloid (thyroglobulin; C) fills the follicles. (From Borysenko, M., and Beringer, T., In "Functional Histology," 2nd Edition, p. 312. Little, Brown, Boston, 1984, by permission of Lippincott Williams & Wilkins.)

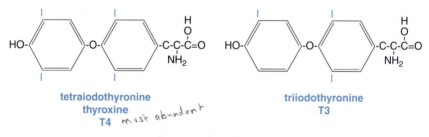

Figure 3 Thyroid hormones.

alanine side chain in the para position on the inner (tyrosyl) ring and a hydroxyl group in the para position in the outer (phenolic) ring. Thyroxine was the first thyroid hormone to be isolated and characterized. Its name derives from thyroid oxyindole, which describes the chemical structure erroneously assigned to it in 1914. Triiodothyronine, a considerably less abundant but three times more potent hormone than thyroxine in most assay systems, was not discovered until 1953. Both hormone molecules are exceptionally rich in iodine, which comprises more than half of their molecular weight. Thyroxine contains four atoms of iodine and thus is referred to as T4; triiodothyronine, which has three atoms of iodine, is referred to as T3.

BIOSYNTHESIS

Several aspects of the production of thyroid hormone are unusual: (1) Thyroid hormones contain large amounts of iodine. Biosynthesis of active hormone requires adequate amounts of this scarce element. This need is met by an efficient energy-dependent mechanism that allows thyroid cells to take up and concentrate iodide. The thyroid gland is also the principal site of storage of this rare dietary constituent. (2) Thyroid hormones are partially synthesized extracellularly at the luminal surface of follicular cells and are stored in an extracellular compartment, the follicular lumen. (3) The hormone therefore is doubly secreted, in that the precursor molecule, thyroglobulin, is released from apical surfaces of follicular cells into the follicular lumen, only to be taken up again by follicular cells and degraded to release T4 and T3, which are then secreted into the blood from the basal surfaces of follicular cells. (4) Thyroxine, the major secretory product, is not the biologically active form of the hormone, but must be transformed to T3 at extrathyroidal sites.

Biosynthesis of thyroid hormones can be considered as the sum of several discrete processes (Figure 4), all of which depend on the products of three genes that are expressed predominantly, if not exclusively, in thyroid follicle cells: the sodium/iodide symporter (NIS), thyroglobulin, and thyroid peroxidase.

IODINE TRAPPING

Under normal circumstances iodide is about 25 to 50 times more concentrated in the cytosol of thyroid follicular cells than in blood plasma, and during periods of active stimulation, it may be as high as 250 times that of plasma. Iodine is accumulated against a steep concentration gradient by the action of an electrogenic "iodide pump" located in the basolateral membranes. The pump is actually a sodium iodide symporter that couples the transfer of two ions of sodium with each ion of iodide. Iodide is thus transported against its concentration gradient driven by the favorable electrochemical gradient for sodium. Energy is expended by the sodium/potassium ATPase (the sodium pump), which then extrudes three ions of sodium in exchange for two ions of potassium to maintain the electrochemical gradient for sodium. Outward diffusion of potassium maintains the membrane potential. Like other transporters, the sodium iodide symporter has a finite capacity and can be saturated. Consequently, other anions, e.g., perchlorate, pertechnetate, and thiocyanate, which compete for binding sites on the sodium

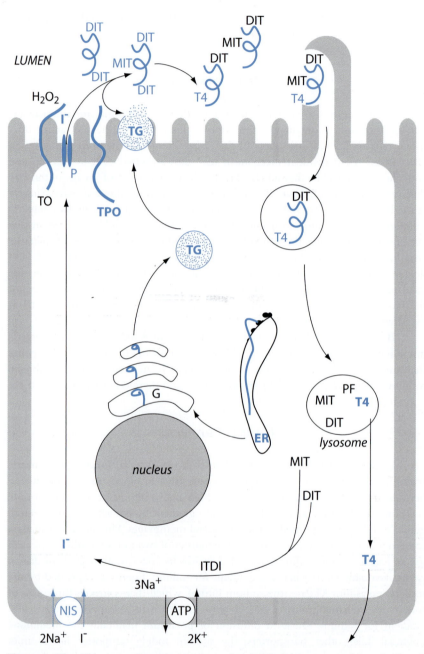

Figure 4 Thyroid hormone biosynthesis and secretion. Iodide (I⁻) is transported into the thyroid follicular cell by the sodium/iodide symporter (NIS) in the basal membrane and diffuses passively into

iodide symporter, can block the uptake of iodide. This property can be exploited for diagnostic or therapeutic purposes.

Thyroglobulin Synthesis

Thyroglobulin is the other major component needed for synthesis of thyroxine and triiodothyronine. Thyroglobulin is the matrix for thyroid hormone synthesis and is the form in which hormone is stored in the gland. It is a large glycoprotein that forms a stable dimer with a molecular mass of about 660,000 Da. Like other secretory proteins, thyroglobulin is synthesized on ribosomes, glycosylated in the cisternae of the endoplasmic reticulum, translocated to the Golgi apparatus, and packaged in secretory vesicles, which discharge it from the apical surface into the lumen. Because thyroglobulin secretion into the lumen is coupled with its synthesis, follicular cells do not have the extensive accumulation of secretory granules characteristic of protein-secreting cells. Iodination to form mature thyroglobulin does not take place until after the thyroglobulin is discharged into the lumen.

Incorporation of Iodine

Iodide that enters at the basolateral surfaces of the follicular cell must be delivered to the follicular lumen, where hormone biosynthesis takes place. Iodide diffuses throughout the follicular cell and exits from the apical membrane by way of a sodium-independent iodide transporter called *pendrin*, which is also expressed in brain and kidney. In order for iodide to be incorporated into tyrosine residues in thyroglobulin, it must first be converted to some higher oxidized state. This step is catalyzed by the thyroid-specific thyroperoxidase in the presence of hydrogen peroxide, whose formation may be rate-limiting. Hydrogen peroxide is generated by the catalytic activity of a calcium-dependent NADPH oxidase that is present in the brush border. Thyroperoxidase is the key enzyme in thyroid hormone

the lumen through the iodide channel called pendrin (P). Thyroglobulin (TG) is synthesized by the rough endoplasmic reticulum (ER); it is then processed in the ER and in the Golgi (G), where it is packaged into secretory granules and released into the follicular lumen. In the presence of hydrogen peroxide (H_2O_2) produced in the lumenal membrane by thyroid oxidase (TO), the thyroid peroxidase (TPO) oxidizes iodide, which reacts with tyrosine residues in thyroglobulin to produce monoiodotyrosyl (MIT) and diiodotyrosyl (DIT) residues within the TG. The TPO reaction also catalyzes the coupling of iodotyrosines to form thyroxine (T4) and some triiodothyronine (T3, not shown) residues within the thyroglobulin. Secretion of T4 begins with phagocytosis of thyroglobulin, fusion of TG-laden endosomes with lysosomes, and proteolytic digestion to peptide fragments (PF), MIT, DIT, and T4, which exits the cell at the basal membrane. MIT and DIT are deiodinated by iodotyrosine deiodinase (ITDI) and recycled.

formation and is thought to catalyze the iodination and coupling reactions described below, in addition to activation of iodide. Thyroperoxidase spans the brush border membrane on the apical surface of follicular cells and is oriented with its catalytic domain facing the follicular lumen.

Addition of iodine molecules to tyrosine residues in thyroglobulin, is called *organification*. Thyroglobulin is iodinated at the apical surface of follicular cells as it is extruded into the follicular lumen. Iodide acceptor sites in thyroglobulin are in sufficient excess over the availability of iodide that no free iodide accumulates in the follicular lumen. Although posttranslational conformational changes orchestrated by endoplasmic reticular proteins organize the configuration of thyroglobulin to increase its ability to be iodinated, iodination and hormone formation do not appear to be particularly efficient. Tyrosine is not especially abundant in thyroglobulin and comprises only about 1 in 40 residues of the peptide chain. Only about 10% of the 132 tyrosine residues in each thyroglobulin dimer appear to be in positions favorable for iodination. The initial products formed are monoiodotyrosine (MIT) and diiodotyrosine (DIT), and they remain in peptide linkage within the thyroglobulin molecules. Normally more DIT is formed than MIT, but when iodine is scarce there is less iodination and the ratio of MIT to DIT is reversed.

Coupling

The final stage of thyroxine biosynthesis is the coupling of two molecules of DIT to form T4 within the peptide chain. This reaction is also catalyzed by thyroperoxidase. Only about 20% of iodinated tyrosine residues undergo coupling, with the rest remaining as MIT and DIT. After coupling is complete, each thyroglobulin molecule normally contains one to three molecules of T4. T3 is considerably scarcer, with one molecule being present in only 20–30% of thyroglobulin molecules. T3 may be formed by deiodination of T4 or coupling of one residue of DIT with one of MIT.

Exactly how coupling is achieved is not known. One possible mechanism involves joining two iodotyrosine residues that are in close proximity to each other, either on two separate strands of thyroglobulin or on adjacent folds of the same strand. Free radicals formed by the action of thyroperoxidase react to form the ether linkage at the heart of the thyronine nucleus, leaving behind in one of the peptide chains the serine or alanine residue that was once attached to the phenyl group that now comprises the outer ring of T4 (Figure 5). An alternative mechanism involves coupling a free diiodophenylpyruvate (deaminated DIT) with a molecule of DIT in peptide linkage within the thyroglobulin molecule by a similar reaction sequence. Regardless of which model proves correct, it is sufficient to recognize the central importance of thyroperoxidase for formation of the thyronine nucleus as well as iodination of tyrosine residues. In addition, the mature

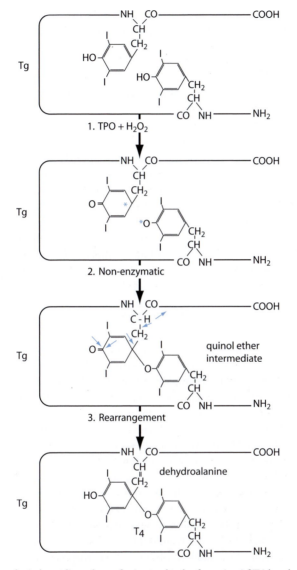

Figure 5 Hypothetical coupling scheme for intramolecular formation of T4 based on model reaction with purified thyroid peroxidase. Tg, Thyroglobulin. (From Taurog, A., In "Werner and Ingbar's The Thyroid," 8th Ed., p. 71, by permission of Lippincott Williams & Wilkins, Philadelphia, 2000.)

hormone is formed while in peptide linkage within the thyroglobulin molecule, and remains a part of that large storage molecule until lysosomal enzymes set it free during the secretory process.

STORAGE

The thyroid is unique among endocrine glands in that it stores its product extracellularly, in follicular lumens, as large precursor molecules. In the normal individual, approximately 30% of the mass of the thyroid gland is thyroglobulin, which corresponds to about a 2 to 3 months' supply of hormone. Mature thyroglobulin is a high-molecular-weight (660,000 da) molecule, probably a dimer of the thyroglobulin precursor peptide, and contains about 10% carbohydrate and about 0.5% iodine. The tyrosine residues that are situated just a few amino acids away from the C and N termini are the principal sites of iodothyronine formation. MIT and DIT at other sites in thyroglobulin comprise an important reservoir for iodine and constitute about 90% of the total pool of iodine in the body.

SECRETION

Thyroglobulin stored within follicular lumens is separated from extracellular fluid and the capillary endothelium by a virtually impenetrable layer of follicular cells. For secretion to occur, thyroglobulin must be brought back into follicular cells by a process of endocytosis. On acute stimulation with TSH, long strands of protoplasm (pseudopodia) reach out from the apical surfaces of follicular cells to surround chunks of thyroglobulin, which are taken up in endocytic vesicles (Figure 6). In chronic situations uptake is probably less dramatic than shown in Figure 6, but nevertheless requires an ongoing endocytic process. The endocytic vesicles migrate toward the basal portion of the cells and fuse with lysosomes, which simultaneously migrate from the basal to the apical region of the cells to meet the incoming endocytic vesicles. As fused lyso-endosomes migrate toward the basement membrane, thyroglobulin is degraded to peptide fragments and free amino acids, including T4, T3, MIT, and DIT. Of these, only T4 and T3 are released into the bloodstream, in a ratio of about 20:1, perhaps by a process of simple diffusion down a concentration gradient.

Monoiodotyrosine and DIT cannot be utilized for synthesis of thyroglobulin and are rapidly deiodinated by a specific microsomal deiodinase. Virtually all of the iodide released from iodotyrosines is recycled into thyroglobulin. Deiodination of iodotyrosine provides about twice as much iodide for hormone synthesis as the iodide pump and is therefore of great significance in hormone biosynthesis. Patients who are genetically deficient in thyroidal tyrosine deiodinase readily

Figure 6 Scanning electron micrographs of the luminal microvilli of dog thyroid follicular cells. Top: TSH secretion suppressed by feeding thyroid hormone (×36,000). Bottom: At 1 hour after TSH (×16,500). (From Balasse, P. D., Rodesch, F. R., Neve, P. E., *et al.*, *C.R. Acad. Sci. [D] (Paris)* **274**, 2332, 1972, by permission of The Academy of Sciences in Paris.)

suffer symptoms of iodine deficiency and excrete MIT and DIT in their urine. Normally, virtually no MIT or DIT escapes from the gland.

Synthesis of thyroglobulin and its export in vesicles into the follicular lumen are ongoing processes that take place simultaneously with uptake of thyroglobulin in other vesicles moving in the opposite direction. These opposite processes, involving vesicles laden with thyroglobulin moving into and out of the cells, are somehow regulated so that under normal circumstances thyroglobulin neither accumulates in follicular cells nor are the lumens depleted. The physiological mechanisms for such traffic control are not yet understood.

REGULATION OF THYROID FUNCTION

EFFECTS OF THYROID-STIMULATING HORMONE

Although the thyroid gland can carry out all the steps of hormone biosynthesis, storage, and secretion in the absence of any external signals, autonomous function is far too sluggish to meet bodily needs for thyroid hormone. The principal regulator of thyroid function is thyroid-stimulating hormone, TSH, which is secreted by thyrotropes in the pituitary gland (see Chapter 2). It may be recalled that TSH consists of two glycosylated peptide subunits, including the same alpha subunit that is also found in FSH and LH (Chapter 2). The beta subunit is the part of the hormone that confers thyroid-specific stimulating activity, but free beta subunits are inactive and stimulate the thyroid only when linked to alpha subunits in a complex three-dimensional configuration.

Thyroid-stimulating hormone binds to a single class of heptihelical G-protein-coupled receptors (see Chapter 1) in the basolateral surface membranes of thyroid follicular cells. The TSH receptor is the product of a single gene, but it is composed of two subunits held together by a disulfide bond. It appears that after the molecule has been properly folded and its disulfide bonds formed, a loop of about 50 amino acids is excised proteolytically from the extracellular portion of the receptor. The alpha subunit includes about 300 residues at the amino terminus and contains most of the TSH binding surfaces. The beta subunit contains the seven membrane-spanning α-helices and the short carboxyl-terminal tail in the cytoplasm. Reduction of the disulfide bond may lead to release of the alpha subunit into the extracellular fluid, and may have important implications for the development of antibodies to the TSH receptor and thyroid disease (see below). Binding of TSH to the receptor results in activation of both adenylyl cyclase, through $G\alpha_s$, and phospholipase C, through $G\alpha_q$, and leads to increases in both the cyclic AMP and the diacylglycerol/IP_3 second messenger pathways (see Chapter 1). Activation of the cyclic AMP pathway appears to be the more important transduction mechanism, because all of the known effects of TSH can be duplicated by cyclic AMP. Because TSH increases cyclic AMP production at much lower concentrations than are needed to increase phospholipid turnover, it is likely that IP_3 and DAG are redundant mediators that reinforce the effects of cyclic AMP at times of intense stimulation, but it is also possible that these second messengers signal some unique responses. Increased turnover of phospholipid is associated with release of arachidonic acid and the consequent increased production of prostaglandins that also follows TSH stimulation of the thyroid.

In addition to regulating all aspects of hormone biosynthesis and secretion, TSH increases blood flow to the thyroid, and with prolonged stimulation also increases the height of the follicular epithelium (hypertrophy), and can stimulate division of follicular cells (hyperplasia). Stimulation of thyroid follicular cells by TSH is a good example of a pleiotropic effect of a hormone in which there are multiple separate but complementary actions that summate to produce an overall response. Each step of hormone biosynthesis, storage, and secretion appears to be stimulated directly by a cyclic AMP-dependent process that is accelerated independently of the preceding or following steps in the pathway. Thus even when increased iodide transport is blocked with a drug that specifically affects the iodide pump, TSH nevertheless accelerates the remaining steps in the synthetic and secretory process. Similarly, when iodination of tyrosine is blocked by a drug specific for the organification process, TSH still stimulates iodide transport and thyroglobulin synthesis.

Most of the responses to TSH depend on activation of protein kinase A and the resultant phosphorylation of proteins, including transcription factors such as CREB (cyclic AMP response element binding protein; see Chapter 1). TSH increases expression of genes for the sodium iodide symporter, thyroglobulin, thyroid oxidase, and thyroid peroxidase. These effects are exerted through cooperative interactions of TSH-activated nuclear proteins with thyroid-specific transcription factors, the expression of which is also enhanced by TSH. TSH appears to increase blood flow by activating the gene for nitric oxide synthase, which increases production of the potent vasodilator, nitric oxide, and by inducing expression of paracrine factors that promote capillary growth (angiogenesis). Precisely how TSH increases thyroid growth is not understood, but it is apparent that synthesis and secretion of a variety of local growth factors is induced.

EFFECTS OF THYROID-STIMULATING IMMUNOGLOBULINS

Overproduction of thyroid hormone, hyperthyroidism, which is also known as Graves' disease, is usually accompanied by extremely low concentrations of TSH in blood plasma, yet the thyroid gland gives every indication of being under intense stimulation. This paradox was resolved when it was found that blood plasma of affected individuals contains a substance that stimulates the thyroid gland to produce and secrete thyroid hormone. This substance is an immunoglobulin secreted by lymphocytes and is almost certainly an antibody to the TSH receptor. Thyroid-stimulating immunoglobulin (TSI) can be found in the serum of virtually all patients with Graves' disease, suggesting an autoimmune etiology to this disorder. It is of interest to note that when reacting with the TSH receptor, antibodies trigger the same sequence of responses that are produced when TSH interacts with the receptor. This fact indicates that all the information needed to produce the characteristic cellular response to TSH resides in the receptor rather than in the hormone. The role of the hormone therefore must be limited to activation of the receptor. Similar effects have also been seen with antibodies to receptors for other hormones.

AUTOREGULATION OF THYROID HORMONE SYNTHESIS

Although production of thyroid hormones is severely impaired when too little iodide is available, iodide uptake and hormone biosynthesis are temporarily blocked when the concentration of iodide in blood plasma becomes too high. This effect of iodide has been exploited clinically to produce short-term suppression of thyroid hormone secretion. This inhibitory effect of iodide apparently depends on its being incorporated into some organic molecule and is thought to represent an autoregulatory phenomenon that protects against overproduction of thyroxine. Blockade of thyroid hormone production is short-lived, and the gland eventually "escapes" from the inhibitory effects of iodide by mechanisms that include down-regulation of the sodium iodide symporter.

Biosynthetic activity of the thyroid gland may also be regulated by the thyroglobulin that accumulates in the follicular lumen. Evidence has been presented that thyroglobulin, acting through a receptor on the apical surface of follicular cells, decreases expression of thyroid-specific transcription factors and thereby decreases expression of the genes for thyroglobulin, the thyroid peroxidase, the sodium iodide symporter, and the TSH receptor. Further effects of thyroglobulin include increased transcription of pendrin, which delivers iodide from the follicular cell to the lumen. Thus thyroglobulin may have significant effects in regulating its own synthesis and may temper the stimulatory effects of TSH, which remains the primary and most important regulator of thyroid function.

THYROID HORMONES IN BLOOD

More than 99% of thyroid hormone circulating in blood is firmly bound to three plasma proteins. They are thyroxine-binding globulin (TBG), transthyretin (TTR), and albumin. Of these, TBG is quantitatively the most important and accounts for more than 70% of the total protein-bound hormone (both T4 and T3). About 10–15% of circulating T4 and 10% of circulating T3 is bound to TTR and nearly equal amounts are bound to albumin. TBG carries the bulk of the hormone, even though its concentration in plasma is only 6% of that of TTR and is less than 0.1% of that of albumin; this is because its affinity for both T4 and T3 is so much higher than that of the other proteins. All three thyroid hormone binding proteins bind T4 at least 10 times more avidly, compared to T3. All are large enough to escape filtration by the renal glomerular membranes, and very little protein crosses the capillary endothelium. The less than 1% of hormone present in free solution is in equilibrium with bound hormone and is the only hormone that can escape from capillaries to produce biological activity or be acted on by tissue enzymes.

The total amount of thyroid hormone bound to plasma proteins represents about three times as much hormone as is secreted and degraded in the course of a single day. Thus plasma proteins provide a substantial reservoir of extrathyroidal hormone. We should therefore not expect acute increases or decreases in the rate of secretion of thyroid hormones to bring about large or rapid changes in circulating concentrations of thyroid hormones. For example, if the rate of thyroxine secretion were doubled for 1 day, we could expect its concentration in blood to increase by no more than 30%, even if there were no accompanying increase in the rate of hormone degradation. A 10-fold increase in the rate of secretion lasting for 60 minutes would give only a 12% increase in total circulating thyroxine, and if thyroxine secretion stopped completely for 1 hour, its concentration would decrease by only 1%. Furthermore, because the binding capacity of plasma proteins for thyroid hormones is far from saturated, an even massive increase in secretion rate would have little effect on the percentage of hormone that is unbound. These considerations seem to rule out changes in thyroid hormone secretion as effectors of minute-to-minute regulation of any homeostatic process. On the other hand, because so much of the circulating hormone is bound to plasma binding proteins, we might expect that the total amount of T4 and T3 in the circulation would be affected significantly by decreases in the concentration of plasma binding proteins, as might occur with liver or kidney disease.

METABOLISM OF THYROID HORMONES

Because T4 is bound much more tightly by plasma proteins, compared to T3, a greater fraction of T3 is free to diffuse out of the vascular compartment and into cells, where it can produce its biological effects or be degraded. Consequently, it is not surprising that the half-time for disappearance of an administered dose of ^{125}I-labeled T3 is only one-sixth of that for T4, or that the lag time needed to observe effects of T3 is considerably shorter than that needed for T4. However, because of

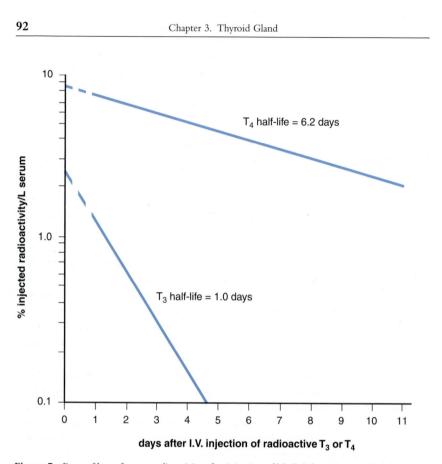

Figure 7   Rate of loss of serum radioactivity after injection of labeled thyroxine or triiodothyronine into human subjects. (Plotted from data of Nicoloff, J. D., Low, J. C., Dussault, J. H., *et al.*, *J. Clin. Invest.* **51**, 473, 1972.)

the binding proteins, both T4 and T3 have unusually long half-lives in plasma, measured in days rather than seconds or minutes (Figure 7). It is noteworthy that the half-lives of T3 and T4 are increased with thyroid deficiency and shortened with hyperthyroidism.

Although T4 is the main secretory product of the thyroid gland and the major form of thyroid hormone present in the circulating plasma reservoir, abundant evidence indicates that it is T3 and not T4 that binds to the thyroid hormone receptor (see below). In fact, T4 can be considered to be a prohormone that serves as the precursor for extrathyroidal formation of T3. Observations in human patients confirm that T3 is actually formed extrathyroidally and can account for most of the biological activity of the thyroid gland. Thyroidectomized subjects given pure T4 in physiological amounts have normal amounts of T3 in their circulation. Furthermore, the rate of metabolism of T3 in normal subjects is such that about 30 µg of T3 is replaced daily, even though the thyroid gland secretes only 5 µg each day. Thus nearly 85% of the T3 that turns over each day must be formed by deiodination of T4 in extrathyroidal tissues. This extrathyroidal formation of T3 consumes about 35% of the T4 secreted each day. The remainder is degraded to inactive metabolites.

Extrathyroidal metabolism of T4 centers around selective and sequential removal of iodine from the thyronine nucleus, catalyzed by three different enzymes called deiodinases (Figure 8). The type I deiodinase is expressed mainly in the liver and kidney, but is also found in the central nervous system, the anterior pituitary gland, and the thyroid gland. The type I deiodinase is a membrane-bound enzyme with its catalytic domain oriented to face the cytoplasm. Despite its intracellular location, however, T3 formed by deiodination, especially in the liver and kidney, readily escapes into the circulation and accounts for about 80% of the T3 in blood. The type I deiodinase can remove an iodine molecule either from the outer (phenolic) ring of T4, or from the inner (tyrosyl) ring. Iodines in the phenolic ring are designated $3'$ and $5'$, whereas iodines in the inner ring are designated simply 3 and 5. The 3 and 5 positions on either ring are chemically equivalent, but there are profound functional consequences of removing an iodine from the inner or outer rings of thyroxine. Removing an iodine from the outer ring produces $3',3,5$-triiodothyronine, usually designated as T3, and converts thyroxine to the form that binds to the thyroid hormone receptor. Removal of an iodine from the inner ring produces $3',5',3$-triiodothyronine, which is called reverse T3 (rT3). Reverse T3 cannot bind to thyroid hormone receptors and can only be further deiodinated.

The type II deiodinase is absent from the liver but is found in many extrahepatic tissues, including the brain and pituitary gland, where it is thought to produce T3 to meet local tissue demands independently of circulating T3, although these tissues can also take up T3 from the blood. Expression of the type II deiodinase is regulated by other hormones; its expression is highest when blood concentrations of T4 are low. In addition, hormones that act through the cyclic AMP second messenger system (Chapter 1) and growth factors stimulate type II deiodinase expression. These characteristics support the idea that this enzyme may provide T3 to meet local demands.

The type III deiodinase removes an iodine from the tyrosyl ring of T4 or T3, and hence its function is solely degradative. It is widely expressed by many tissues throughout the body. Reverse T3 is produced by both type I and type III deiodinases and may be further deiodinated by the type III deiodinase by removal of the second iodide from inner ring (Figure 8). Reverse T3 is also a favored

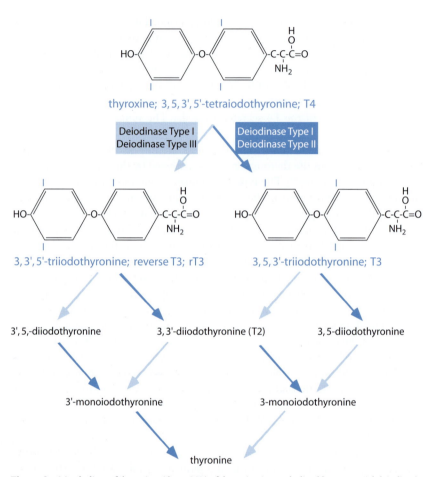

Figure 8   Metabolism of thyroxine. About 90% of thyroxine is metabolized by sequential deiodination catalyzed by deiodinases (types I, II, and III); the first step removes an iodine from either the phenolic or tyrosyl ring, producing an active (T3) or an inactive (rT3) compound. Subsequent deiodinations continue until all of the iodine is recovered from the thyronine nucleus. Dark blue arrows designate deiodination of the phenolic ring and light blue arrows indicate deiodination of the tyrosyl ring. Less than 10% of thyroxine is metabolized by shortening the alanine side chain prior to deiodination.

substrate for the type I deiodinase, and although it and T3 are formed at similar rates, it is degraded much faster as compared to T3. Some rT3 escapes into the bloodstream, where it is avidly bound to TBG and TTR.

All three deiodinases can catalyze the oxidative removal of iodine from partially deiodinated hormone metabolites, and through their joint actions the thyronine nucleus can be completely stripped of iodine. The liberated iodide is then available to be taken up by the thyroid and recycled into hormone. A quantitatively less important route for degradation of thyroid hormones includes shortening of the alanine side chain to produce tetraiodothyroacetic acid (Tetrac) and its subsequent deiodination products. Thyroid hormones are also conjugated with glucuronic acid and excreted intact in the bile. Bacteria in the intestine can split the glucuronide bond, and some of the thyroxine liberated can be taken up from the intestine and can be returned to the general circulation. This cycle of excretion in bile and absorption from the intestine is called the enterohepatic circulation and may be of importance in maintaining normal thyroid economy when thyroid function is marginal or dietary iodide is scarce. Thyroxine is one of the few naturally occurring hormones that is sufficiently resistant to intestinal and hepatic destruction that it can readily be given by mouth.

PHYSIOLOGICAL EFFECTS OF THYROID HORMONES

GROWTH AND MATURATION

Skeletal System

One of the most striking effects of thyroid hormones is on bodily growth (see Chapter 10). Although fetal growth appears to be independent of the thyroid, growth of the neonate and attainment of normal adult stature require optimal amounts of thyroid hormone. Because stature or height is determined by the length of the skeleton, we might anticipate an effect of thyroid hormone on growth of bone. However, there is no evidence that T3 acts directly on cartilage or bone cells to signal increased bone formation. Rather, at the level of bone formation, thyroid hormones appear to act permissively or synergistically with growth hormone, insulin-like growth factor I (see Chapter 10), and other growth factors that promote bone formation. Thyroid hormones also promote bone growth indirectly by actions on the pituitary gland and hypothalamus. Thyroid hormone is required for normal growth hormone synthesis and secretion.

Skeletal maturation is distinct from skeletal growth. Maturation of bone results in the ossification and eventual fusion of the cartilaginous growth plates, which occurs with sufficient predictability in normal development that individuals can be assigned a specific "bone age" from radiological examination of ossification centers. Thyroid hormones profoundly affect skeletal maturation, perhaps by a direct action. Bone age is retarded relative to chronological age in children who are deficient in thyroid hormone and is advanced prematurely in hyperthyroid children. Uncorrected deficiency of thyroid hormone during childhood results in retardation of growth and malformation of facial bones characteristic of juvenile hypothyroidism, or cretinism.

Central Nervous System

The importance of the thyroid hormones for normal development of the nervous system is well established. Thyroid hormones and their receptors are present early in the development of the fetal brain, well before the fetal thyroid gland becomes functional. T4 and T3 present in the fetal brain at this time probably arise in the mother and readily cross the placenta to the fetus. Some evidence suggests that maternal hypothyroidism may lead to deficiencies in postnatal neural development, but direct effects of thyroid deficiency on the fetal brain have not been established. However, babies with failure of thyroid gland development who are born to mothers with normal thyroid function have normal brain development if properly treated with thyroid hormones after birth. Maturation of the nervous system during the perinatal period has an absolute dependence on thyroid hormone. During this critical period thyroid hormone must be present for normal development of the brain. In rats made hypothyroid at birth, cerebral and cerebellar growth and nerve myelination are severely delayed. Overall size of the brain is reduced along with its vascularity, particularly at the capillary level. The decrease in size may be partially accounted for by a decrease in axonal density and dendritic branching. Thyroid hormone deficiency also leads to specific defects in cell migration and differentiation. In human infants the absence or deficiency of thyroid hormone during this period is catastrophic and results in permanent, irreversible mental retardation, even if large doses of hormone are given later in childhood (Figure 9). If replacement therapy is instituted early in postnatal life, however, the tragic consequences of neonatal hypothyroidism can be averted. Mandatory neonatal screening for hypothyroidism has therefore been instituted throughout the United States and other countries. Precisely what thyroid hormones do during the critical period, how they do it, and why the opportunity for intervention is so brief are subjects of active research.

Effects of T3 and T4 on the central nervous system are not limited to the perinatal period of life. In the adult, hyperthyroidism produces hyperexcitability, irritability, restlessness, and exaggerated responses to environmental stimuli. Emotional instability that can lead to full-blown psychosis may also occur. Conversely, decreased thyroid hormone results in listlessness, lack of energy, slowness of speech, decreased sensory capacity, impaired memory, and somnolence. Mental capacity is dulled, and psychosis (myxedema madness) may occur. Conduction velocity in peripheral nerves is slowed and reflex time is increased in hypothyroid individuals. The underlying mechanisms for these changes are not understood.

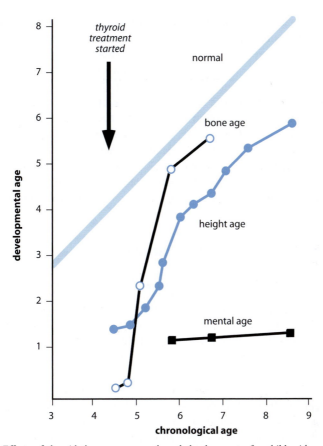

Figure 9 Effects of thyroid therapy on growth and development of a child with no functional thyroid tissue. Daily treatment with thyroid extract began at 4.5 years of age (vertical arrow). Bone age rapidly returned toward normal, and the rate of growth (height age) paralleled the normal curve. Mental development, however, remained infantile. (From Wilkins, L., "The Diagnosis and Treatment of Endocrine Disorders in Childhood and Adolescence." Charles C. Thomas, Springfield, Illinois, 1965, with permission.)

AUTONOMIC NERVOUS SYSTEM

Interactions between thyroid hormones and the autonomic nervous system, particularly the sympathetic branch, are important throughout life. Increased secretion of thyroid hormone exaggerates many of the responses that are mediated by the neurotransmitters norepinephrine and epinephrine, which are released from sympathetic neurons and the adrenal medulla (see Chapter 4). In fact, many symptoms of hyperthyroidism, including tachycardia (rapid heart rate) and increased cardiac output, resemble increased activity of the sympathetic nervous system. Thyroid hormones increase the number of receptors for epinephrine and norepinephrine (β-adrenergic receptors) in the myocardium and some other tissues. Thyroid hormones may also increase expression of the stimulatory G-protein ($G\alpha_s$) associated with adrenergic receptors and down-regulate the inhibitory G-protein ($G\alpha_i$). Either of these effects results in greater production of cyclic AMP (Chapter 1). Furthermore, through the agency of cyclic AMP, sympathetic stimulation activates the type II deiodinase, which accelerates local conversion of T4 to T3. Because thyroid hormones exaggerate a variety of responses mediated by β-adrenergic receptors, pharmacological blockade of these receptors is useful for reducing some of the symptoms of hyperthyroidism. Conversely, the diverse functions of the sympathetic nervous system are compromised in hypothyroid states.

METABOLISM

Oxidative Metabolism and Thermogenesis

More than a century has passed since it was recognized that the thyroid gland exerts profound effects on oxidative metabolism in humans. The so-called basal metabolic rate (BMR), which is a measure of oxygen consumption under defined resting conditions, is highly sensitive to thyroid status. A decrease in oxygen consumption results from a deficiency of thyroid hormones, and excessive thyroid hormone increases BMR. Oxygen consumption in all tissues except brain, testis, and spleen is sensitive to the thyroid status and increases in response to thyroid hormone (Figure 10). Even though the dose of thyroid hormone given to hypothyroid animals in the experiment shown in Figure 10 was large, there was a delay of many hours before effects were observable. In fact, the rate of oxygen consumption in the whole animal did not reach its maximum until 4 days after a single dose of hormone. The underlying mechanisms for increased oxygen consumption are incompletely understood.

Oxygen consumption ultimately reflects activity of mitochondria and is coupled with formation of high-energy bonds in ATP. Physiologically, oxygen consumption is proportional to energy utilization. Thus if there is increased consumption of oxygen, there must be increased utilization of energy or the efficiency of coupling ATP production with oxygen consumption must be altered. T3 appears to accelerate ATP-dependent processes, including activity of the sodium/potassium ATPase that maintains ionic integrity of all cells, and to decrease efficiency of oxygen utilization. In normal individuals activity of the sodium/potassium ATPase is thought to account for about 20% of the resting oxygen consumption. Activity of this enzyme is decreased in hypothyroid individuals, and its synthesis is accelerated by thyroid hormone. A variety of other metabolic reactions are also accelerated by

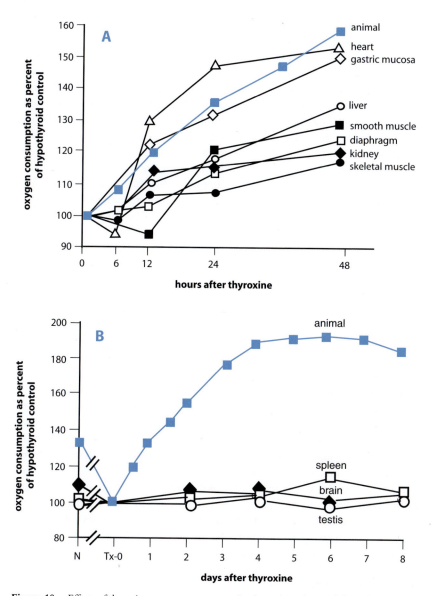

Figure 10   Effects of thyroxine on oxygen consumption by various tissues of thyroidectomized rats. Note in A that the abscissa is in units of hours and in B the units are days. (B) N, Normal; Tx-0, thyroidectomized just prior to thyroxine. (Redrawn from Barker, S. B., and Klitgaard, H. M., *Am. J. Physiol.* **170**, 81, 1952, with permission.)

T3, and the accompanying increased turnover of ATP contributes to the increase in oxygen consumption.

Phosphorylation of ADP to form ATP is driven by the proton gradient generated across the inner mitochondrial membrane by the electron transport system, which delivers protons to oxygen to form water. Thus ATP synthesis is coupled to oxygen consumption. Leakage of protons across the inner mitochondrial membrane "uncouples" oxygen consumption from ATP production by partially dissipating the gradient. As a result, oxygen consumption proceeds at a faster rate than ATP generation, and the extra energy derived is dissipated as heat. Leakage of protons into the mitochondria depends on the presence of special uncoupling proteins (UCPs) in the inner mitochondrial membrane. To date three proteins thought to have uncoupling activity have been identified in mitochondrial membranes of various tissues. All three appear to be up-regulated by T3. Although the physiological importance of UCP-1 seems firmly established (see below), the physiological roles of UCP-2 and UCP-3 remain controversial.

Splitting of ATP not only energizes cellular processes but also results in heat production. Thyroid hormones are said to be "calorigenic" because they promote heat production. It is therefore not surprising that one of the classical signs of hypothyroidism is decreased tolerance to cold, whereas excessive heat production and sweating are seen in hyperthyroidism. Effects of thyroid hormone on oxidative metabolism are seen only in animals that maintain a constant body temperature, consistent with the idea that calorigenic effects may be related to thermoregulation. Thyroidectomized animals have severely reduced ability to survive cold temperature. T3 contributes to both heat production and heat conservation.

Individuals exposed to a cold environment maintain constant body temperature by increasing heat production by at least two mechanisms: (1) shivering, which is a rapid increase in involuntary activity of skeletal muscle, and (2) the so-called nonshivering thermogenesis seen in cold-acclimated individuals. Details of the underlying mechanisms for each of these responses are still not understood. As we have seen, the metabolic effects of T3 have a long lag time and hence increased production of T3 cannot be of much use for making rapid adjustments to cold temperatures. The role of T3 in the shivering response is probably limited to maintenance of tissue sensitivity to sympathetic stimulation. In this context, the importance of T3 derives from actions that were established before exposure to cold temperature. Maintenance of sensitivity to sympathetic stimulation permits efficient mobilization of stored carbohydrate and fat, needed to fuel the shivering response and to make circulatory adjustments for increased activity of skeletal muscle. It may be also recalled that the sympathetic nervous system regulates heat conservation by decreasing blood flow through the skin. Piloerection in animals increases the thickness of the insulating layer of fur. These responses are likely to be of importance in both acute and chronic responses to cold exposure.

Chronic nonshivering thermogenesis appears to require increased production of T3, which acts in concert with the sympathetic nervous system to increase heat production and conservation. Some data indicate that norepinephrine may increase permeability of brown fat and skeletal muscle cells to sodium. Increased activity of the sodium pump could account for increased oxygen consumption and heat production in the cold-acclimatized individual. In muscles of cold-acclimated rats, activity of the sodium/potassium ATPase is increased in a manner that appears to depend on thyroid hormone. Some experimental results support a similar effect on calcium pumps.

Brown fat is an important source of heat in newborn humans and throughout life in small mammals. This form of adipose tissue is especially rich in mitochondria, which give it its unique brown color. Mitochondria in this tissue contain UCP-1, sometimes called *thermogenin*, which allows mitochondria to oxidize relatively large amounts of fatty acids and to produce heat unfettered by limitations in availability of ADP. Although both T3 and the sympathetic neurotransmitter norepinephrine can each induce the synthesis of UCP-1, their cooperative interaction results in production of three to four times as much of this mitochondrial protein as the sum of their independent actions. In addition, T3 increases the efficacy of norepinephrine to release fatty acids from stored triglycerides and thus provides fuel for heat production. Brown adipose tissue increases synthesis of the type II deiodinase in response to sympathetic stimulation, and produces abundant T3 locally to meet its needs. Adult humans have little brown fat, and may increase heat production through similar effects of UCP-2 and UCP-3 in white fat and muscle, but supporting evidence for this possibility is not available.

In rodents and other experimental animals, exposure to cold temperatures is an important stimulus for increased TSH secretion from the pituitary and the resultant increase in T4 and T3 secretion from the thyroid gland. Cold exposure does not increase TSH section in humans except in the newborn. In humans and experimental animals, however, exposure to cold temperatures increases conversion of T4 to T3, probably as a result of increased sympathetic nervous activity that leads to increased cyclic AMP production in various tissues. It may be recalled that expression of the type II deiodinase is activated by cyclic AMP.

Carbohydrate Metabolism

T3 accelerates virtually all aspects of metabolism, including carbohydrate utilization. It increases glucose absorption from the digestive tract, glycogenolysis and gluconeogenesis in hepatocytes, and glucose oxidation in liver, fat, and muscle cells. No single or unique reaction in any pathway of carbohydrate metabolism has been identified as the rate-determining target of T3 action. Rather, carbohydrate degradation appears to be driven by other factors, such as increased demand for ATP,

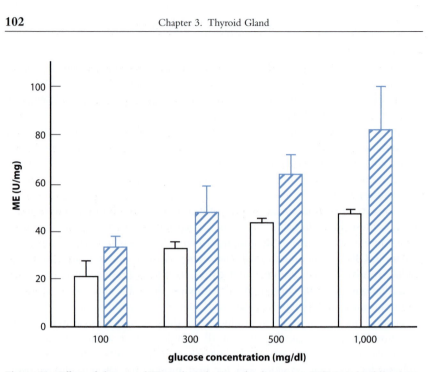

Figure 11   Effects of glucose and T3 on the induction of malic enzyme (ME) in isolated hepatocyte cultures. Note that the amount of enzyme present in tissues was increased by growing cells in higher and higher concentrations of glucose. Open bars show effects of glucose in the presence of a low (10^{-10} M) concentration of T3. Hatched bars indicate that the effects of glucose were exaggerated when cells were grown in a high concentration of T3 (10^{-8} M). (From Mariash, G. N., and Oppenheimer, J. H., *Fed. Proc., Fed. Soc., Exp. Biol.* **41**, 2674, 1982. Reproduced with permission of the Federation of American Societies for Experimental Biology.)

the content of carbohydrate in the diet, or the nutritional state. Although T3 may induce synthesis of specific enzymes of carbohydrate and lipid metabolism, e.g., the malic enzyme, glucose 6-phosphate dehydrogenase, and 6-phosphogluconate dehydrogenase, it appears principally to behave as an amplifier or gain control working in conjunction with other signals (Figure 11). In the example shown in Figure 11, induction of the malic enzyme in hepatocytes was dependent both on the concentration of glucose in the culture medium and on the concentration of T3. T3 had little effect on enzyme induction when there was no glucose but amplified the effectiveness of glucose as an inducer of genetic expression. This experiment provides a good example of how T3 can amplify the readout of genetic information.

Lipid Metabolism

Because glucose is the major precursor for fatty acid synthesis in both liver and fat cells, it should not be surprising that optimal amounts of thyroid hormone are necessary for lipogenesis in these cells. Once again the primary determinant of lipogenesis is not T3, but rather the amount of available carbohydrate or insulin (see Chapter 5), with thyroid hormone acting as a gain control. Similarly, mobilization of fatty acids from storage depots in adipocytes is compromised in the thyroid-deficient subject and is increased above normal when thyroid hormones are present in excess. Once again, T3 amplifies physiological signals for fat mobilization without itself acting as such a signal.

Increased blood cholesterol (hypercholesterolemia) is typically found in hypothyroidism. Thyroid hormones reduce cholesterol in the plasma of normal subjects and restore blood concentrations of cholesterol to normal in hypothyroid subjects. Hypercholesterolemia in hypothyroid subjects results from decreased ability to excrete cholesterol in bile, rather than overproduction of cholesterol. In fact, cholesterol synthesis is impaired in the hypothyroid individual. T3 may facilitate hepatic excretion of cholesterol by increasing the abundance of low-density lipoprotein (LDL) receptors in hepatocyte membranes, thereby enhancing uptake of cholesterol from the blood.

Nitrogen Metabolism

Body proteins are constantly being degraded and resynthesized. Both synthesis and degradation of protein are slowed in the absence of thyroid hormones; conversely, both are accelerated by thyroid hormone. In the presence of excess T4 or T3, the effects of degradation predominate, and often there is severe catabolism of muscle. In hyperthyroid subjects body protein mass decreases despite increased appetite and ingestion of dietary proteins. With thyroid deficiency there is a characteristic accumulation of a mucus-like material consisting of protein complexed with hyaluronic acid and chondroitin sulfate in extracellular spaces, particularly in the skin. Because of its osmotic effect, this material causes water to accumulate in these spaces, giving rise to the edema typically seen in hypothyroid individuals and to the name *myxedema* for hypothyroidism.

REGULATION OF THYROID HORMONE SECRETION

As already indicated, secretion of thyroid hormones depends on stimulation of thyroid follicular cells by TSH, which bears primary responsibility for integrating thyroid function with bodily needs (Chapter 2). In the absence of TSH, thyroid cells are quiescent and atrophy, and, as we have seen, administration of TSH increases both synthesis and secretion of T4 and T3. Secretion of TSH by the pituitary gland is governed by positive input from the hypothalamic hormone thyrotropin-releasing hormone and negative input from thyroid hormones.

Little TSH is produced by the pituitary gland when it is removed from contact with the hypothalamus and transplanted to some extrahypothalamic site, and disruption of the TRH gene reduces the TSH content of mouse pituitaries to less than half that of wild-type litter mate controls. Positive input for thyroid hormone secretion thus originates in the central nervous system by way of TRH and the anterior pituitary gland. TRH increases expression of the genes for both the alpha and the beta subunits of TSH, and increases the posttranslational incorporation of carbohydrate that is required for normal potency of TSH, but these processes can go on at a reduced level in the absence of TRH. Blood levels of thyroid hormones in mice lacking a functional TRH gene are less than half of normal, but the mice grow, develop, and reproduce almost normally, indicating that their hypothyroidism is relatively mild.

Maintaining constant levels of thyroid hormones in blood depends on negative feedback effects of T4 and T3, which inhibit synthesis and secretion of TSH (Figure 12). The contribution of free T4 in blood is quite significant in this regard. Because thyrotropes are rich in type II deiodinase, they can convert this more abundant form of thyroid hormone to T3 and thereby monitor the overall amount of free hormone in blood. High concentrations of thyroid hormones may shut off TSH secretion completely and, when maintained over time, produce atrophy of the thyroid gland. Measurement of relative concentrations of TSH and thyroid hormones in the blood provide critically important information for diagnosing thyroid disease. For instance, low blood concentrations of free T3 and T4 in the presence of elevated levels of TSH signal a primary defect in the thyroid gland, whereas high concentrations of free T3 and T4 accompanied by high concentrations of TSH reflect a defect in the pituitary or hypothalamus. As already noted, the high concentrations of T4 and T3 seen in Graves' disease are accompanied by very low concentrations of TSH in blood as a result of negative feedback inhibition of TSH secretion.

Negative feedback inhibition of TSH secretion results from actions of thyroid hormones exerted both on TRH neurons in the paraventricular nuclei of the hypothalamus and on thyrotropes in the pituitary. Results of animal studies indicate that T3 and T4 inhibit TRH synthesis and secretion. Events thought to occur within the thyrotropes are illustrated in Figure 13. TRH binds its G-protein-coupled heptihelical receptors (Chapter 1) on the surface of thyrotropes. The resulting activation of phospholipase C generates the second messengers inositol trisphosphate (IP_3) and diacylglycerol (DAG). IP_3 promotes calcium mobilization, and DAG activates protein kinase C, both of which rapidly stimulate release of stored hormone. This effect is augmented by influx of extracellular calcium following activation of membrane calcium channels. In addition, transcription of genes for both subunits of TSH is increased. TRH also promotes processing of the carbohydrate components of TSH necessary for maximum biological activity. Meanwhile, both T4 and T3 enter the cell at a rate

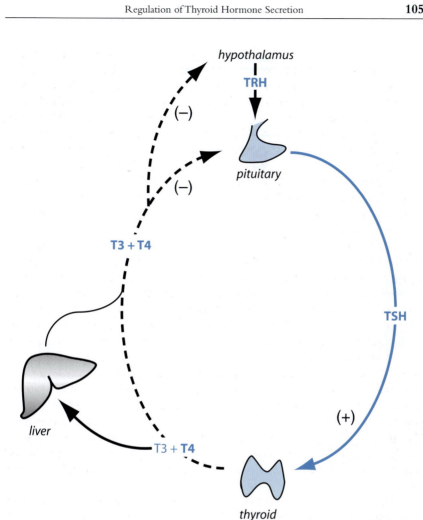

Figure 12   Feedback regulation of thyroid hormone secretion. TRH, Thyrotropin-releasing hormone; TSH, thyroid-stimulating hormone; (+) stimulation; (–) inhibition.

determined by their free concentrations in blood plasma, and T4 is deiodinated to T3 in the cytoplasm. T3 enters the nucleus, binds to its receptors, and down-regulates transcription of the genes for both the alpha and the beta subunits of TSH and for TRH receptors. In addition, T3 inhibits release of stored hormone and accelerates TRH receptor degradation. The net consequence of these actions of T3 is a reduction in the sensitivity of the thyrotropes to TRH (Figure 14).

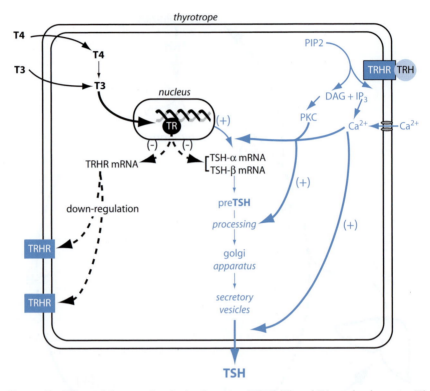

Figure 13 Effects of thyrotropoin–releasing hormone (TRH), T3, and T4 on the thyrotrope. T3 down-regulates expression of genes for TRH receptors (TRHR) and both subunits of TSH (dashed arrows). TRH effects (shown in blue) include up-regulation of TSH gene expression, enhanced TSH glycosylation, and accelerated secretion; (+), increase; (−), decrease.

MECHANISM OF THYROID HORMONE ACTION

As must already be obvious, virtually all cells appear to require optimal amounts of thyroid hormone for normal operation, even though different aspects of function may be affected in different cells. Thyroid hormones are quite hydrophobic and may either diffuse across the cell membrane or enter target cells by a carrier-mediated transport process. T3 formed within the target cell by deiodination of T4 appears to mix freely with T3 taken up from the plasma and to enter the nucleus, where it binds to specific receptors (see Chapter 1). Thyroid hormone receptors are members of the large family of nuclear hormone receptors and bind to specific nucleotide sequences (thyroid response elements, or TREs) in the genes they regulate. Unlike most other nuclear receptors, thyroid hormone receptors

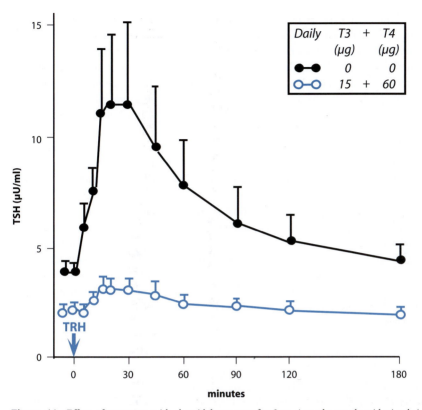

Figure 14 Effect of treatment with thyroid hormones for 3 to 4 weeks on thyroid-stimulating hormone (TSH) secretion in normal young men in response to an intravenous injection of thyrotropin-releasing hormone (TRH). Six normal subjects received 25 mg of TRH, indicated by the arrow. Values are expressed as means ± SEM. (From Snyder, P. J., and Utiger, R. D., *J. Clin. Invest.* **52**, 2077, 1972, with permission.)

bind to their response elements in the absence of hormone. They bind as monomers or as homodimers composed of either of two thyroid hormone receptors, or they may form heterodimers with other nuclear receptor family members, usually the receptor for an isomer of retinoic acid. In the absence of T3, the unoccupied receptor, in conjunction with a corepressor protein, inhibits T3-dependent gene expression by maintaining the DNA in a tightly coiled configuration that bars access of transcription activators or RNA polymerase. On binding T3, the configuration of the receptor is modified in a way that causes it to release the corepressor and bind instead to a coactivator. Although T3 acts in an analogous way to suppress expression of some genes, the underlying mechanism for negative control of gene expression is not understood.

Nuclear receptors for T3 are encoded in two genes, designated *TRα* and *TRβ*. The TRα gene resides on chromosome 17 and gives rise to two isoforms, TRα$_1$ and TRα$_2$, as a result of alternate splicing that deletes the T3 binding site from the TRα$_2$ isoform. The TRα$_2$ isoform, therefore, cannot act as a hormone receptor, but it nevertheless plays a vital physiological role (see below). The TRβ gene maps to chromosome 3 and also gives rise to two alternately spliced products, TRβ$_1$ and TRβ$_2$. TRα$_1$ and TRβ$_1$ are widely distributed throughout the body and are present in different ratios in the nuclei of all target tissues examined, but TRβ$_2$ appears to be expressed primarily in the anterior pituitary gland and the brain.

Efforts to determine which T3 responses are mediated by each form of the T3 receptor have been greatly advanced by the advent of technology that permits disruption or "knockout" of individual genes in mouse embryos. Mice lacking both TRβ isoforms have no developmental deficiencies, are fertile, and exhibit no obvious behavioral abnormalities. However, these animals have abnormally high rates of TSH, T4, and T3 secretion, presumably because TRβ$_2$ mediates the negative feedback action of T3. These symptoms are remarkably similar to those seen in a rare genetic disease that is characterized by resistance to thyroid hormone. Like the knockout mice, patients exhibit few abnormalities but have increased circulating levels of TSH, T4, and T3. They have enlarged thyroid glands (goiter) stemming from increased TSH levels, but exhibit none of the consequences of T4 hypersecretion. This disease typically results from mutations in the TRβ gene.

No effects on life span or fertility result from manipulation of TRα gene so that it encodes only the TRα$_2$ isoform, which cannot bind T3. However, animals that lack the TRα$_1$ isoform have low heart rates and low body temperature. When the TRα gene was knocked out so that neither the α$_1$ or α$_2$ isoform could be expressed, the animals stopped growing after about 2 weeks and died shortly after weaning, with apparent failure of intestinal development. Thus although few symptoms of hypothyroidism result from knockout of any of the TR receptors that are capable of binding T3, loss of the α$_2$ isoform produced devastating effects, suggesting that it plays a critical, though perhaps T3-independent, role in gene transcription. The combined absence of TRα$_1$, TRβ$_1$, and TRβ$_2$ produces more symptoms of hypothyroidism than lack of either TRα$_1$ or TRβ, suggesting that these receptors have redundant or overlapping functions. However, the hypothyroid symptoms are mild compared to those seen when the complement of TRs is normal but thyroid hormone is absent. Unoccupied TRs that repress gene expression may therefore produce harmful effects. Consistent with this idea, a mutation of the TRβ gene that prevents it from binding T3 produced severe defects in neurological development, similar to those seen in hypothyroid mice even though T3 was abundant. Thus at least one of the physiological roles of T3 may be to counteract the consequences of T3 receptors in silencing of some genes.

Although extensive evidence indicates that T3 and T4 produce the majority of their actions through nuclear receptors, extranuclear specific binding proteins for thyroid hormones have also been found in the cytosol and mitochondria. The function, if any, of these proteins is not known. In addition, some rapid effects of T3 and T4 that may not involve the genome have also been described. It is highly likely that T3 and T4 have physiologically important actions that are not dependent on nuclear events, but detailed understanding will require further research.

SUGGESTED READING

Braverman, L. E., and Utiger, R. D. (eds). (2000). "Werner and Ingbar's The Thyroid," 8th Ed. Lippincott Williams and Wilkins, Philadelphia. (This book provides excellent coverage of a broad range of basic and clinical topics.)

De La Vieja, A., Dohan, O., Levy, O., and Carrasco, N. (2000). Molecular analysis of the sodium/iodide symporter: Impact on thyroid and extrathyroid pathophysiology. *Physiol. Rev.* **80**, 1083–1105.

Gershengorn, M. C., and Osman, R. (1996). Molecular and cellular biology of thyrotropin-releasing hormone receptors. *Physiol. Rev.* **76**, 175–191.

Köhrle, J. (1999). Local activation and inactivation of thyroid hormones: The deiodinase family. *Mol. Cell. Endocrinol.* **151**, 103–119.

Koibuchi, N., and Chin, W. W. (2000). Thyroid hormone action and brain development. *Trends Endocrinol. Metab.* **11**, 123–128.

Rapoport, B., Chazenbalk, D., Jaume, J. C., and McLachlan, S. M. (1998). The thyrotropin (TSH) receptor: Interactions with TSH and autoantibodies. *Endocr. Rev.* **19**, 673–716.

Vassart, G., and Dumont, J. (1992). The thyrotropin receptor and the regulation of thyrocyte function and growth. *Endocr. Rev.* **13**, 596–611.

Weiss, R. E., and Refetoff, S. (2000). Resistance to thyroid hormone. *Rev. Endocr. Metab. Disord.* **1**, 97–108.

Yen, P. M. (2001). Physiological and molecular basis of thyroid hormone action. *Physiol. Rev.* **81**, 1097–1142.

CHAPTER 4

Adrenal Glands

OVERVIEW

The adrenal glands are complex polyfunctional organs that secrete hormones that are required for maintenance of life. Without adrenal hormones, deranged electrolyte or carbohydrate metabolism leads to circulatory collapse or hypoglycemic coma and death. The hormones of the outer region, or cortex, are steroids that act at the level of the genome to regulate the expression of genes that govern the operation of fundamental processes in virtually all cells. The inner adrenal gland region, the medulla, is actually a component of the sympathetic nervous system and participates in the wide array of regulatory responses that are characteristic of that branch of the nervous system.

There are three major categories of adrenal steroid hormones: mineralocorticoids, which act to defend the body content of sodium and potassium; glucocorticoids, which affect body fuel metabolism, responses to injury, and general cell function; and androgens, which function in a manner similar to that of the hormone of the male gonads. Secretion of mineralocorticoids is primarily controlled by the kidneys through secretion of renin and the consequent production of angiotensin. Secretion of glucocorticoids and androgens is controlled by the anterior pituitary gland through secretion of ACTH. We focus on actions of these hormones on the limited number of processes that are most thoroughly studied, but it should be kept in mind that adrenal cortical hormones directly or indirectly affect almost every physiological process and hence are central to the maintenance of homeostasis.

The adrenal cortex and the medulla often behave as a functional unit and together confer a remarkable capacity to cope with changes in the internal or external environment. Fast-acting medullary hormones are signals for physiological adjustments, and slower acting cortical hormones maintain or increase sensitivity of tissues to medullary hormones and other signals as well as maintain or enhance the capacity of tissues to respond to such signals. The cortical hormones thus tend to be modulators rather than initiators of responses.

MORPHOLOGY

The adrenal glands are bilateral structures situated above the kidneys. They are composed of an outer region, or cortex, consisting of three zones that normally make up more than three-quarters of the adrenal mass, and an inner region, or medulla (Figure 1). The medulla is a modified sympathetic ganglion that, in response to signals reaching it through cholinergic, preganglionic fibers, releases either or both of its two hormones, epinephrine and norepinephrine, into adrenal venous blood. The cortex arises from mesodermal tissue and produces a class of lipid-soluble hormones derived from cholesterol and called *steroids*. The cortex is subdivided histologically into three zones. Cells in the outer zone, or *zona glomerulosa*, are arranged in clusters (glomeruli) and produce the hormone aldosterone. In the *zona fasciculata*, which comprises the bulk of the cortex, rows of

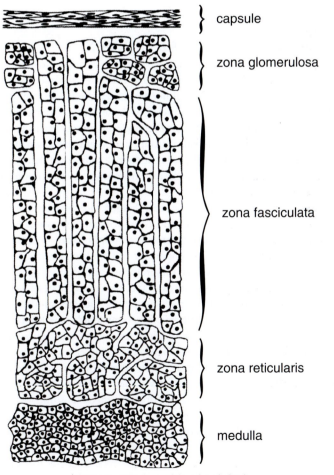

capsule

zona glomerulosa

zona fasciculata

zona reticularis

medulla

Figure 1 Histology of the adrenal gland.

lipid-laden cells are arranged radially in bundles of parallel cords (fasces). The innermost zone of the cortex, the *zona reticularis*, consists of a tangled network of cells. The fasciculata and reticularis, which produce both cortisol and the adrenal androgens, are functionally separate from the zona glomerulosa.

The adrenal glands receive their blood supply from numerous small arteries that branch off the renal arteries or the lumbar portion of the aorta and its various major branches. These arteries penetrate the adrenal capsules and divide to form the subcapsular plexus, from which small arterial branches pass centripetally toward the medulla. The subcapsular plexuses also give rise to long loops of capillaries that pass between the cords of fascicular cells and empty into sinusoids in the reticularis and medulla. Sinusoidal blood collects through venules into a single large central vein in each adrenal and drains into either the renal vein or the inferior vena cava.

ADRENAL CORTEX

In all species thus far studied the adrenal cortex is essential for maintenance of life. Insufficiency of adrenal cortical hormones (Addison's disease) produced by pathological destruction or surgical removal of the adrenal cortices results in death within 1 to 2 weeks unless replacement therapy is instituted. Virtually every organ system goes awry with adrenal cortical insufficiency, but the most likely cause of death appears to be circulatory collapse secondary to sodium depletion. When food intake is inadequate, death may result instead from insufficient amounts of glucose in the blood (hypoglycemia).

Adrenal cortical hormones have been divided into two categories based on their ability to protect against these two causes of death. The so-called mineralo-corticoids are necessary for maintenance of sodium and potassium balance. Aldosterone is the physiologically important mineralocorticoid, although some deoxycorticosterone, another potent mineralocorticoid, is also produced by the normal adrenal gland (Figure 2). Cortisol and, to a lesser extent, corticosterone are the physiologically important glucocorticoids and are so named for their ability to maintain carbohydrate reserves. Glucocorticoids have a variety of other effects as well. At high concentrations, aldosterone may exert glucocorticoid-like activity, and conversely, cortisol and corticosterone may exert some mineralocorti-coid activity (see p. 131). The adrenal cortex also produces androgens, which as their name implies have biological effects similar to those of the male gonadal hormones (see Chapter 11). Adrenal androgens mediate some of the changes that occur at puberty. Adrenal steroid hormones are closely related to steroid hormones produced by the testis and ovary and are synthesized from common precursors. In some abnormal states the adrenals may secrete any of the gonadal steroids.

Glucocorticoids

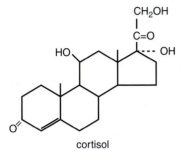

cortisol

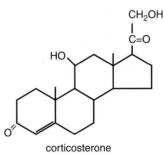

corticosterone

Mineralocorticoids

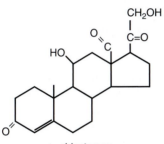

aldosterone

11-deoxycorticosterone (DOC)

Androgens

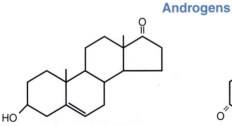

dehydroepiandrosterone

androstenedione

Figure 2 Principal adrenal steroid hormones.

ADRENOCORTICAL HORMONES

All of the adrenal steroids are derivatives of the polycyclic phenanthrene nucleus, which is also present in cholesterol, ovarian and testicular steroids, bile acids, and precursors of vitamin D. Use of some of the standard conventions for

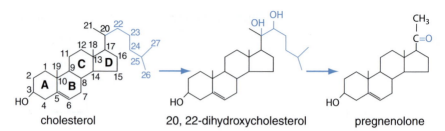

Figure 3 Conversion of cholesterol to pregnenolone. Carbons 20 and 22 are sequentially oxidized (in either order), followed by oxidative cleavage of the bond between them. All three reactions are catalyzed by cytochrome P450scc.

naming the rings and the carbons facilitates discussion of the biosynthesis and metabolism of the steroid hormones. When drawing structures of steroid hormones, carbon atoms are indicated by junctions of lines that represent chemical bonds. The carbons are numbered and the rings lettered as shown in Figure 3. It should be remembered that steroid hormones have complex three-dimensional structures; they are not flat, two-dimensional molecules as we depict them for simplicity. Substituents on the steroid nucleus that project toward the reader are usually designated by the prefix β. Those that project away from the reader are designated by α and are shown diagrammatically with dashed lines. The fully saturated 21-carbon molecule is called pregnane. When a double bond is present in any of the rings, the -*ane* in the ending is changed to -*ene*, or to -*diene* when there are two double bonds, i.e., pregnene or pregnadiene. The location of the double bond is designated by the Greek letter Δ followed by one or more superscripts to indicate the location. The presence of a hydroxyl group (OH) is indicated by the ending -*ol*, and the presence of a keto group (O) by the ending -*one*. Thus the important intermediate in the biosynthetic pathway for steroid hormones shown in Figure 3 has a double bond in the B ring, a keto group on carbon 20, and a hydroxyl group on carbon 3, and hence is called Δ^5-pregnenolone.

The starting material for steroid hormone biosynthesis is cholesterol, most of which arrives at the adrenal cortex in the form of low-density lipoproteins, which are avidly taken up from blood by a process of receptor-mediated endocytosis. Adrenal cortical cells also synthesize cholesterol from carbohydrate or fatty acid precursors. Substantial amounts of cholesterol are stored in steroid hormone-producing cells in the form of fatty acid esters.

Key reactions in the biosynthesis of the adrenal hormones are catalyzed by a particular class of oxidizing enzymes, the cytochromes P450, which includes a large number of hepatic detoxifying enzymes called mixed-function oxidases. These enzymes contain a heme group covalently linked through a sulfur–iron bond and absorb light in the visible range. The name "P450" derives from the property of these pigments to absorb light at 450 nanometers when reduced by carbon monoxide. The P450 enzymes utilize molecular oxygen and electrons donated from NADPH$^+$ to oxidize their substrates. Although they have a single substrate-binding site, some of the P450 enzymes catalyze more than one oxidative step in steroid hormone synthesis.

The rate-limiting step in the biosynthesis of all of the steroid hormones is cleavage of the side chain to convert the 27-carbon cholesterol molecule to the 21-carbon pregnenolone molecule (Figure 3). The enzyme that catalyzes this complex reaction, is called P450scc (for side chain cleavage) and is located on the inner mitochondrial membrane. P450scc catalyzes the oxidation of carbons 20 and 22 and then cleaves the bond between them to shorten the side chain. This initial step in hormone biosynthesis requires a complicated series of molecular events. Cholesterol must first be released from its esterified storage form by the action of an esterase. The free but water-insoluble cholesterol must then be transferred to the mitochondria, perhaps through the agency of a cholesterol-binding protein and participation of cytoskeletal elements. Cholesterol must then enter the mitochondria to gain access to P450scc, whose activity is limited primarily by availability of its substrate. The steroid acute regulatory (StAR) protein plays an indispensable role in presenting cholesterol to P450scc through mechanisms that are still incompletely understood. The StAR protein has a very short half-life, and stimulation of its synthesis appears to be the critical regulated step in steroid hormone biosynthesis. Unlike cholesterol, 21-carbon steroids apparently pass through the mitochondrial membrane rather freely.

Pregnenolone is the common precursor of all steroid hormones produced by the adrenals or the gonads. An early step in hormone biosynthesis is oxidation of the hydroxyl group at carbon 3 to a keto group. This reaction is catalyzed by the enzyme 3β-hydroxysteroid dehydrogenase (3βHSD) and initiates a rearrangement that shifts the double bond from the B ring to the A ring. A ketone group at carbon 3 is found in all biologically important adrenal steroids and appears necessary for physiological activity. Biosynthesis of the various steroid hormones involves oxygenation of carbons 21, 17, 11, and 18, as depicted in Figure 4. The exact sequence of hydroxylations may vary, and some of the reactions may take place in a different order than that presented in the figure. The specific hormone that is ultimately secreted once the cholesterol–pregnenolone roadblock has been passed is determined by the enzymatic makeup of the particular cells involved. For example, there are two different P450 enzymes that catalyze the hydroxylation of carbon 11. One of these is found exclusively in cells of the zona glomerulosa (P450c11AS) and catalyzes the oxidation of both carbon 11 and carbon 18. The other enzyme (P450c11β) is found in cells of the zonae fasciculata and reticularis and can oxidize only carbon 11. At the same time, cells of the zonae fasciculata and reticularis, but not of the zona glomerulosa, express the enzyme P450c17, also called P450 17α-hydroxylase/lyase, which catalyzes the oxidation of the carbon at position 17.

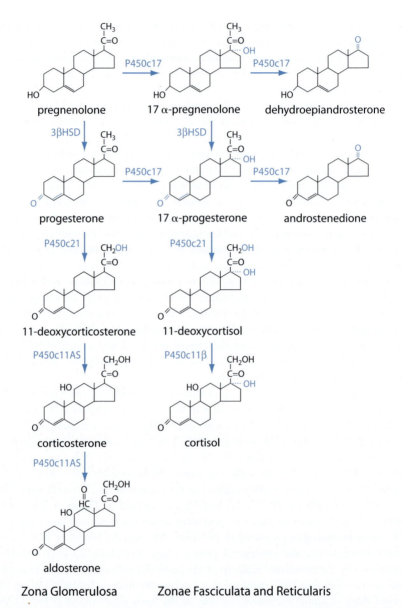

Figure 4 Biosynthesis of adrenal cortical hormones. The pathway on the left represents that found in the zona glomerulosa. The pathways on the right are followed in the zona fasciculata and zona reticularis. Hydroxylation at carbon 17 usually precedes hydroxylation at carbon 11. Note that changes produced in each reaction are shown in blue.

Hence glomerulosa cells can produce corticosterone, aldosterone, and deoxy-corticosterone, but not cortisol, whereas cells of the zonae fasciculata and reticularis can form cortisol and 17α-hydroxyprogesterone. When reducing equivalents are delivered to P450c17 rapidly enough, the reaction continues beyond 17α-progesterone hydroxylation to cleavage of the bond between carbon 17 and carbon 20. Removal of carbons 20 and 21 produces the 19-carbon androgens. Hence androgens can also be produced by these cells but not by glomerulosa cells.

As is probably already apparent, steroid chemistry is complex and can be bewildering; but because these compounds are so important physiologically and therapeutically, some familiarity with their structures is required. We can simplify the task somewhat by noting that steroid hormones can be placed into three major categories: those that contain 21 carbon atoms, those that contain 19 carbon atoms, and those that contain 18 carbon atoms. In addition, there are relatively few sites where modification of the steroid nucleus determines its physiological activity.

The physiologically important steroid hormones of the 21-carbon series are as follows:

1. Progesterone. Progesterone has the simplest structure and can serve as a precursor molecule for all of the other steroid hormones. Note that the only modifications to the basic carbon skeleton of the 21-carbon steroid nucleus are keto groups at positions 3 and 20. Normal adrenal cortical cells convert progesterone to other products so rapidly that none escapes into adrenal venous blood. Progesterone is a major secretory product of the ovaries and the placenta.

2. 11-Deoxycorticosterone. Addition of a hydroxyl group to carbon 21 of progesterone is the minimal change required for adrenal corticoid activity. This addition produces 11-deoxycorticosterone, a potent mineralocorticoid that is virtually devoid of glucocorticoid activity. Deoxycorticosterone is only a minor secretory product of the normal adrenal gland but may become important in some disease states.

3. Corticosterone. A hydroxyl group at carbon 11 is found in all gluco-corticoids. Adding the hydroxyl group at carbon 11 confers glucocorticoid activity to deoxycorticosterone and reduces its mineralocorticoid activity 10-fold. This compound is corticosterone and can be produced in cells of all three zones of the adrenal cortex. Corticosterone is the major glucocorticoid in the rat but is of only secondary importance in humans.

4. Aldosterone. Corticosterone is a precursor of aldosterone, which is pro-duced in cells of the zona glomerulosa by oxidation of carbon 18 to an aldehyde. The oxygen at carbon 18 increases the mineralocorticoid potency of cortico-sterone by a factor of 200 and only slightly decreases glucocorticoid activity.

5. Cortisol. Cortisol differs from corticosterone only by the presence of a hydroxyl group at carbon 17. Cortisol is the most potent of the naturally occurring glucocorticoids. It has 10 times as much glucocorticoid activity as aldosterone, but less than 0.25% of aldosterone's mineralocorticoid activity in normal human subjects. Synthetic glucocorticoids with even greater potency than cortisol are available for therapeutic use.

6. Cortisone. Cortisone differs from cortisol only in that the substituent on carbon 11 is a keto group rather than a hydroxyl group. Cortisone is produced from cortisol at extraadrenal sites by oxidation of the hydroxyl group on carbon 11 and circulates in blood at about one-fifth the concentration of cortisol. Oxidation of cortisol to cortisone profoundly lowers its affinity for adrenal steroid hormone receptors and hence inactivates it. Cortisol is oxidized to cortisone in mineralo-corticoid target cells, and cortisone can be reduced to cortisol in the liver and other glucocorticoid target tissues. This so-called cortisol/cortisone shuttle is catalyzed by two enzyme forms of 11-hydroxysteroid dehydrogenase, HSD I and HSD II, which are products of different genes and are expressed in different tissues. HSD I can catalyze the reaction in either direction, and hence may activate or inactivate the hormone. HSD II, which is expressed in all mineralocorticoid target tissues, catalyzes only the oxidation of cortisol to cortisone (Figure 5).

Steroids in the 19-carbon series usually have androgenic activity and are precursors of the estrogens (female hormones). Hydroxylation of either pregnenolone or progesterone at carbon 17 is the critical prerequisite for cleavage of the $C_{20,21}$ side chain to yield the adrenal androgens dehydroepiandrosterone or androstenedione (see Figure 4). These compounds are also called 17-ketosteroids. The principal testicular androgen is testosterone, which has a hydroxyl group rather than a keto group at carbon 17. Although 19-carbon androgens are products of the same enzyme that catalyzes 17α-hydroxylation in the adrenals and the gonads, cleavage of the bond linking carbons 17 and 20 in the adrenals normally occurs to

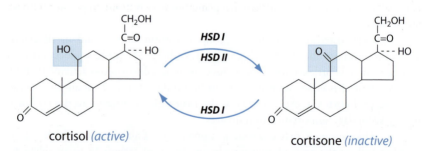

cortisol *(active)*   cortisone *(inactive)*

Figure 5 The cortisol/cortisone shuttle. Two isoenzymes of 11-hyrdoxysteroid dehydrogenase (HSD I and HSD II) catalyze the inactivating conversion of cortisol to cortisone. HSD I can also catalyze conversion of the inactive cortisone to cortisol.

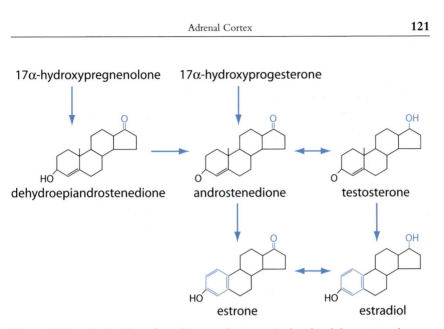

17α-hydroxypregnenolone   17α-hydroxyprogesterone

dehydroepiandrostenedione   androstenedione   testosterone

estrone   estradiol

Figure 6 Biosynthetic pathway for androgens and estrogens. In the adrenal the sequence does not usually proceed all the way to testosterone and to the estrogens, which are the gonadal hormones. Because the cells of the zona glomerulosa lack 17α-hydroxylase, these reactions can occur only in the inner zones.

a significant extent only after puberty, and then is confined largely to cells of the zona reticularis.

Steroids of the 18-carbon series usually have estrogenic activity. Estrogens characteristically have an unsaturated A ring. Oxidation of the A ring (a process called *aromatization*) results in loss of the methyl carbon at position 19 (Figure 6). This reaction, which takes place principally in ovaries and placenta, can also occur in a variety of nonendocrine tissues; aromatization of the A ring of either testicular or adrenal androgens comprises the principal source of estrogens in men and postmenopausal women.

Regulation of Adrenocorticol Hormone Synthesis

Effects of ACTH

Adrenocorticotropic hormone maintains normal secretory activity of the inner zones of the adrenal cortex. After removal of the pituitary gland, little or no steroidogenesis occurs in the zona fasciculata or reticularis, but the zona glomerulosa continues to function. In cells of all three zones, ACTH interacts with a specific G-protein-coupled membrane receptor and triggers production of cyclic AMP by activating adenylyl cyclase (see Chapter 1). Cyclic AMP activates protein kinase A, which catalyzes the phosphorylation of a variety of proteins and thereby modifies their activity. In the zones fasciculata and reticularis this results in accelerated deesterification of cholesterol esters, increased transport of cholesterol to the mitochondria, and increased synthesis of StAR protein. Early studies of how ACTH increases steroidogenesis revealed a puzzling requirement for protein synthesis that ultimately led to the discovery of the StAR protein. It appears that ACTH increases StAR protein by stimulating its synthesis on preexisting mRNA templates and by promoting gene transcription (Figure 7). Thus the immediate actions of ACTH accelerate the delivery of cholesterol to P450scc to form pregnenolone. Once pregnenolone is formed, remaining steps in steroid biosynthesis can proceed without further intervention from ACTH, although some evidence suggests that ACTH may also speed up some later reactions in the biosynthetic sequence. With continued stimulation, ACTH, acting through cyclic AMP and protein kinase A, also stimulates transcription of the genes encoding the P450 enzymes (P450scc, P450c21, P450c17, and P450c11β) and the LDL receptor responsible for uptake of cholesterol.

ACTH is the only hormone known to control synthesis of the adrenal androgens, the 19-carbon steroids produced primarily in the zona reticularis. Their production is limited first by the rate of conversion of cholesterol to pregnenolone and subsequently by cleavage of the bond between carbons 17 and 20. As already mentioned, P450c17, the same enzyme that catalyzes α-hydroxylation of carbon 17 of cortisol, also catalyzes the second oxidative reaction at carbon 17 (17,20-lyase) that removes the $C_{20,21}$ side chain. Some evidence suggests that the lyase reaction is increased by phosphorylation of the P450c17, and other studies suggest that androgen production is driven by the capacity of reticularis cells to deliver reducing equivalents to the reaction. Little or no androgen is produced in young children whose adrenal glands contain only a rudimentary zona reticularis. The reticularis, with its unique complement of enzymes, develops shortly before puberty. The arrival of puberty is heralded by a dramatic increase in production of the adrenal androgens, principally dehydroepiandrosterone sulfate (DHEAS), which are responsible for growth of pubic and axillary hair. Secretion of DHEAS gradually rises to reach a maximum by age 20 to 25, and thereafter declines. This pattern of androgen secretion is quite different from the pattern of cortisol secretion and therefore appears to be governed by factors other than simply the ACTH-dependent rate of pregnenolone formation. These findings have led some investigators to propose separate control of androgen production, possibly by another, as yet unidentified, pituitary hormone, but to date no such hormone has been found. It is important to emphasize that increased stimulation of both the fasciculata and the reticularis by ACTH can profoundly increase adrenal androgen production.

Effects of ACTH on the adrenal cortex are not limited to accelerating the rate-determining step in steroid hormone production. ACTH either directly or

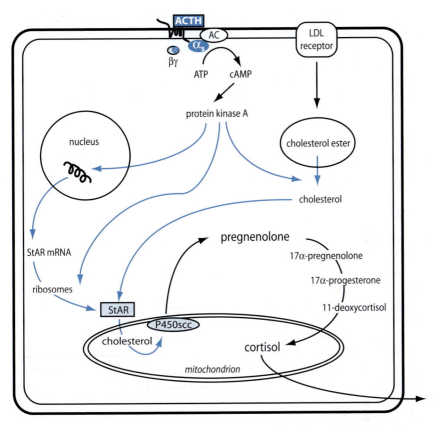

Figure 7 Stimulation of steroidogenesis by ACTH in a zona fasciculata cell of the adrenal cortex. The rate-determining reaction is the conversion of cholesterol to pregnenolone, which requires mobilization of cholesterol from its storage droplet and transfer to the P450scc (side chain cleavage enzyme) on the inner mitochondrial membrane. See text for discussion. ACTH may also increase cholesterol uptake by increasing the number or affinity of low-density lipoprotein (LDL) receptors. α_s, Stimulatory α subunit of the guanine nucleotide-binding protein; AC, adenylate cyclase; $\beta\gamma$, $\beta\gamma$ subunits of the guanine nucleotide-binding protein; StAR, steroid acute regulatory protein.

indirectly also increases blood flow to the adrenal glands and thereby provides not only the needed oxygen and metabolic fuels but also increased capacity to deliver newly secreted hormone to the general circulation. ACTH maintains the functional integrity of the inner zones of the adrenal cortex: absence of ACTH leads to atrophy of these two zones, and chronic stimulation increases their mass.

Stimulation with ACTH increases steroid hormone secretion within 1 to 2 minutes, and peak rates of secretion are seen in about 15 minutes. Unlike other glandular cells, steroid-producing cells do not store hormones, and hence biosynthesis and secretion are components of a single process regulated at the step of cholesterol conversion to pregnenolone. Because steroid hormones are lipid soluble they can diffuse through the plasma membrane and enter the circulation through simple diffusion down a concentration gradient. Even under basal conditions, cortisol concentrations are more than 100 times higher in fasciculata cells than in plasma. It is not surprising, therefore, that biosynthetic intermediates may escape into the circulation during intense stimulation. Human adrenal glands normally produce about 20 mg of cortisol, about 2 mg of corticosterone, 10–15 mg of androgens, and about 150 µg of aldosterone each day, but with sustained stimulation they can increase this output manyfold.

Regulation of Aldosterone Synthesis

The regulation of aldosterone synthesis is considerably more complex than that of the glucocorticoids and is not completely understood. Although cells of the zona glomerulosa express ACTH receptors and ACTH is required for optimal secretion, ACTH is not an important regulator of aldosterone production in most species. Angiotensin II, an octapeptide whose production is regulated by the kidney (see below and Chapter 7), is the hormonal signal for increased production of aldosterone (Figure 8). The cellular events entrained by angiotensin II are not as well established as those described for ACTH. Like ACTH, angiotensin II reacts with specific heptahelical membrane receptors, but angiotensin II does not activate adenylyl cyclase or use cyclic AMP as its second messenger. Instead, it acts through IP_3 and calcium to promote the formation of pregnenolone from cholesterol. The ligand-bound angiotensin receptor associates with $G\alpha_q$ and activates phospholipase C to release IP_3 (see Chapter 1) and diacylglycerol from membrane phosphatidylinositol bisphosphate. $G\alpha_q$ may also interact directly with potassium channels and cause them to close. The resulting depolarization of the membrane opens voltage-gated calcium channels and allows calcium to enter. Simultaneously, the beta/gamma subunits directly activate these channels, further promoting calcium entry. Intracellular calcium concentrations are also increased by interaction of IP_3 with its receptor in the endoplasmic reticulum to release stored calcium. Increased intracellular calcium activates a calmodulin-dependent protein kinase that promotes transfer of cholesterol into the mitochondria by increasing synthesis of the StAR protein in the same manner as described for protein kinase A in fasciculata cells. The increase in cytosolic calcium in turn raises the intramitochondrial calcium concentration and stimulates P450c11AS, which catalyzes the critical final reactions in aldosterone synthesis. The DAG that is released by activation of phospholipase C activates protein kinase C, but the role of this enzyme is uncertain. It may augment synthesis of StAR protein, and it may participate in calcium channel activation. Protein kinase C may also play an important role in mediating

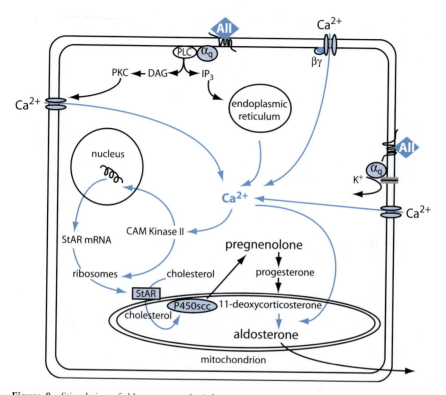

Figure 8 Stimulation of aldosterone synthesis by angiotensin II (A_{II}). A_{II} accelerates the conversion of cholesterol to pregnenolone and 11-deoxycorticosterone to aldosterone. α_q, $\beta\gamma$, Subunits of the guanine nucleotide-binding protein; PLC, phospholipase C; DAG, diacylglycerol; IP_3, inositol trisphosphate; PKC, protein kinase C; CAM kinase II, calcium/calmodulin-dependent protein kinase II; StAR, steroid acute regulatory protein. High concentrations of potassium (K^+) in blood also increase aldosterone synthesis and secretion by increasing intracellular calcium secondary to partial membrane depolarization.

the hypertrophy of the zona glomerulosa seen after prolonged stimulation of the adrenal glands with angiotensin II.

Cells of the zona glomerulosa are exquisitely sensitive to changes in concentration of potassium in the extracellular fluid and adjust aldosterone synthesis accordingly. An increase of as little as 0.1 mM in the concentration of potassium, which corresponds to a change of only about 2–3%, may increase aldosterone production by as much as 35%. Increased extracellular potassium depolarizes the plasma membrane and activates voltage-gated calcium channels. The resulting increase in intracellular calcium stimulates aldosterone synthesis as already described. The rate of aldosterone secretion can also be affected by the concentration of sodium in the extracellular fluid, but relatively large changes are required. A decline in sodium concentration increases sensitivity to angiotensin II and potassium, but direct effects of sodium on glomerulosa cells are relatively unimportant except in extreme cases of sodium depletion. However, sodium profoundly affects aldosterone synthesis indirectly through its influence on production of angiotensin II, as described below and in Chapter 7.

Synthesis and secretion of aldosterone are also negatively regulated by the atrial natriuretic factor (ANF) secreted primarily by the cardiac atria. This hormone and its role in normal physiology are discussed in Chapter 7. ANF receptors have intrinsic guanylyl cyclase activity, and when bound to ANF, catalyze the conversion of guanosine triphosphate to cyclic guanosine monophosphate (cyclic GMP). Precisely how an increase in cyclic GMP antagonizes the actions of angiotensin and potassium on aldosterone synthesis has not been established. Cyclic GMP is known to activate the enzyme cyclic AMP phosphodiesterase, and may thereby lower basal levels of cyclic AMP, or it may act through stimulating cyclic GMP-dependent protein kinase.

Adrenocorticol Hormones in Blood

Adrenal cortical hormones are transported in blood bound to the specific plasma protein transcortin, or corticosteroid-binding globulin (CBG), and to a lesser extent to albumin. Like albumin, CBG is synthesized and secreted by the liver, but its concentration of ~1 μM in plasma is only about 1000th that of albumin. CBG is a glycoprotein with a molecular weight of about 58,000, and is a member of the serine proteinase inhibitor (SERPIN) superfamily of proteins. It has a single steroid hormone binding site whose affinity for glucocorticoids is nearly 20 times higher than for aldosterone. About 95% of the glucocorticoids and about 60% of the aldosterone in blood are bound to protein. Under normal circumstances the concentration of free or unbound cortisol in plasma is about 100 times that of aldosterone. Probably because they circulate bound to plasma proteins, adrenal steroids have a relatively long half-life in blood: 1.5 to 2 hours for cortisol, and about 15 minutes for aldosterone.

Metabolism and Excretion of Adrenocorticol Hormones

Because mammals cannot degrade the steroid nucleus, elimination of steroid hormones is achieved by inactivation through metabolic changes that make them unrecognizable to their receptors. Inactivation of glucocorticoids occurs mainly in liver and is achieved primarily by reduction of the A ring and its keto group at position 3. Conjugation of the resulting hydroxyl group on carbon 3 with glucuronic acid or sulfate increases water solubility and decreases binding to CBG so the steroid can now pass through renal glomerular capillaries and be excreted in the urine. The major urinary products of steroid hormone degradation are glucuronide esters of 17-hydroxycorticosteroids (17-OHCS) derived from cortisol, and 17-ketosteroids (17-KS) derived from glucocorticoids and androgens. Because recognizable hormonal products can be identified in the urine, it is possible to estimate the daily secretory rate of steroid hormones by the noninvasive technique of analyzing urinary excretory products.

PHYSIOLOGY OF THE MINERALOCORTICOIDS

Although several naturally occurring adrenal cortical hormones, including glucocorticoids, can produce mineralocorticoid effects, aldosterone is by far the most important mineralocorticoid physiologically. In its absence there is a progressive loss of sodium by the kidney, which results secondarily in a loss of extracellular fluid (see Chapter 7). It may be recalled that the kidney adjusts the composition of the extracellular fluid by processes that involve formation of an ultrafiltrate of plasma, followed by secretion or selective reabsorption of solutes and water. With severe loss of blood volume (hypovolemia), water is retained in an effort to restore volume, and the concentration of sodium in blood plasma may gradually fall (hyponatremia) from the normal value of 140 mEq/liter to 120 mEq/liter or even lower in extreme cases. With the decrease in concentration of sodium, the principal cation of extracellular fluid, there is a net transfer of water from extracellular to intracellular space, further aggravating hypovolemia. Diarrhea is frequently seen, and it, too, worsens hypovolemia. Loss of plasma volume increases the hematocrit and the viscosity of blood (hemoconcentration). Simultaneous with the loss of sodium, the ability to excrete potassium is impaired, and with continued dietary intake, plasma concentrations of potassium may increase from the normal value of 4 mEq/liter to 8–10 mEq/liter (hyperkalemia). Increased concentrations of potassium in blood, and therefore in extracellular fluid, result in partial depolarization of plasma membranes of all cells, leading to cardiac arrhythmia and weakness of muscles, including the heart. Blood pressure falls from the combined effects of decreased vascular volume, decreased cardiac contractility, and decreased responsiveness of vascular smooth muscle to vasoconstrictor agents caused by hyponatremia. Mild acidosis is seen with mineralocorticoid deficiency, partly as a result of deranged potassium balance and partly from lack of the direct effects of aldosterone on hydrogen ion excretion.

All of these life-threatening changes can be reversed by administration of aldosterone and can be traced to the ability of aldosterone to promote inward transport of sodium across epithelial cells of kidney tubules and the outward transport of potassium and hydrogen ions into the urine. It has been estimated that aldosterone is required for the reabsorption of only about 2% of the sodium filtered at the renal glomeruli; even in its absence, about 98% of the filtered sodium is reabsorbed. However, 2% of the sodium filtered each day corresponds to the amount present in about 3.5 liters of extracellular fluid. Aldosterone also promotes sodium and potassium transport by the sweat glands, the colon, and the salivary glands. Of these target tissues, the kidney is by far the most important.

Effects of Aldosterone on the Kidney

Initial insights into the action of aldosterone on the kidney were obtained from observations of the effects of hormone deprivation or administration on the composition of the urine. Mineralocorticoids decrease the ratio of urinary sodium to potassium concentrations; in the absence of mineralocorticoids, the ratio increases. However, although aldosterone promotes both sodium conservation and potassium excretion, the two effects are not tightly coupled, and sodium is not simply exchanged for potassium. Indeed, the same amount of aldosterone that increased both sodium retention and potassium excretion when given to adrenalectomized dogs stimulated only potassium excretion in normal, sodium-replete dogs. Similarly, when normal human subjects were given aldosterone for 25 days, the sodium-retaining effects lasted only for the first 15 days, but increased excretion of potassium persisted for as long as the hormone was given (Figure 9). Renal handling of sodium and potassium is complex, and compensatory mechanisms exerted at aldosterone-insensitive loci within the kidney can offset sustained effects of aldosterone on sodium absorption when measured in the otherwise normal subject (see Chapter 7).

Renal tubular epithelial cells are polarized. Permeability properties of the membrane that faces the lumen are different from those of the basolateral membranes that face the interstitium. Reabsorption of sodium depends on entry through channels in the luminal membrane followed by extrusion by the sodium/potassium-dependent ATPase in the basolateral membranes. This enzyme is energized by cleavage of ATP and exchanges three sodium ions for two potassium ions. Potassium, which would otherwise accumulate within the cells, can then passively diffuse out through channels located in both the luminal and basolateral membranes (Figure 10). Consequently, movement of sodium from the lumen to the interstitium is not necessarily accompanied by equivalent movement of potassium in the opposite direction. The proportion of potassium that back-diffuses into the interstitium depends on the relative strengths of the electrochemical gradients across the luminal and basolateral membranes, and these in turn are determined by the ionic composition of the interstitial fluid and the urine.

The principal cells in the connecting segments and the cortical collecting ducts are the most important targets for aldosterone (see Chapter 7). Aldosterone increases the sodium permeability of the luminal membrane. This action of aldosterone requires a lag period of at least 30 minutes, is sensitive to inhibitors of RNA

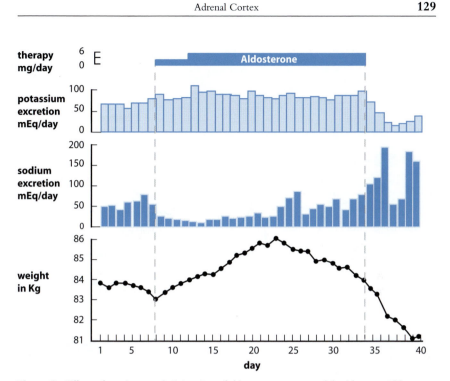

Figure 9 Effects of continuous administration of aldosterone to a normal, healthy man. Aldosterone (3–6 mg/day) increased potassium excretion and sodium retention, represented here as a decrease in urinary sodium. The increased retention of sodium, which continued for 2 weeks, caused fluid retention and hence an increase in body weight. The subject "escaped" from the sodium-retaining effects but continued to excrete increased amounts of potassium for as long as aldosterone was given. (From August, J. T., Nelson, D. H., and Thorn, G. W., *J. Clin. Invest.*, **37**, 1549–1559, 1958, with permission.)

and protein synthesis, and is mediated by transcriptional events initiated by nuclear receptors (see Chapter 1). After binding aldosterone, the mineralocorticoid receptor, in conjunction with other nuclear regulatory proteins, regulates transcription of certain genes. Aldosterone induces transcription of sodium channel genes, and may increase expression of proteins that transform nonconductive sodium channels to a functionally active state and perhaps insert sequestered channels into the luminal membrane. Aldosterone also increases expression of sodium/potassium ATPase in the basolateral membrane by inducing its synthesis and membrane insertion. At the same time, aldosterone increases the capacity for ATP generation by promoting synthesis of some mitochondrial citric acid cycle enzymes, such as isocitrate dehydrogenase.

Aldosterone also promotes hydrogen ion excretion through actions exerted on the intercalated cells of the distal tubule and collecting duct. The underlying

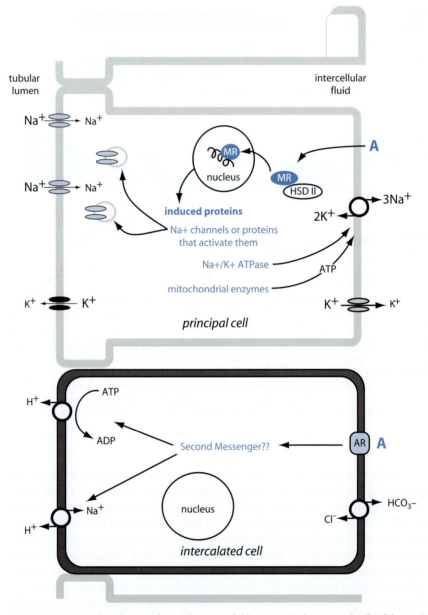

Figure 10   Proposed mechanisms for renal actions of aldosterone. In the principal cells of the cortical collecting ducts, aldosterone (A) induces transcription and translation of new proteins. Note that entry of sodium into the principal cell and increased activity of the sodium potassium ATPase increase the electrochemical gradient that provides the driving force for outflow of potassium into the lumen

mechanisms have not been firmly established, but aldosterone appears to stimulate an electrogenic proton pump (H^+-ATPase) in the luminal membrane. Proton secretion is accompanied by exchange of chloride for bicarbonate in the basolateral membrane. There is increasing evidence that aldosterone may also activate sodium:proton exchange. This effect is too rapid to depend on transcription and may be mediated by membrane-bound rather than nuclear receptors. Because sodium ions are almost a million times more abundant than hydrogen ions in extracellular fluid, secretion of protons in exchange for sodium accounts for only a minuscule fraction of the sodium that is retained. It is possible that transcriptional changes in the intercalated cells also contribute to the effects of aldosterone on acid excretion.

Similar responses to aldosterone are seen in extrarenal tissues. Aldosterone also promotes the absorption of sodium and secretion of potassium in the colon and decreases the ratio of sodium to potassium concentrations in sweat and salivary secretions. Formation of sweat and saliva is analogous to formation of urine. Initial secretions are ultrafiltrates of blood plasma whose ionic composition is modified by epithelial cells of the duct that carries the fluid from its site of generation to its site of release. Under the influence of aldosterone, sodium is reabsorbed in exchange for potassium probably by the same cellular mechanism as described for the cells of the cortical collecting duct (Figure 10). The effect of aldosterone on these secretions is not subject to the escape phenomenon. Because perspiration can be an important avenue for sodium loss, minimizing sodium loss in sweat is physiologically significant. Persons suffering from adrenal insufficiency are especially sensitive to extended exposure to a hot environment and may become severely dehydrated.

The Cortisol/Cortisone Shuttle and the Mechanism of Mineralocorticoid Specificity

Receptors for adrenal steroids were originally classified based on their affinity and selectivity for mineralocorticoids or glucocorticoids. The mineralocorticoid receptors, also called type I receptors, have a high and nearly equal affinity for aldosterone and cortisol. The type II, or glucocorticoid receptors, have a considerably greater affinity for cortisol than for aldosterone. Expression of mineralocorticoid receptors is confined to aldosterone target tissues and the brain, whereas glucocorticoid receptor expression is widely disseminated. Both receptors

---

or back into the interstitium. MR, Mineralocorticoid receptor; HSD II, 11-hydroxysteroid dehydrogenase II. In the intercalated cells, aldosterone promotes the secretion of protons by a mechanism that bypasses the nucleus and probably involves an aldosterone receptor (AR) on the cell surface and activation of some second messenger.

reside in the cytosol bound to other proteins in the unstimulated state, and on binding hormone, release their associated proteins and migrate as dimers to the nucleus, where they activate gene expression (see Chapter 1).

Because the mineralocorticoid receptor binds aldosterone and cortisol with equal affinity, it cannot distinguish between the two classes of steroid hormones. Nevertheless, even though the concentration of cortisol in blood is about 1000 times higher than that of aldosterone, mineralocorticoid responses normally reflect only the availability of aldosterone. This is due in part to differences in plasma protein binding; only 3–4% of cortisol is in free solution, compared to nearly 40% of the aldosterone. Although hormone binding lowers the discrepancy in the available hormone concentrations by 10-fold, free cortisol is 100 times as abundant as free aldosterone, and readily diffuses into mineralocorticoid target cells. Access to mineralocorticoid receptors, however, is guarded by the enzyme HSD II, which colocalizes with the receptors and defends mineralocorticoid specificity. The high efficiency of this enzyme inactivates cortisol by converting it to cortisone, which is released into the blood (see Figure 5). Consequently, the kidneys, which are the major targets for aldosterone, are the major source of circulating cortisone. Persons with a genetic defect in HSD II suffer from symptoms of mineralocorticoid excess (hypertension and hypokalemia) as a result of constant saturation of the mineralocorticoid receptor by cortisol. An acquired form of the same ailment is seen after ingestion of excessive amounts of licorice, which contains the potent inhibitors of HSD, glycyrrhizic acid and its metabolite glycyrrhetinic acid.

Regulation of Aldosterone Secretion

Angiotensin II is the primary stimulus for aldosterone secretion, although ACTH and high concentrations of potassium are also potent stimuli. Angiotensin II is formed in blood by a two-step process that depends on proteolytic cleavage of the plasma protein angiotensinogen, by the enzyme renin, to release the inactive decapeptide angiotensin I. Angiotensin I is then converted to angiotensin II by the ubiquitous angiotensin-converting enzyme. Control of angiotensin II production is achieved by regulating the secretion of renin from smooth muscle cells of the afferent glomerular arterioles. The principal stimulus for renin secretion is a decrease in the vascular volume, and the principal physiological role of aldosterone is to defend the vascular volume. Aldosterone secretion is regulated by negative feedback, with vascular volume, and not the concentration of aldosterone, as the controlled variable. Reabsorption of sodium is accompanied by a proportionate reabsorption of water, and because sodium remains extracellular, sodium retention expands the extracellular volume and hence blood volume. Expansion of the blood volume, which is the ultimate result of sodium retention, provides the negative feedback signal for regulation of renin and aldosterone secretion (Figure 11). Although preservation of body sodium is central to aldosterone action, the

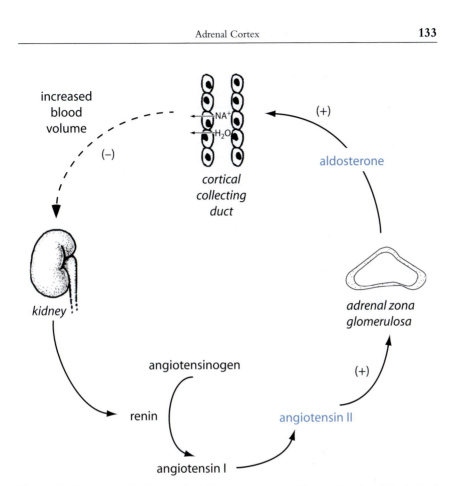

Figure 11 Negative feedback control of aldosterone secretion. The monitored variable is blood volume. (+), Stimulation; (−), inhibition.

concentration of sodium in blood does not appear to be monitored directly, and fluctuations in plasma concentrations have little direct effect on the secretion of renin. This topic is considered in detail in Chapter 7.

PHYSIOLOGY OF THE GLUCOCORTICOIDS

Although named for their critical role in maintaining carbohydrate reserves, glucocorticoids produce diverse physiological actions, many of which are still not well understood and therefore can be considered only phenomenologically. Virtually every tissue of the body is affected by an excess or deficiency of

Table 1

Some Effects of Glucocorticoids

Tissue	Effects
Central nervous system	Taste, hearing, and smell ↑ in acuity with adrenal cortical insuffiency and ↓ in Cushing's disease ↓ Corticotropin releasing hormone (see text) ↓ ADH secretion
Cardiovascular system	Maintain sensitivity to epinephrine and norepinephrine ↑ Sensitivity to vasoconstrictor agents Maintain microcirculation
Gastrointestinal tract	↑ Gastric acid secretion ↓ Gastric mucosal cell proliferation
Liver	↑ Gluconeogenesis
Lungs	↑ Maturation and surfactant production during fetal development
Pituitary	↓ ACTH secretion and synthesis
Kidney	↑ GFR Needed to excrete dilute urine
Bone	↑ Resorption ↓ Formation
Muscle	↓ Fatigue (probably secondary to cardiovascular actions) ↑ Protein catabolism ↓ Glucose oxidation ↓ Insulin sensitivity ↓ Protein synthesis
Immune system (see text)	↓ Mass of thymus and lymph nodes ↓ Blood concentrations of eosinophils, basophils, and lymphocytes ↓ Cellular immunity
Connective tissue	↓ Activity of fibroblasts ↓ Collagen synthesis

glucocorticoids (Table 1). If any simple phrase could describe the role of glucocorticoids, it would be "coping with adversity." Even if sodium balance could be preserved and carbohydrate intake were adequate to meet energy needs, individuals suffering from adrenal insufficiency would still teeter on the brink of disaster when faced with a threatening environment. We shall consider here only the most thoroughly studied actions of glucocorticoids.

Effects on Energy Metabolism

Ability to maintain and draw on metabolic fuel reserves is ensured by actions and interactions of many hormones and is critically dependent on normal function of the adrenal cortex (Figure 12). Although we speak of maintaining carbohydrate reserves as the hallmark of glucocorticoid activity, it must be understood that metabolism of carbohydrate, protein, and lipid are inseparable components of overall energy balance. This complex topic is considered further in Chapter 9.

In the absence of adrenal function, even relatively short periods of fasting may produce a catastrophic decrease in blood sugar (hypoglycemia) accompanied by depletion of muscle and liver glycogen. A drastically compromised ability to produce sugar from nonglucose precursors (gluconeogenesis) forces these individuals to rely almost exclusively on dietary sugars to meet their carbohydrate needs. Their metabolic problems are further complicated by decreased ability to utilize alternate substrates such as fatty acids and protein. Glucocorticoids promote gluconeogenesis by complementary mechanisms.

1. Extrahepatic actions provide substrate. Glucocorticoids promote proteolysis and inhibit protein synthesis in muscle and lymphoid tissues, thereby causing amino acids to be released into the blood. In addition, they increase blood glycerol concentrations by acting with other hormones to increase lipolysis in adipose tissue.

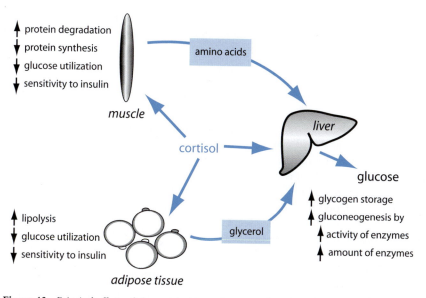

Figure 12   Principal effects of glucocorticoids on glucose production and metabolism of body fuels (↑, increased; ↓, decreased).

2. Hepatic actions enhance the flow of glucose precursors through existing enzymatic machinery and induce the synthesis of additional gluconeogenic and glycogen-forming enzymes along with enzymes needed to convert amino acids to usable precursors of carbohydrate.

Nitrogen excretion during fasting is lower than normal with adrenal insufficiency, reflecting decreased conversion of amino acids to glucose. High concentrations of glucocorticoids, as seen in states of adrenal hyperfunction, inhibit protein synthesis and promote rapid breakdown of muscle and lymphoid tissues that serve as repositories for stored protein. These effects result in increased blood urea nitrogen (BUN) and enhanced nitrogen excretion. Individuals with hyperfunction of the adrenal cortex (Cushing's disease) characteristically have spindly arms and legs, reflecting increased breakdown of their muscle protein. Protein wasting in these patients may also extend to skin and connective tissue, and it contributes to their propensity to bruise easily.

Glucocorticoids defend against hypoglycemia in yet another way. In experimental animals glucocorticoids decrease utilization of glucose by muscle and adipose tissue and lower the responsiveness of these tissues to insulin. Prolonged exposure to high levels of glucocorticoids often leads to diabetes mellitus (Chapter 5); about 20% of patients with Cushing's disease are also diabetic, and virtually all of the remainder have some milder impairment of glucose metabolism. Despite the relative decrease in insulin sensitivity and increased tendency for fat mobilization seen in experimental animals, patients with Cushing's disease paradoxically accumulate fat in the face (moon face), between the shoulders (buffalo hump), and in the abdomen (truncal obesity). As yet there is no explanation for this redistribution of body fat.

Effects on Water Balance

In the absence of the adrenal glands, renal plasma flow and glomerular filtration are reduced and it is difficult to produce either concentrated or dilute urine. One of the diagnostic tests for adrenal cortical insufficiency is the rapidity with which a water load can be excreted. Glucocorticoids facilitate excretion of free water and are more important in this regard than are mineralocorticoids. The mechanism for this effect is still debated. It has been suggested that glucocorticoids may maintain normal rates of glomerular filtration by acting directly on glomeruli or glomerular blood flow, or indirectly by facilitating the action or production of the atrial natriuretic hormone (see Chapter 7). In addition, in the absence of glucocorticoids antidiuretic hormone (vasopressin) secretion is increased.

Effects on Lung Development

One of the dramatic physiological changes that must be accommodated in the newborn infant is the shift from the placenta to the lungs as source of oxygen and carbon dioxide exchange. Preparation of fetal lungs for this transition includes thinning of the alveolar walls to diminish the diffusion barrier and the production of surfactant. Surfactant consists largely of phospholipids associated with specific proteins. It reduces alveolar surface tension, increases lung compliance, and allows even distribution of inspired air. Glucocorticoids play a crucial role in maturation of alveoli and production of surfactant. A major problem of preterm delivery is a condition known as respiratory distress syndrome, which is caused by impaired pulmonary mechanics resulting from incomplete development of pulmonary alveoli and production of surfactant. Although the fetal adrenal gland is capable of secreting some glucocorticoids by about week 24 of pregnancy, the major secretory products of the fetal adrenal gland are androgens, which serve as precursors for placental estrogen synthesis (see Chapter 13). Administration of large doses of glucocorticoid to mothers carrying fetuses at risk for premature delivery diminishes the incidence of respiratory distress syndrome. Steady-state levels of mRNAs for surfactant proteins may be reached within about 15 hours after exposure of fetuses to glucocorticoids.

Glucocorticoids and Responses to Injury

One of the most remarkable effects of glucocorticoids was discovered almost by chance during the late 1940s, when it was observed that glucocorticoids dramatically reduced the severity of disease in patients suffering from rheumatoid arthritis. This observation, in addition to leading to the award of a Nobel Prize, called attention to the antiinflammatory effects of glucocorticoids. Because supraphysiological concentrations are needed to produce these effects therapeutically, some investigators consider them to be pharmacological side effects. As understanding of the antiinflammatory actions has increased, however, it is apparent that glucocorticoids are physiological modulators of the inflammatory response through actions mediated by the activated glucocorticoid receptor. It is likely that free cortisol concentrations are higher at local sites of tissue injury than in the general circulation. Partial degradation of corticosteroid-binding globulin by the proteolytic enzyme, elastase, secreted by activated mononuclear leukocytes, decreases its affinity for cortisol and raises the concentration of free cortisol at the site of inflammation. In addition, inflammatory mediators up-regulate 11β-hydroxysteroid dehydrogenase I in some cells and cause localized conversion of circulating cortisone to cortisol. As might be anticipated, glucocorticoids and related compounds devised by the pharmaceutical industry are exceedingly important therapeutic agents for treating such diverse conditions as poison ivy, asthma, a host of inflammatory conditions, and various autoimmune diseases. The latter reflects their related ability to diminish the immune response.

Antiinflammatory Effects

Inflammation is the term used to encompass the various responses of tissues to injury. It is characterized by redness, heat, swelling, pain, and loss of function. Redness and heat are manifestations of increased blood flow and result from vasodilation. Swelling is due to formation of a protein-rich exudate that collects because capillaries and venules become leaky to proteins. Pain is caused by chemical products of cellular injury and sometimes by mechanical injury to nerve endings. Loss of function may be a direct consequence of injury or secondary to the pain and swelling that injury evokes. An intimately related component of the early response to tissue injury is the migration of white blood cells to the injured area and the subsequent recruitment of the immune system.

The initial pattern of the inflammatory response is independent of the injurious agent or causal event. This response is presumably defensive and may be a necessary antecedent of the repair process. Increased blood flow accelerates delivery of the white blood cells that combat invading foreign substances or organisms and clean up the debris of injured and dead cells. Increased blood flow also facilitates dissemination of chemoattractants to white blood cells and promotes their migration to the site of injury. In addition, increased blood flow provides more oxygen and nutrients to cells at the site of damage and facilitates removal of toxins and wastes. Increased permeability of the microvasculature allows fluid to accumulate in the extravascular space in the vicinity of the injury and thus dilute noxious agents.

Although we are accustomed to thinking of physiological responses as having beneficial effects, it is apparent that some aspects of inflammation may actually cause or magnify tissue damage. Lysosomal hydrolases released during phagocytosis of cellular debris or invading organisms may damage nearby cells that were not harmed by the initial insult. Loss of fluid from the microvasculature at the site of injury may increase viscosity of the blood, slowing its flow, and even leaving some capillaries clogged with stagnant red blood cells. Decreased perfusion may cause further cell damage. In addition, massive disseminated fluid loss into the extravascular space sometimes compromises cardiovascular function. Consequently, long-term survival demands that checks and balances be in place to prevent the defensive and positive aspects of the inflammatory response from becoming destructive. We may regard the physiological role of the glucocorticoids to modulate inflammatory responses as a major component of such checks and balances. Exaggeration of such physiological modulation with supraphysiological amounts of glucocorticoids upsets the balance in favor of suppression of inflammation and provides the therapeutic efficacy of pharmacological treatment.

Inflammation is initiated, sustained, and amplified by the release of a large number of chemical mediators derived from multiple sources. Cytokines are a diverse group of peptides that range in size from about 8 to about 40 kDa and are produced mainly by cells of the hematopoietic and immune systems, but can be synthesized and secreted by virtually any cell. Cytokines may promote or antagonize development of inflammation, or may have a mixture of pro- and antiinflammatory effects, depending on the particular cells involved. Prostaglandins and leukotrienes are released principally from vascular endothelial cells and macrophages, but virtually all cell types can produce and release them. They may also produce either pro- or antiinflammatory effects, depending on the particular compound formed and the cells on which they act. Histamine and serotonin are released from mast cells and platelets. Enzymes and superoxides released from dead or dying cells or from cells that remove debris by phagocytosis contribute directly and indirectly to the spread of inflammation by activating other mediators (e.g., bradykinin) and leukocyte attractants that arise from humoral precursors associated with the immune and clotting systems.

Glucocorticoids and the Metabolites of Arachidonic Acid

Prostaglandins and the closely related leukotrienes are derived from the polyunsaturated essential fatty acid arachidonic acid (Figure 13). Because of their 20-carbon backbone they are also sometimes referred to collectively as *eicosanoids*. These compounds play a central role in the inflammatory response. They generally act locally on cells in the immediate vicinity of their production, including the cells that produced them, but some also survive in blood long enough to act on distant tissues. Prostaglandins act directly on blood vessels to cause vasodilation and indirectly increase vascular permeability by potentiating the actions of histamine and bradykinin. Prostaglandins sensitize nerve endings of pain fibers to other mediators of inflammation, such as histamine, serotonin, bradykinin, and substance P, thereby producing increased sensitivity to touch (hyperalgesia). The leukotrienes stimulate production of cytokines and act directly on the microvasculature to increase permeability. Leukotrienes also attract white blood cells to the site of injury and increase their stickiness to vascular endothelium. The physiology of arachidonate metabolites is complex, and a thorough discussion is not possible here. There are a large number of these compounds with different biological activities. Although some eicosanoids have antiinflammatory actions that may limit the overall inflammatory response, arachidonic acid derivatives are major contributors to inflammation.

Arachidonic acid is released from membrane phospholipids by phospholipase A_2 (PLA_2; see Chapter 1), which is activated by injury, phagocytosis, or a variety of other stimuli in responsive cells. Activation is mediated by a cytosolic PLA_2-activating protein that closely resembles a protein in bee venom called *mellitin*. In addition, PLA_2 activity also increases as a result of an increased enzyme synthesis. The first step in the production of prostaglandins from arachidonate is catalyzed by a cytosolic enzyme, cyclooxygenase (COX). One isoform of this enzyme,

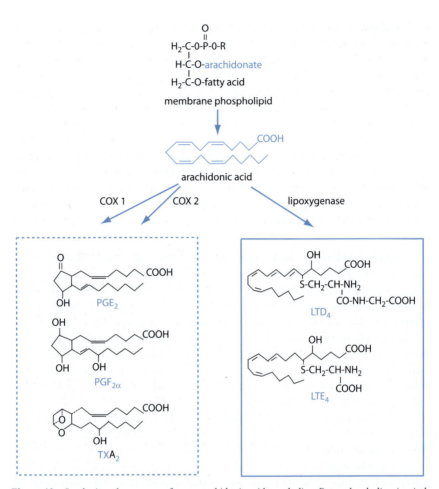

Figure 13   Synthesis and structures of some arachidonic acid metabolites; R may be choline, inositol, serine, or ethanolamine. PG, Prostaglandin; LT, leukotriene. The terminal designations E_2 or $F_{2\alpha}$ refer to substituents on the ring structure of the PG. The designations D_4 and E_4 refer to glutathione derivatives in thioester linkage at carbon 6 of LT. TXA_2, thromboxane.

COX 1, is constitutively expressed. A second form, COX 2, is induced by the inflammatory response. Glucocorticoids suppress the formation of prostaglandins by inhibiting synthesis of COX 2 and probably also by inducing expression of a protein that inhibits PLA_2. Nonsteroidal antiinflammatory drugs such as indomethacin and aspirin also block the cyclooxygenase reaction catalyzed by both COX 1 and COX 2. Some of the newer antiinflammatory drugs specifically block COX 2 and hence may target inflammation more specifically.

Glucocorticoids and Cytokines The large number of compounds designated as cytokines include one or more isoforms of the interleukins (IL-1 through IL-18), tumor necrosis factor (TNF), the interferons (IFN-α, -β, and -γ), colony-stimulating factor (CSF), granulocyte/macrophage colony-stimulating factor (GM-CSF), transforming growth factor (TGF), leukemia inhibiting factor (LIF), oncostatin, and a variety of cell- or tissue-specific growth factors. It is not clear just how many of these hormone-like molecules are produced, and not all have a role in inflammation. Two of these factors, IL-1 and TNFα, are particularly important in the development of inflammation. The intracellular signaling pathways and biological actions of these two cytokines are remarkably similar. They enhance each other's actions in the inflammatory response and differ only in the respect that TNFα may promote cell death (apoptosis) whereas IL-1 does not.

IL-1 is produced primarily by macrophages and to a lesser extent by other connective tissue elements, skin, and endothelial cells. Its release from macrophages is stimulated by interaction with immune complexes, activated lymphocytes, and metabolites of arachidonic acid, especially leukotrienes. IL-1 is not stored in its cells of origin but is synthesized and secreted within hours of stimulation in a response mediated by increased intracellular calcium and protein kinase C (see Chapter 1). IL-1 acts on many cells to produce a variety of responses (Figure 14) all of which are components of the inflammatory/immune response. Many of the consequences of these actions can be recognized from personal experience as nonspecific symptoms of viral infection. TNFα is also produced in macrophages and other cells in response to injury and immune complexes, and can act on many cells, including those that secrete it. Secretions of both IL-1 and TNFα and their receptors are increased by some of the cytokines and other mediators of inflammation whose production they increase, so that an amplifying positive feedback cascade is set in motion. Some products of these cytokines also feed back on their production in a negative way to modulate the inflammatory response. Glucocorticoids play an important role as negative modulators of IL-1 and TNFα by (1) inhibiting their production, (2) interfering with signaling pathways, and (3) inhibiting the actions of their products. Glucocorticoids also interfere with the production and release of other proinflammatory cytokines as well, including IFN-γ, IL-2, IL-6, and IL-8.

Production of IL-1 and TNFα and many of their effects on target cells are mediated by activation of genes by the transcription factor called nuclear factor kappa B (NF-κB). In the unactivated state NF-κB resides in the cytoplasm bound to the NF-κB inhibitor (I-κB). Activation of the signaling cascade by some tissue insult or by the binding of IL-1 and TNFα to their respective receptors is initiated by activation of a kinase (I-κK), which phosphorylates I-κB, causing it to dissociate from NF-κB and to be degraded. Free NF-κB is then able to translocate to the nucleus, where it binds to response elements in genes that it regulates, including genes for the cytokines IL-1, TNFα, IL-6, and IL-8 and for enzymes

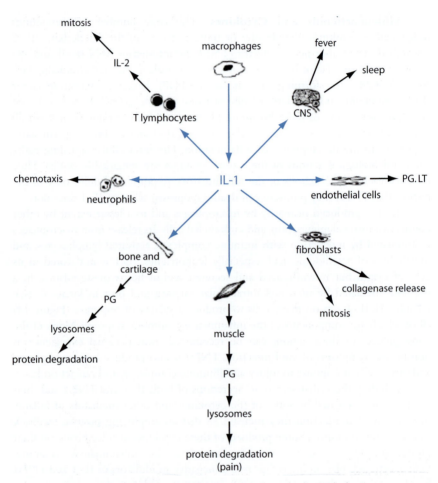

Figure 14 Effects of interleukin-1 (IL-1). PG, Prostaglandin; LT, leukotriene.

such as PLA$_2$, COX 2, and nitric oxide synthase (Figure 15). IL-6 is an important proinflammatory cytokine that acts on the hypothalamus, liver, and other tissues, and IL-8 plays an important role as a leukocyte attractant. Nitric oxide is important as a vasodilator and may have other effects as well.

  Glucocorticoids interfere with the actions of IL-1 and TNFα by promoting the synthesis of I-κB, which traps NF-κB in the cytosol, and by interfering with the ability of the NF-κB that enters the nucleus to activate target genes. The mechanism for interference with gene activation is thought to invoke protein:protein interaction between the liganded glucocorticoid receptor and NF-κB. Glucocorticoids also appear to interfere with IL-1- or TNFα-dependent activation of other genes by the activator protein (AP-1) transcription complex. In addition, cortisol induces expression of a protein that inhibits PLA_2 and destabilizes the mRNA for COX 2. It is noteworthy that many of the responses attributed to IL-1 may be mediated by prostaglandins or other arachidonate metabolites. For example, IL-1, which is identical with what was once called endogenous pyrogen,

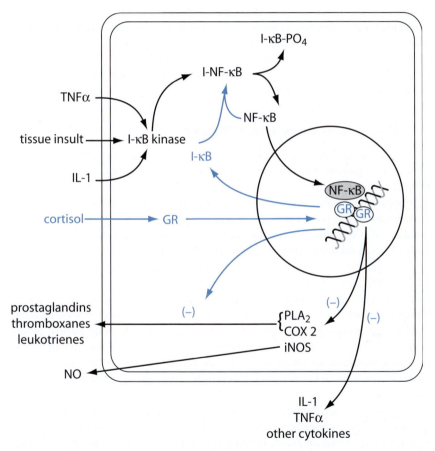

Figure 15 Antiinflammatory actions of cortisol. Cortisol induces the formation of the nuclear factor κB inhibitor (I-κB), which binds to nuclear factor κB (NF-κB) and prevents it from entering the nucleus and activating target genes. The activated glucocorticoid receptor (GR) also interferes with NF-κB binding to its response elements in DNA, thus preventing induction of phospholipase A_2 (PLA_2), cyclooxygenase 2 (COX 2), and inducible nitric oxide synthase (iNOS). TNFα, Tumor necrosis factor-α; IL-1, interleukin-1; NO, nitric oxide.

may cause fever by inducing the formation of prostaglandins in the thermoregula-
tory center of the hypothalamus. Glucocorticoids might therefore exert their
antipyretic effect at two levels: at the level of the macrophage, by inhibiting
IL-1 production, and at the level of the hypothalamus, by interfering with
prostaglandin synthesis.

Glucocorticoids and the Release of Other Inflammatory Mediators

Granulocytes, mast cells, and macrophages contain vesicles filled with serotonin,
histamine, or degradative enzymes, all of which contribute to the inflammatory
response. These mediators and lysosomal enzymes are released in response to
arachidonate metabolites, cellular injury, reaction with antibodies, or during
phagocytosis of invading pathogens. Glucocorticoids protect against the release
of all these compounds by inhibiting cellular degranulation. It has also been suggested that glucocorticoids inhibit histamine formation and stabilize lysosomal
membranes, but the molecular mechanisms for these effects are unknown.

Glucocorticoids and the Immune Response

The immune system, which functions to destroy and eliminate foreign substances or organisms, has two major components: the B lymphocytes, which are
formed in bone marrow and develop in liver or spleen, and the thymus-derived
T lymphocytes. Humoral immunity is the province of B lymphocytes, which, on
differentiation into plasma cells, are responsible for production of antibodies. Large
numbers of B lymphocytes circulate in blood or reside in lymph nodes. Reaction
with a foreign substance (antigen) stimulates B cells to divide and produce a clone
of cells capable of recognizing the antigen and producing antibodies to it.
Such proliferation depends on cytokines released from the macrophages and
helper T cells. Antibodies, which are circulating immunoglobulins, bind to
foreign substances and thus mark them for destruction. Glucocorticoids inhibit
cytokine production by macrophages and T cells and thus decrease normal proliferation of B cells and reduce circulating concentrations of immunoglobulins.
At high concentrations, glucocorticoids may also act directly on B cells to inhibit
antibody synthesis and may even kill B cells by activating apoptosis (programmed
cell death).

The T cells are responsible for cellular immunity, and participate in destruction of invading pathogens or cells that express foreign surface antigens, as might
follow viral infection or transformation into tumor cells. IL-1 stimulates T lymphocytes to produce IL-2, which promotes proliferation of T lymphocytes that
have been activated by coming in contact with antigens. Antigenic stimulation
triggers the temporary expression of IL-2 receptors only in those T cells that
recognize the antigen. Consequently, only certain clones of T cells are stimulated
to divide because there are no receptors for IL-2 on the surface membranes of T

lymphocytes until they interact with their specific antigens. Glucocorticoids block the production of, but probably not the response to, IL-2 and thereby inhibit proliferation of T lymphocytes. IL-2 also stimulates T lymphocytes to produce IFN-γ, which participates in destruction of virus-infected or tumor cells and also stimulates macrophages to produce IL-1. Macrophages, T lymphocytes, and secretory products are thus arranged in a positive feedback relationship and produce a self-amplifying cascade of responses. Glucocorticoids restrain the cycle by suppressing production of each of the mediators. Glucocorticoids also activate apoptosis in some T lymphocytes.

The physiological implications of the suppressive effects of glucocorticoids on humoral and cellular immunity are incompletely understood. It has been suggested that suppression of the immune response might prevent development of autoimmunity that might otherwise follow from the release of fragments of injured cells. However, it must be pointed out that much of the immunosuppression by glucocorticoids requires concentrations that may never be reached under physiological conditions. High doses of glucocorticoids can so impair immune responses that relatively innocuous infections with some organisms can become overwhelming and cause death. Thus, excessive antiimmune or antiinflammatory influences are just as damaging as unchecked immune or inflammatory responses. Under normal physiological circumstances, these influences are balanced and protective. Nevertheless, the immunosuppressive property of glucocorticoids is immensely important therapeutically, and high doses of glucocorticoids are often administered to combat rejection of transplanted tissues and to suppress various immune and allergic responses.

Other Effects of Glucocorticoids on Lymphoid Tissues

Sustained high concentrations of glucocorticoids produce a dramatic reduction in the mass of all lymphoid tissues, including thymus, spleen, and lymph nodes. The thymus contains germinal centers for lymphocytes, and large numbers of T lymphocytes are formed and mature within it. Lymph nodes contain large numbers of both T and B lymphocytes. Immature lymphocytes of both lineages have glucocorticoid receptors and respond to hormonal stimulation by the same series of events as seen in other steroid-responsive cells, except that the DNA transcribed contains the program for apoptosis. Loss in mass of thymus and lymph nodes can be accounted for by the destruction of lymphocytes rather than the stromal or supporting elements. Mature lymphocytes and germinal centers seem to be unresponsive to this action of glucocorticoids.

Glucocorticoids also decrease circulating levels of lymphocytes and particularly a class of white blood cells known as eosinophils (for their cytological staining properties). This decrease is partly due to apoptosis and partly to sequestration in the spleen and lungs. Curiously, the total white blood cell count does not decrease because glucocorticoids also induce a substantial mobilization of neutrophils from bone marrow.

Maintenance of Vascular Responsiveness to Catecholamines

A final action of glucocorticoids relevant to inflammation and the response to injury is maintenance of sensitivity of vascular smooth muscle to vasoconstrictor effects of norepinephrine released from autonomic nerve endings or the adrenal medulla. By counteracting local vasodilator effects of inflammatory mediators, norepinephrine decreases blood flow and limits the availability of fluid to form the inflammatory exudate. In addition, arteriolar constriction decreases capillary and venular pressure and favors reabsorption of extracellular fluid, thereby reducing swelling. The vasoconstrictor action of norepinephrine is compromised in the absence of glucocorticoids. The mechanism for this action is not known, but at high concentrations glucocorticoids may block inactivation of norepinephrine.

Adrenocortical Function during Stress

During the mid-1930s the Canadian endocrinologist Hans Selye observed that animals respond to a variety of seemingly unrelated threatening or noxious circumstances with a characteristic pattern of changes, including an increase in size of the adrenal glands, involution of the thymus, and a decrease in the mass of all lymphoid tissues. He inferred that the adrenal glands are stimulated whenever an animal is exposed to any unfavorable circumstance, which he called "stress." Stress does not directly affect adrenal cortical function, but rather increases the output of ACTH from the pituitary gland (see below). In fact, stress is now defined operationally by endocrinologists as any of the variety of conditions that increase ACTH secretion.

Although it is clear that relatively benign changes in the internal or external environment may become lethal in the absence of the adrenal glands, we understand little more than Selye did about what cortisol might be doing to protect against stress. The favored experimental model used to investigate this problem was the adrenalectomized animal, which might have further complicated an already complex experimental question.

It appears that many cellular functions require glucocorticoids either directly or indirectly for their maintenance, suggesting that these steroid hormones govern some process that is fundamental to normal operation of most cells. Consequently, without replacement therapy many systems are functioning only marginally even before the imposition of stress. Any insult may therefore prove overwhelming. It further became apparent that glucocorticoids are required for normal responses to other hormones or to drugs, even though steroids do not initiate similar responses in the absence of these agents.

Treatment of adrenalectomized animals with a constant basal amount of glucocorticoid prior to and during a stressful incident prevented the devastating effects of stress and permitted expression of expected responses to stimuli. This finding introduced the idea that glucocorticoids act in a normalizing, or *permissive*, way. That is, by maintaining normal operation of cells, glucocorticoids permit normal regulatory mechanisms to act. Because it was not necessary to increase the amounts of adrenal corticoids to ensure survival of stressed adrenalectomized animals, it was concluded that increased secretion of glucocorticoids was not required to combat stress. However, this conclusion is not consistent with clinical experience. Persons suffering from pituitary insufficiency or who have undergone hypophysectomy have severe difficulty withstanding stressful situations, even though at other times they get along reasonably well on the small amounts of glucocorticoids produced by their adrenals in the absence of ACTH. Patients suffering from adrenal insufficiency are routinely given increased doses of glucocorticoids before undergoing surgery or other stressful procedures. We have already seen that glucocorticoids suppress the inflammatory response. It is also known that these hormones increase the sensitivity of various tissues to epinephrine and norepinephrine, which are also secreted in response to stress (see below). Although we still do not understand the role of increased concentrations of glucocorticoids in the physiological response to stress, it appears likely that they are beneficial. The question remains open, however, and will not be resolved until a better understanding of glucocorticoid actions is obtained.

Mechanism of Action of Glucocorticoids

With few exceptions, the physiological actions of cortisol at the molecular level fit the general pattern of steroid hormone action described in Chapter 1. The gene for the glucocorticoid receptor gives rise to two isoforms as a result of alternate splicing of RNA. The alpha isoform binds glucocorticoids, sheds its associated proteins, and migrates to the nucleus, where it can form homodimers that bind to response elements in target genes. The beta isoform cannot bind hormone, is constitutively located in the nucleus, and apparently cannot bind to DNA. The beta isoform, however, can dimerize with the alpha isoform and diminish or block the ability of the alpha isoform to activate transcription. Some evidence suggests that formation of the beta isoform may be a regulated process that modulates glucocorticoid responsiveness.

Glucocorticoids act on a great variety of cells and produce a wide range of effects that depend on activating or suppressing transcription of specific genes. The ability to regulate different genes in different tissues presumably reflects differing accessibility of the activated glucocorticoid receptor to glucocorticoid-responsive genes in each differentiated cell type, and presumably reflects the presence or absence of different coactivators and corepressors. Glucocorticoids also inhibit expression of some genes that lack glucocorticoid response elements. Such inhibitory effects are thought to be the result of protein:protein interactions between the glucocorticoid receptor and other transcription factors, to modify their ability to activate gene transcription. The mechanisms for such interference are the subject of active research. The glucocorticoid receptor can be phosphorylated to various degrees on serine residues. Phosphorylation may modulate the affinity of the receptor for hormone, or DNA, or may modify its ability to interact with other proteins.

Regulation of Glucocorticoid Secretion

Secretion of glucocorticoids is regulated by the anterior pituitary gland through the hormone ACTH, whose effects on the inner zones of the adrenal cortex have already been described (see above). In the absence of ACTH the concentration of cortisol in blood decreases to very low values, and the inner zones of the adrenal cortex atrophy. Regulation of ACTH secretion requires vascular contact between the hypothalamus and the anterior lobe of the pituitary gland, and is driven primarily by corticotropin-releasing hormone (CRH). CRH-containing neurons are widely distributed in the forebrain and brain stem but are heavily concentrated in the paraventricular nuclei in close association with vasopressin-secreting neurons. They stimulate the pituitary to secrete ACTH by releasing CRH into the hypophyseal portal capillaries (Chapter 2). Arginine vasopressin (AVP) also exerts an important influence on ACTH secretion by augmenting the response to CRH. AVP is cosecreted with CRH, particularly in response to stress. It should be noted that the AVP that is secreted into the hypophyseal portal vessels along with CRH arises in a population of paraventricular neurons different from those that produce the AVP that is secreted by the posterior lobe of the pituitary in response to changes in blood osmolality or volume.

CRH binds to G-protein-coupled receptors in the corticotrope membrane and activates adenylyl cyclase. The resulting increase in cyclic AMP activates protein kinase A, which directly or indirectly inhibits potassium outflow through at least two classes of potassium channels. Buildup of positive charge within the corticotrope decreases the membrane potential, and results in calcium influx through activation of voltage-sensitive calcium channels. Direct phosphorylation of these channels may enhance calcium entry by lowering their threshold for activation. Increased intracellular calcium and perhaps additional effects of protein kinase A on secretory vesicle trafficking trigger ACTH secretion. Protein kinase A also phosphorylates CREB, which initiates production of the AP-1 nuclear factor that activates POMC transcription. AVP binds to its G-protein-coupled receptor and activates phospholipase C, to cause the release of DAG and IP_3. This action of AVP has little effect on CRH secretion in the absence of CRH, but in its presence amplifies the effects of CRH on ACTH secretion without affecting synthesis.

As described in Chapter 1, IP_3 stimulates release of calcium from intracellular stores, and DAG activates protein kinase C, although the role of this enzyme in ACTH secretion is unknown These effects are summarized in Figure 16.

On stimulation with ACTH, the adrenal cortex secretes cortisol, which inhibits further secretion of ACTH in a typical negative feedback arrangement (Figure 17). Cortisol exerts its inhibitory effects both on CRH neurons in the hypothalamus and on corticotropes in the anterior pituitary. These effects are mediated by the glucocorticoid receptor. The negative feedback effects on secretion depend on transcription of genes that code for proteins that either activate potassium channels or block the effects of PKA-catalyzed phosphorylation on these channels and may also act at the level of secretory vesicle trafficking. Initial actions of glucocorticoids suppress secretion of CRH and ACTH from storage granules. Subsequent actions of glucocorticoids result from inhibition of transcription of the genes for CRH and POMC in hypothalamic neurons and corticotropes, perhaps by direct interaction of the glucocorticoid receptor with transcription factors that regulate synthesis of CRH and POMC. This feedback system closely resembles the one described earlier for regulation of thyroid hormone secretion, even though the adrenal ACTH system is much more dynamic and subject to episodic changes.

The relative importance of the pituitary and the CRH-producing neurons of the paraventricular nucleus for negative feedback regulation of ACTH secretion has been explored in mice that were made deficient in CRH by disruption of the CRH gene. These CRH knockout mice secrete normal basal levels of ACTH and glucocorticoid, and their corticotropes express normal levels of mRNA for POMC. In normal mice, disruption of negative feedback by surgical removal of the adrenal glands results in a prompt increase both in POMC gene expression and in ACTH secretion. Adrenalectomy of CRH knockout mice produces no increase in ACTH secretion, although POMC mRNA increases normally. These animals also suffer a severe impairment, but not total lack of ACTH secretion in response to stress. Thus it seems that basal function of the pituitary/adrenal negative feedback system does not require CRH, but that CRH is crucial for increasing ACTH secretion above basal levels. Further, it appears that transcription of the POMC gene is inhibited by glucocorticoids even under basal conditions.

It was pointed out earlier that negative feedback systems ensure constancy of the controlled variable. However, even in the absence of stress, ACTH and cortisol concentrations in blood plasma are not constant but oscillate with a 24-hour periodicity. This so-called circadian rhythm is sensitive to the daily pattern of physical activity. For all but those who work the night shift, hormone levels are highest in the early morning hours just before arousal and lowest in the evening (Figure 18). This rhythmic pattern of ACTH secretion is consistent with the negative feedback model shown in Figure 17 and is sensitive to glucocorticoid input throughout the day. In the negative feedback system, the positive limb (CRH and ACTH

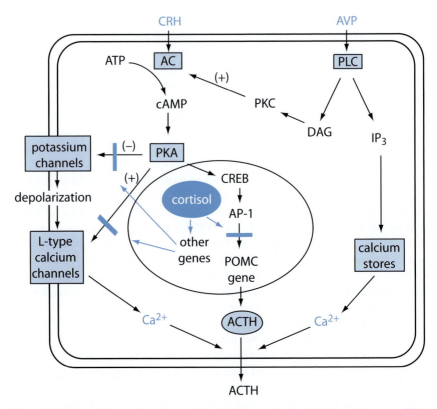

Figure 16 Hormonal interactions that regulate ACTH secretion by pituitary corticotropes. CRH, Corticotropin-releasing hormone; AVP, arginine vasopressin; AC, adenylyl cyclase; PLC, phospholipase C; ATP, adenosine triphosphate; cAMP, cyclic adenosine monophosphate; PKC, protein kinase C; DAG, diacylglycerol; IP_3, inositol trisphosphate; PKA, protein kinase A; CREB, cyclic AMP response element binding protein; AP-1, activator protein-1; POMC, proopiomelanocortin. Inhibitory actions of cortisol are shown in dark blue.

secretion) is inhibited when the negative limb (cortisol concentration in blood) reaches some set point. For basal ACTH secretion, the set point of the corticotropes and the CRH-secreting cells is thought to vary in its sensitivity to cortisol at different times of day. Decreased sensitivity to inhibitory effects of cortisol in the early morning results in increased output of CRH, ACTH, and cortisol. As the day progresses, sensitivity to cortisol increases, and there is a decrease in the output of CRH and consequently of ACTH and cortisol. The cellular mechanisms underlying the periodic changes in set point are not understood, but although they vary with time of day, cortisol concentrations in blood are precisely controlled throughout the day.

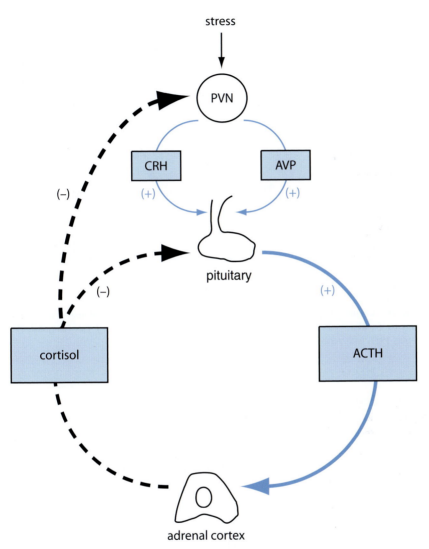

Figure 17 Negative feedback control of glucocorticoid secretion. PVN, Paraventricular nuclei; CRH, corticotropin-releasing hormone; AVP, arginine vasopressin; (+), stimulation; (-), inhibition.

Negative feedback also governs the response of the pituitary–adrenal axis to most stressful stimuli. Different mechanisms appear to apply at different stages of the response. With the imposition of a stressful stimulus, there is a sharp increase in ACTH secretion driven by CRH and AVP. The rate of ACTH secretion is

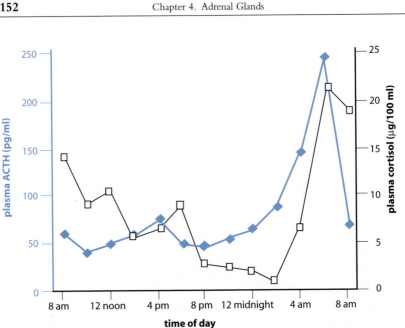

Figure 18   Variations in plasma concentrations of ACTH and cortisol at different times of day. (From Matsukura, S., West, C. D., Ichikawa, Y., Jubiz, W., Harada, G., and Tyler, F. H., *J. Lab. Clin. Med.* **77**, 490–500, 1971, with permission.)

determined by both the intensity of the stimulus to CRH-secreting neurons and the negative feedback influence of cortisol. In the initial moments of the stress response, pituitary corticotropes and CRH neurons monitor the rate of change rather than the absolute concentration of cortisol and decrease their output accordingly. After about 2 hours, negative feedback seems to be proportional to the total amount of cortisol secreted during the stressful episode. With chronic stress a new steady state is reached, and the negative feedback system again seems to monitor the concentration of cortisol in blood, but with the set point readjusted at a higher level.

    Each phase of negative feedback involves different cellular mechanisms. During the first few minutes the inhibitory effects of cortisol occur without a lag period and are expressed too rapidly to be mediated by altered genomic expression. Indeed, the rapid inhibitory action of cortisol is unaffected by inhibitors of protein synthesis. Its molecular basis is unknown, but it may be mediated by nongenomic responses of receptors in neuronal membranes. The negative feedback effect of cortisol in the subsequent interval occurs after a lag period and seems to require RNA and protein synthesis, typical of the steroid actions discussed earlier. In this phase cortisol restrains secretion of CRH and ACTH but not their synthesis. At this time, corticotropes are less sensitive to CRH. With chronic administration of glucocorticoids or with chronic stress, negative feedback is also exerted at the level of POMC gene transcription and translation.

Regulation of ACTH secretion includes the following major features:

1. Basal secretion of ACTH follows a diurnal rhythm driven by CRH and perhaps intrinsic rhymicity of the corticotropes.
2. Stress increases CRH and AVP secretion through neural pathways.
3. ACTH secretion is subject to negative feedback control under basal conditions and during the response to most stressful stimuli.
4. Cortisol inhibits secretion of both CRH and ACTH.

Some observations suggest that cytokines produced by cells of the immune system may directly affect secretion by the hypothalamic–pituitary–adrenal axis. In particular, IL-1, IL-2, and IL-6 stimulate CRH secretion, and may also act directly on the pituitary to increase ACTH secretion. IL-2 and IL-6 may also stimulate cortisol secretion by a direct action on the adrenal gland. In addition, lymphocytes express ACTH and related products of the POMC gene and are responsive to the stimulatory effects of CRH and the inhibitory effects of glucocorticoids. Because glucocorticoids inhibit cytokine production, there is another negative feedback relationship between the immune system and the adrenals (Figure 19). It has been suggested that this communication between the endocrine and immune systems provides a mechanism to alert the body to the presence of invading organisms or antigens.

In our discussion of the regulation of cortisol and ACTH secretion we have ignored other members of the ACTH family that reside in the same secretory granule and are released along with ACTH. Endocrinologists have focused their attention on the physiological implications of increased secretion of ACTH and glucocorticoids in response to stress. Recent observations suggest that other peptides, such as β-endorphin and α-melanocyte-stimulating hormone, whose concentrations in blood increase in parallel with ACTH, may exert antiinflammatory actions.

Understanding of the negative feedback relation between the adrenal and pituitary glands has important diagnostic and therapeutic applications. Normal adrenocortical function can be suppressed by injection of large doses of glucocorticoids. For these tests a potent synthetic glucocorticoid, usually dexamethasone, is administered, and at a predetermined time later the natural steroids or their metabolites are measured in blood or urine. If the hypothalamo–pituitary–adrenal system is intact, production of cortisol is suppressed and its concentration in blood is low. If, on the other hand, cortisol concentration remains high, an autonomous adrenal or ACTH-producing tumor may be present.

Another clinical application is treatment of the adrenogenital syndrome. As pointed out earlier, adrenal glands produce androgenic steroids by extension of

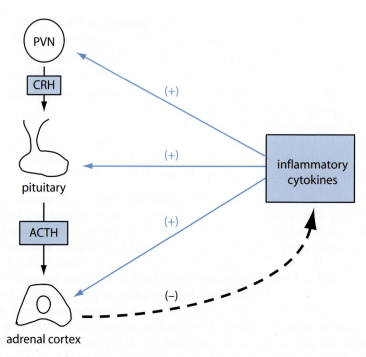

Figure 19   Negative feedback regulation of the hypothalamic–pituitary–adrenal axis by inflammatory cytokines. PVN, Paraventricular nuclei; CRH, corticotropin-releasing hormone; (+), stimulation; (−), inhibition.

the synthetic pathway for glucocorticoids (Figure 4). Defects in production of glucocorticoids, particularly in enzymes responsible for hydroxylation of carbons 21 or 11, may lead to increased production of adrenal androgens. Overproduction of androgens in female patients leads to masculinization, which is manifest, for example, by enlargement of the clitoris, increased muscular development, and growth of facial hair. Severe defects may lead to masculinization of the genitalia of female infants, and in male babies produce the supermasculinized "infant Hercules." Milder defects may show up simply as growth of excessive facial hair (hirsutism) in women. Overproduction of androgens occurs in the following way: Stimulation of the adrenal cortex by ACTH increases pregnenolone production (see above), most of which is normally converted to cortisol, which exerts negative feedback inhibition of ACTH secretion. With a partial block in cortisol production, much of the pregnenolone is diverted to androgens, which have no inhibitory effect on ACTH secretion. ACTH secretion therefore remains high and stimulates more pregnenolone production and causes adrenal hyperplasia (Figure 20). Eventually, the hyperactive adrenals produce enough cortisol for

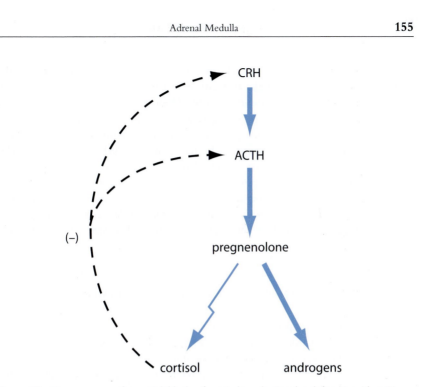

Figure 20 Consequences of a partial block of cortisol production by defects in either 11- or 21-hydroxylase. Pregnenolone is diverted to androgens, which exert no feedback activity on ACTH secretion. The thickness of the arrows connotes relative amounts. Broken arrows indicate impairment is in the inhibitory limb of the feedback system. Administration of glucocorticoids shuts down androgen production by inhibiting ACTH secretion.

negative feedback to be operative, but at the expense of maintaining a high rate of androgen production. The whole system can be brought into proper balance by giving sufficient glucocorticoids to decrease ACTH secretion and therefore remove the stimulus for androgen production.

ADRENAL MEDULLA

The adrenal medulla accounts for about 10% of the mass of the adrenal gland and is embryologically and physiologically distinct from the cortex, although cortical and medullary hormones often act in a complementary manner. Cells of the adrenal medulla have an affinity for chromium salts in histological preparations and hence are called chromaffin cells. They arise from neuroectoderm and are innervated by neurons whose cell bodies lie in the intermediolateral cell column in the thoracolumbar portion of the spinal cord. Axons of these cells pass through the paravertebral sympathetic ganglia to form the splanchnic nerves. Chromaffin cells are thus modified postganglionic neurons. Their principal secretory products, epinephrine and norepinephrine, are derivatives of the amino acid tyrosine and belong to a class of compounds called *catecholamines*. About 5 to 6 mg of catecholamines are stored in membrane-bound granules within chromaffin cells. Epinephrine is about five times as abundant in the human adrenal medulla as norepinephrine, but only norepinephrine is found in postganglionic sympathetic neurons and extra-adrenal chromaffin tissue. Although medullary hormones affect virtually every tissue of the body and play a crucial role in the acute response to stress, the adrenal medulla is not required for survival so long as the rest of the sympathetic nervous system is intact.

BIOSYNTHESIS OF MEDULLARY HORMONES

The biosynthetic pathway for epinephrine and norepinephrine is shown in Figure 21. Hydroxylation of tyrosine to form dihydroxyphenylalanine (DOPA) is the rate-determining reaction and is catalyzed by the enzyme tyrosine hydroxylase. Activity of this enzyme is inhibited by catecholamines (product inhibition) and stimulated by phosphorylation. In this way, regulatory adjustments are made rapidly and are closely tied to bursts of secretion. A protracted increase in secretory activity induces synthesis of additional enzyme after a lag time of about 12 hours.

Tyrosine hydroxylase and DOPA decarboxylase are cytosolic enzymes, but the enzyme that catalyzes the β-hydroxylation of dopamine to form norepinephrine resides within the secretory granule. Dopamine is pumped into the granule by an energy-dependent, stereospecific process. For sympathetic nerve endings and those adrenomedullary cells that produce norepinephrine, synthesis is complete with the formation of norepinephrine, and the hormone remains in the granule until it is secreted. Synthesis of epinephrine, however, requires that norepinephrine reenter the cytosol for the final methylation reaction. The enzyme required for this reaction, phenylethanolamine-N-methyltransferase (PNMT), is at least partly inducible by glucocorticoids. Induction requires concentrations of cortisol that are considerably higher than those found in peripheral blood. The vascular arrangement in the adrenals is such that interstitial fluid surrounding cells of the medulla can equilibrate with venous blood that drains the cortex and therefore has a much higher content of glucocorticoids than arterial blood. Glucocorticoids may thus determine the ratio of epinephrine to norepinephrine production. Once methylated, epinephrine is pumped back into the storage granule, whose membrane protects stored catecholamines from oxidation by cytosolic enzymes.

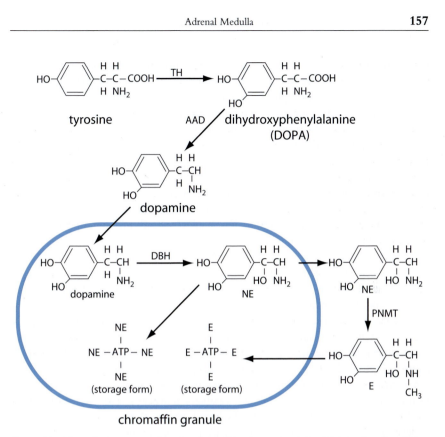

Figure 21   Biosynthetic sequence for epinephrine (E) and norepinephrine (NE) in adrenal medullary cells. TH, Tyrosine hydroxylase; AAD, aromatic L-amino acid decarboxylase (also called DOPA decarboxylase); DBH, dopamine betahydroxylase; PNMT, phenylethanolamine-*N*-methyltransferase.

STORAGE, RELEASE, AND METABOLISM OF MEDULLARY HORMONES

Catecholamines are stored in secretory granules in close association with ATP and at a molar ratio of 4:1, suggesting some hydrostatic interaction between the positively charged amines and the four negative charges on ATP. Some opioid peptides, including the enkephalins, β-endorphin, and their precursors, are also found in these granules. Acetylcholine released during neuronal stimulation increases sodium conductance of the chromaffin cell membrane. The resulting influx of sodium ions depolarizes the plasma membrane, leading to an influx of calcium through voltage-sensitive channels. Calcium is required for catecholamine secretion. Increased cytosolic concentrations of calcium promote phosphorylation of microtubules and the consequent translocation of secretory granules to the cell surface. Secretion occurs when membranes of the chromaffin granules fuse with plasma membranes and the granular contents are extruded into the extracellular space. Fusion of the granular membrane with the cell membrane may also require calcium. ATP, opioid peptides, and other contents of the granules are released along with epinephrine and norepinephrine. As yet, the physiological significance of opioid secretion by the adrenals is not known, but it has been suggested that the analgesic effects of these compounds may be of importance in the stress response.

All the epinephrine in blood originates in the adrenal glands. However, norepinephrine may reach the blood by either adrenal secretion or diffusion from sympathetic synapses. The half-lives of medullary hormones in the peripheral circulation have been estimated to be less than 10 seconds for epinephrine and less than 15 seconds for norepinephrine. Up to 90% of the catecholamines is removed in a single passage through most capillary beds. Clearance from the blood requires uptake by both neuronal and nonneuronal tissues. Significant amounts of norepinephrine are taken up by sympathetic nerve endings and incorporated into secretory granules for release at a later time. Epinephrine and norepinephrine that are taken up in excess of storage capacity are degraded in neuronal cytosol principally by the enzyme monoamine oxidase (MAO). This enzyme catalyzes oxidative deamination of epinephrine, norepinephrine, and other biologically important amines (Figure 22). Catecholamines taken up by endothelium, heart, liver, and other tissues are also inactivated enzymatically, principally by catecholamine-O-methyltransferase (COMT), which catalyzes transfer of a methyl group from S-adenosylmethionine to one of the hydroxyl groups. Both of these enzymes are widely distributed and can act sequentially in either order on both epinephrine and norepinephrine. A number of pharmaceutical agents have been developed to modify the actions of these enzymes and thus modify sympathetic responses. Inactivated catecholamines, chiefly vanillylmandelic acid (VMA) and 3-methoxy-4-hydroxyphenylglycol (MHPG), are conjugated with sulfate or glucuronide and excreted in urine. As with steroid hormones, measurement of urinary metabolites of catecholamines is a useful, noninvasive source of diagnostic information.

PHYSIOLOGICAL ACTIONS OF MEDULLARY HORMONES

The sympathetic nervous system and adrenal medullary hormones, like the cortical hormones, act on a wide variety of tissues to maintain the integrity of the internal environment, both at rest and in the face of internal and external challenges. Catecholamines enable us to cope with emergencies and equip us for what Cannon called "fright, fight, or flight." Responsive tissues make no distinctions between bloodborne catecholamines and those released locally from nerve endings. In contrast to adrenal cortical hormones, effects of catecholamines are expressed within seconds and dissipate as rapidly when the hormone is removed. Medullary hormones are thus

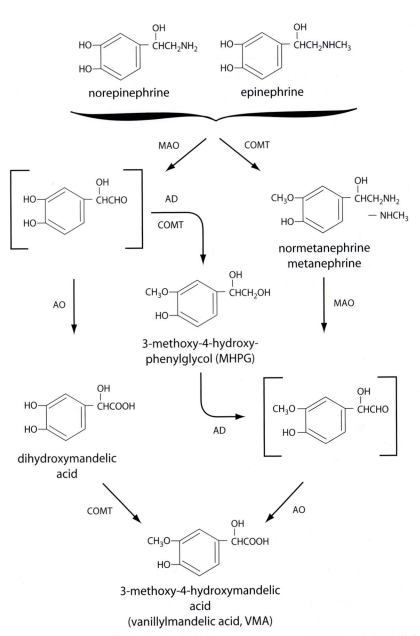

Figure 22 Catecholamine degradation. MAO, Monoamine oxidase; COMT, catechol–O–methyl-transferase; AD, alcohol dehydrogenase; AO, aldehyde oxidase. (From Cryer, *In* "Endocrinology and Metabolism," 3rd Ed., p. 716. McGraw-Hill, New York, 1995, by permission of The McGraw-Hill Companies.)

Table 2

Typical Responses to Stimulation of the Adrenal Medulla

Target	Responses
Cardiovascular system	
Heart	↑ Force and rate of contraction
	↑ Conduction
	↑ Blood flow (dilation of coronary arterioles)
	↑ Glycogenolysis
Arterioles	
Skin	Constriction
Mucosae	Constriction
Skeletal muscle	Constriction, dilation
Metabolism	
Fat	↑ Lipolysis, ↑ blood FFA and glycerol
Liver	↑ Glycogenolysis and gluconeogenesis,
	↑ Blood sugar
Muscle	↑ Glycogenolysis, ↑ Lactate and pyruvate release
Bronchial muscle	Relaxation
Stomach and intestines	↑ Motility, ↑ sphincter contraction
Urinary bladder	↑ Sphincter contraction
Skin	↑ Sweating
Eyes	Contraction of radial muscle of the iris
Salivary gland	↑ Amylase secretion, ↑ watery secretion
Kidney	↑ Renin secretion
Skeletal muscle	↑ Tension generation, ↑ neuromuscular transmission (defatiguing effect)

ideally suited for making the rapid short-term adjustments demanded by a changing environment, whereas cortical hormones, which act only after a lag period of at least 30 minutes, are of little use at the onset of stress. The cortex and medulla together, however, provide an effective "one–two punch," with cortical hormones maintaining and even amplifying the effectiveness of medullary hormones.

Cells in virtually all tissues of the body express G-protein-coupled receptors for epinephrine and norepinephrine on their surface membranes (see Chapter 1). These so-called adrenergic receptors were originally divided into two categories, α and β, based on their activation or inhibition by various drugs. Subsequently, the α and β receptors were further subdivided into α_1, α_2, β_1, β_2, and β_3 receptors. All these receptors recognize both epinephrine and norepinephrine at least to some extent, and a given cell may have more than one class of adrenergic receptor.

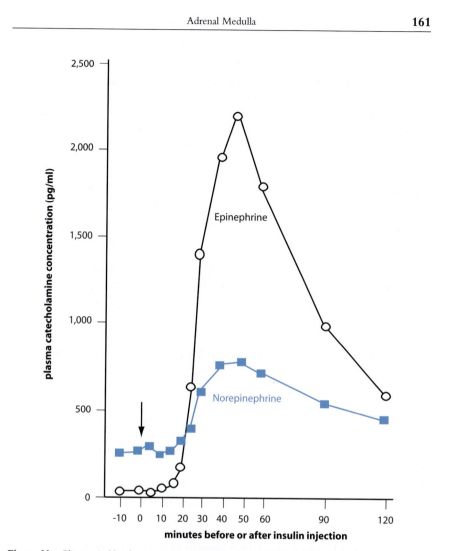

Figure 23 Changes in blood concentrations of epinephrine and norepinephrine in response to hypoglycemia. Insulin, which produces hypoglycemia, was injected at the time indicated by the arrow. (From Garber, A. J., Bier, D. M., Cryer, P. E., and Pagliara, A. S., *J. Clin. Invest.* **58**, 7–15, 1976, with permission.)

Biochemical mechanisms of signal transduction follow the pharmacological subdivisions of the adrenergic receptors. All of the β-adrenergic receptors communicate with adenylyl cyclase through the stimulatory G-protein, (G_s) (Chapter 1) and activate adenylyl cyclase, but subtle differences distinguish them. The β_3 receptors may couple to both G_s and G_i heterotrimeric proteins, and hence give a less robust response than do β_1 and β_2 receptors. From a physiological perspective, the only difference between β_1 and β_2 receptors is the low sensitivity of the β_2

receptors to norepinephrine. Stimulation of α_2 receptors inhibits adenylyl cyclase and may block the increase in cyclic AMP produced by other agents. For α_2 effects, the receptor communicates with adenylyl cyclase through the inhibitory G protein (G_i). Responses initiated by the α_1 receptor, which couples with G_q, are mediated by the inositol trisphosphate–diacylglycerol mechanism (Chapter 1).

Some of the physiological effects of catecholamines are listed in Table 2. Although these actions may seem diverse, in actuality they constitute a magnificently coordinated set of responses that Cannon aptly called "the wisdom of the body." When producing their effects, catecholamines maximize the contributions of each of the various tissues to resolve the challenges to survival. On the whole, cardiovascular effects maximize cardiac output and ensure perfusion of the brain and working muscles. Metabolic effects ensure an adequate supply of energy-rich substrate. Relaxation of bronchial muscles facilitates pulmonary ventilation. Ocular effects increase visual acuity. Effects on skeletal muscle and transmitter release from motor neurons increase muscular performance, and quiescence of the gut permits diversion of blood flow, oxygen, and fuel to reinforce these effects.

REGULATION OF ADRENAL MEDULLARY FUNCTION

The sympathetic nervous system, including its adrenal medullary component, is activated by any actual or threatened change in the internal or external environment. It responds to physical changes, emotional inputs, and anticipation of increased physical activity. Input reaches the adrenal medulla through its sympathetic innervation. Signals arising in the hypothalamus and other integrating centers activate both the neural and hormonal components of the sympathetic nervous system, but not necessarily in an all-or-none fashion. Activation may be general or selectively limited to discrete targets. The adrenals can be preferentially stimulated, and it is even possible that norepinephrine- or epinephrine-secreting cells may be selectively activated, as shown in Figure 23. In response to hypoglycemia detected by glucose-monitoring cells in the central nervous system, the concentration of norepinephrine in blood increased threefold, whereas that of epinephrine, which tends to be a more effective hyperglycemic agent, increased 50-fold. Metabolic actions of epinephrine are discussed further in Chapter 9.

SUGGESTED READING

Blalock, J. E. (1999). Proopiomelanocortin and the immune–neuroendocrine connection. *Ann. N.Y. Acad. Sci.* **885**, 161–172.

Clark, A. J. L., and Weber, A. (1998). Adrenocorticotropin insensitivity syndromes. *Endocr. Rev.* **19**, 828–843.

Dallman, M. F., and Bhatnagar, S. (2001). Chronic stress and energy balance: Role of the hypothal-amo–pituitary–adrenal axis. *In* "Handbook of Physiology, Section 7: Endocrinology, Volume IV: Coping with the Environment: Neural and Endocrine Mechanisms" (B. S. McEwen, ed.), pp. 179–210. American Physiological Society and Oxford University Press, New York.

Funder, J. W. (1991). Steroids, receptors, and response elements: The limits of signal specificity. *Recent Prog. Horm. Res.* **47**, 191–210.

Keller-Wood, M. E., and Dallman, M. F. (1984). Corticosteroid inhibition of ACTH secretion. *Endocr. Rev.* **5**, 1–24.

McKay, L. I., and Cidlowski, J. A. (1999). Molecular control of immune/Inflammatory responses: Interactions between nuclear factor-[{kappa}] B and steroid receptor-signaling pathways. *Endocr. Rev.* **20**, 435–459.

Needleman, P., Turk, J., Jakschik, B. A., Morrison, A. R., and Lefkowith, J. B. (1986). Arachidonic acid metabolism. *Annu. Rev. Biochem.* **55**, 69–102.

Orth, D. N. (1992). Corticotropin-releasing hormone in humans. *Endocr. Rev.* **13**, 164–191.

Rogerson, F. M., and Fuller, P. J. (2000). Mineralocorticoid action. *Steroids* **65**, 61–73.

Sapolsky, R. M., Romero, L. M., and Munck, A. U. (2000). How do glucocorticoids influence stress responses? Integrating permissive, suppressive, stimulatory, and preparative actions. *Endocr. Rev.* **21**, 55–89.

Stocco, D. M. (2001). StAR protein and the regulation of steroid hormone biosynthesis. *Annu. Rev. Physiol.* **63**, 193–213.

Ungar, A., and Phillips, J. H. (1983). Regulation of the adrenal medulla. *Physiol. Rev.* **63**, 787–843.

Weninger, S. C., and Majzoub, J. A. (2001). Regulation and actions of corticotropin releasing hormone. *In* "Handbook of Physiology, Section 7: Endocrinology, Volume IV: Coping with the Environment: Neural and Endocrine Mechanisms" (B. S. McEwen, ed.), pp. 103–124. American Physiological Society and Oxford University Press, New York.

White, P. C., Mune, T., and Agarwal, A. K. (1997). 11β-Hydroxysteroid dehydrogenase and the syndrome of apparent mineralocorticoid excess. *Endocr. Rev.* **18**, 135–156.

Young, J. B., and Landsberg, L. (2001). Synthesis, storage and secretion of adrenal medullary hormones: Physiology and pathophysiology. *In* "Handbook of Physiology, Section 7: Endocrinology, Volume IV: Coping with the Environment: Neural and Endocrine Mechanisms" (B. S. McEwen, ed.), pp. 3–20. American Physiological Society and Oxford University Press, New York.

CHAPTER 5

The Pancreatic Islets

OVERVIEW

The principal pancreatic hormones are insulin and glucagon, whose opposing effects on the liver regulate hepatic storage, production, and release of energy-rich fuels. Insulin is an anabolic hormone that promotes sequestration of carbohydrate, fat, and protein in storage depots throughout the body. Its powerful actions are exerted principally on skeletal muscle, liver, and adipose tissue, whereas those of glucagon are restricted to the liver, which responds by forming and secreting energy-rich water-soluble fuels: glucose, acetoacetic acid, and β-hydroxybutyric acid. Interplay of these two hormones contributes to constancy in the availability of metabolic fuels to all cells. Somatostatin is a third islet hormone, but a physiological role for pancreatic somatostatin has not been established. A fourth substance, pancreatic polypeptide, is even less understood.

Glucagon acts in concert with other fuel-mobilizing hormones to counterbalance the fuel-storing effects of insulin. Because compensatory changes in secretion of all of these hormones are readily made, states of glucagon excess or deficiency rarely lead to overt human disease. Insulin, on the other hand, acts alone, and prolonged survival is not possible in its absence. Inadequacy of insulin due either to insufficient production [diabetes mellitus type I; insulin-dependent diabetes mellitus (IDDM)] or end-organ unresponsiveness [diabetes mellitus type II; noninsulin-dependent diabetes mellitus (NIDDM)] results in one of the most common of the endocrine diseases affecting more than 3% of the American population.

MORPHOLOGY OF THE ENDOCRINE PANCREAS

The 1–2 million islets of the human pancreas range in size from about 50 to about 500 mm in diameter and contain from 50 to 300 endocrine cells. Collectively the islets comprise only 1–2% of the pancreatic mass. They are highly vascular, with each cell seemingly in direct contact with a capillary. Blood is supplied by the pancreatic artery and drains into the portal vein, which thus delivers the entire output of pancreatic hormones to the liver. The islets are also richly innervated with both sympathetic and parasympathetic fibers that terminate on or near the secretory cells.

Histologically, the islets consist of three cell types. Beta cells, which synthesize and secrete insulin, make up about 60–75% of a typical islet. Alpha cells are the source of glucagon and comprise perhaps as much as 20% of islet tissue. Delta cells, which are considerably less abundant, produce somatostatin. An additional but rarer cell type, the F cell, may also appear in the exocrine part of the pancreas.

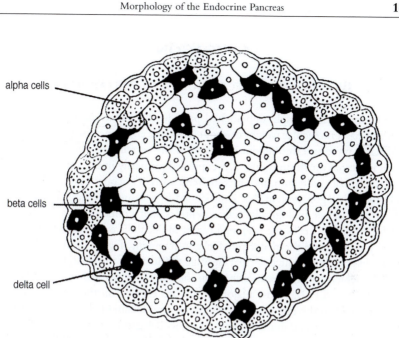

Figure 1 Arrangement of cells in a typical islet. The clear cells in the center of the islet are the beta (insulin-secreting) cells. The stippled cells in the periphery are the alpha (glucagon-secreting) cells, and the solid black cells are the delta (somatostatin-secreting) cells. (From Orci, L., and Unger, R. H., *Lancet* **2**, 1243–1244, 1975, with permission.)

It contains and secretes a compound of called pancreatic polypeptide, which inhibits pancreatic exocrine cell functions.

Beta cells occupy the central region of the islet or microlobules within islets, whereas alpha cells occupy the outer rim. Delta cells are interposed between alpha and beta cells and are thus in contact with both types (Figure 1). Gap junctions link alpha cells to each other, beta cells to each other, and alpha cells to beta cells. Although experimental proof is lacking, this arrangement may account for synchronous secretory activity. There are also tight junctions between various islet cells. These sites of close apposition or actual fusion of plasma membranes of adjacent cells may affect diffusion of substances into or out of intercellular spaces. This arrangement could either facilitate or hinder paracrine communication between alpha, beta, and delta cells. Blood flows through an anastomosing network of capillaries from the center of the islet toward the periphery. This arrangement favors intraislet delivery of insulin from the centrally located beta cells to the peripherally located alpha cells. The physiological consequences of these complex anatomical specializations are not understood.

GLUCAGON

BIOSYNTHESIS, SECRETION, AND METABOLISM

Glucagon is a simple unbranched peptide chain that consists of 29 amino acids and has a molecular weight of about 3,500. Its amino acid sequence has been remarkably preserved throughout evolution of the vertebrates. The glucagon gene, which is located on chromosome 2, is expressed primarily in the alpha cells, L-cells of the intestinal epithelium, and discrete brain areas. It encodes a large, 158-amino-acid preproglucagon protein that is processed in a tissue-specific manner to give rise to at least six biologically active peptides that contain similar amino acid sequences. In alpha cells the preproglucagon molecule is enzymatically cleaved to release glucagon and the major proglucagon fragment (Figure 2). In the intestine and the hypothalamus the principal cleavage products are the *glucagon-like peptide 1* (GLP-1), which has important effects on islet function (see below), and *glucagon-like peptide 2* (GLP-2), whose major actions are exerted on the intestine. GLP-1 may also regulate the rate of gastric emptying and feeding behavior. Proglucagon is a member of a superfamily of genes that encodes gastrointestinal hormones and neuropeptides, including secretin, vasoactive inhibitory peptide (VIP), pituitary adenylyl cyclase-activating peptide (PACAP), glucose-dependent insulinotropic peptide (GIP), and the growth hormone-releasing hormone (GHRH). Glucagon is packaged, stored in membrane-bound granules, and secreted by exocytosis like other peptide hormones.

Glucagon circulates without binding to carrier proteins and has a half-life in blood of about 5 minutes. Its concentrations in peripheral blood are considerably

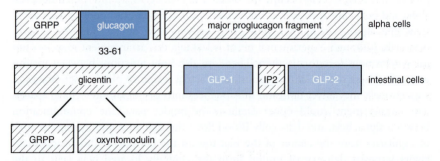

Figure 2   Cell-specific posttranslational processing of preproglucagon. GRPP, Glicentin-related pancreatic peptide; GLP-1, glucagon-like peptide 1; GLP-2, glucagon-like peptide 2; IP2, intervening peptide 2. Intervening peptide 1 is the small fragment between glucagon and the major proglucagon fragment at the top of the figure.

lower than in portal venous blood. This difference reflects not only greater dilution in the general circulation but also the fact that about 25% of the secreted glucagon is destroyed during passage through the liver. The kidney is another important site of degradation, and a considerable fraction of circulating glucagon is destroyed by plasma peptidases.

PHYSIOLOGICAL ACTIONS OF GLUCAGON

The physiological role of glucagon is to stimulate hepatic production and secretion of glucose and, to a lesser extent, ketone bodies that are derived from fatty acids. Under normal circumstances, liver and possibly pancreatic beta cells are the only targets of glucagon action. A number of other tissues, including fat and heart, express glucagon receptors, and can respond to glucagon experimentally, but considerably higher concentrations of glucagon are needed than are normally found in peripheral blood. Glucagon stimulates the liver to release glucose and produces a prompt increase in blood glucose concentration. Glucose that is released from the liver is obtained from breakdown of stored glycogen (*glycogenolysis*) and from new synthesis (*gluconeogenesis*). Because the principal precursors for gluconeogenesis are amino acids, especially alanine, glucagon also increases hepatic production of urea (*ureogenesis*) from the amino groups. Glucagon also increases production of ketone bodies (*ketogenesis*) by directing metabolism of long-chain fatty acids toward oxidation and away from esterification and export as lipoproteins. Concomitantly, glucagon may also promote breakdown of hepatic triglycerides to yield long-chain fatty acids, which, along with fatty acids that reach the liver from peripheral fat depots, provide the substrate for ketogenesis.

All of the effects of glucagon appear to be mediated by cyclic AMP (see Chapter 1). In fact, it was studies of the glycogenolytic action of glucagon that led to the discovery of cyclic AMP and its role as a second messenger. Activation of protein kinase A by cyclic AMP results in phosphorylation of enzymes, which increases or decreases enzymatic activity, or phosphorylation of the transcription factor CREB, which usually increases transcription of target genes. Glucagon may also increase intracellular concentrations of calcium by a mechanism that depends on activation of protein kinase A, and the increased calcium may reinforce some actions of glucagon, particularly on glycogenolysis.

Glucose Production

To understand how glucagon stimulates the hepatocyte to release glucose, we must first consider some of the biochemical reactions that govern glucose metabolism in the liver. Biochemical pathways that link these reactions are illustrated in

Figure 3. It is important to recognize that not all enzymatic reactions are freely reversible under conditions that prevail in living cells. Phosphorylation and dephosphorylation of substrate usually require separate enzymes. This sets up substrate cycles that would spin futilely in the absence of some regulatory influence exerted on either or both opposing reactions. These reactions are often strategically situated at or near branch points in metabolic pathways and can therefore direct flow of substrates toward one fate or another. Regulation is achieved both by modulating the activity of enzymes already present in cells and by increasing or decreasing rates of enzyme synthesis and therefore amounts of enzyme molecules. Enzyme activity can be regulated allosterically by changes in conformation produced by substrates or cofactors, or covalently by phosphorylation and dephosphorylation of regulatory sites in the enzymes. Changing the activity of an enzyme requires only seconds, whereas many minutes or even hours are needed to change the amount of an enzyme.

Glycogenolysis

Cyclic AMP formed in response to the interaction of glucagon with its G-protein-coupled receptors on the surface of the hepatocyte (see Chapter 1) activates protein kinase A, which catalyzes phosphorylation, and hence activation, of an enzyme called *phosphorylase kinase* (Figure 4). This enzyme, in turn, catalyzes phosphorylation of another enzyme, *glycogen phosphorylase*, which cleaves glycogen stepwise to release glucose-1-phosphate. Glucose-1-phosphate is the substrate for *glycogen synthase*, which catalyzes the incorporation of glucose into glycogen. Glycogen synthase is also a substrate for protein kinase A and is inactivated when phosphorylated. Thus by increasing the formation of cyclic AMP, glucagon simultaneously promotes glycogen breakdown and prevents recycling of glucose to glycogen. Cyclic AMP-dependent phosphorylation of enzymes that regulate the glycolytic pathway at the level of phosphofructokinase and acetyl coenzyme A (CoA) carboxylase (see below) prevents consumption of glucose-6-phosphate by the hepatocyte, leaving dephosphorylation and diffusion into the blood as the only pathway open to newly depolymerized glucose.

Gluconeogenesis

Precursors of glucose enter the gluconeogenic pathway as three- or four-carbon compounds. Glucagon directs their conversion to glucose by accelerating their condensation to fructose-6-phosphate while simultaneously blocking their escape from the gluconeogenic pathway (cycles III and IV in Figure 3). Cyclic AMP controls production of a potent allosteric regulator of metabolism called *fructose-2,6-bisphosphate*. This compound, when present even in tiny amounts, activates *phosphofructokinase* and inhibits *fructose-1,6-bisphosphatase*, thereby directing

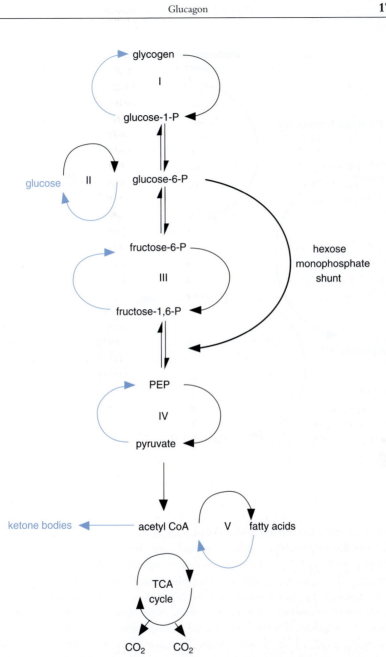

Figure 3 Biochemical pathways of glucose metabolism in hepatocytes. Reactions that are accelerated in the presence of glucagon are shown in blue. PEP, Phosphoenolpyruvate; TCA, tricarboxylic acid.

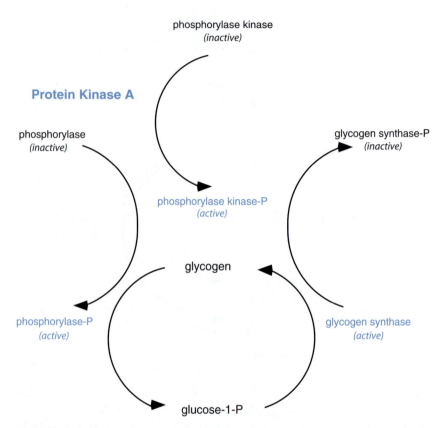

Figure 4   Role of protein kinase A (cyclic AMP-dependent protein kinase) in glycogen metabolism.

flow of substrate toward glucose breakdown rather than glucose formation (Figure 5). Fructose-2,6-bisphosphate, which should not be confused with fructose-1, 6-bisphosphate, is formed from fructose-6-phosphate by the action of an unusual bifunctional enzyme that catalyzes either phosphorylation of fructose-6-phosphate to fructose-2,6-bisphosphate or dephosphorylation of fructose-2,6-bisphosphate to fructose-6-phosphate, depending on its own state of phosphorylation. This enzyme is a substrate for protein kinase A and behaves as a phosphatase when it is phosphorylated. Its activity in the presence of cyclic AMP rapidly depletes the hepatocyte of fructose-2,6-bisphosphate, and substrate therefore flows toward glucose production.

The other important regulatory step in gluconeogenesis is phosphorylation and dephosphorylation of pyruvate (cycle IV in Figure 3). It is here that three- and

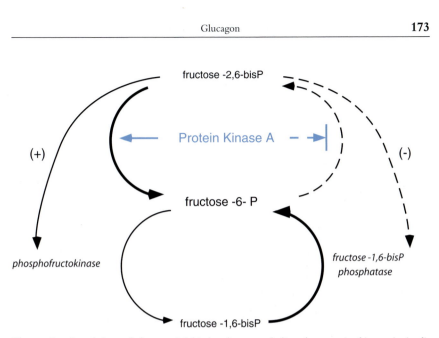

Figure 5 Regulation of fructose-1,6-bisphosphate metabolism by protein kinase A (cyclic AMP-dependent protein kinase) and fructose-2,6-bisphosphate. Protein kinase A is required for formation of fructose-2,6-bisphosphate, which activates (+) phosphofructokinase and inhibits (−) fructose-1,6-bisphosphatase.

four-carbon fragments enter or escape from the gluconeogenic pathway. The cytosolic enzyme that catalyzes dephosphorylation of phosphoenol pyruvate (PEP) was inappropriately named pyruvate kinase before it was recognized that direct phosphorylation of pyruvate does not occur under physiological conditions and that this enzyme acts only in the direction of dephosphorylation (Figure 6). Regulation of this enzyme is complex. Pyruvate kinase is another substrate for protein kinase A and is powerfully inhibited when phosphorylated, but the inhibition can be overcome by fructose-1,6-bisphosphate. Thus, activation of protein kinase A has the dual effect of decreasing pyruvate kinase activity directly and of decreasing the abundance of its activator, fructose-1,6-bisphosphate, by reactions shown in Figure 5. Inhibiting pyruvate kinase may be the single most important effect of glucagon on the gluconeogenic pathway. On a longer time scale, glucagon inhibits the synthesis of pyruvate kinase. Phosphorylation of pyruvate requires a complex series of reactions in which pyruvate must first enter mitochondria, where it is carboxylated to form oxaloacetate. Entry of pyruvate across the mitochondrial membrane is accelerated by glucagon, but the mechanism for this effect is not known. Oxaloacetate is converted to cytosolic PEP by the catalytic activity of *PEP carboxykinase*. Synthesis of this enzyme is accelerated by increased cyclic AMP.

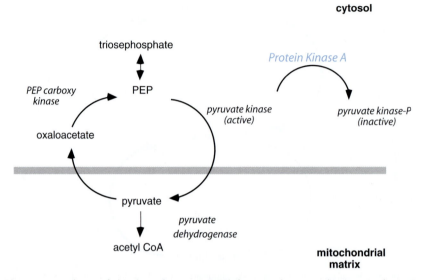

Figure 6 Regulation of phosphoenol pyruvate (PEP) formation by protein kinase A (cyclic AMP-dependent protein kinase). Protein kinase A catalyzes the phosphorylation and, hence, inactivation of pyruvate kinase, whose activity limits the conversion of PEP to pyruvate.

Lipogenesis and Ketogenesis

An alternate fate of pyruvate in mitochondria is decarboxylation to form acetyl CoA (Figure 6). This two-carbon acetyl unit is the building block of fatty acids and eventually finds its way back to the cytosol where fatty acid synthesis (*lipogenesis*) takes place. Lipogenesis is the principal competitor of gluconeogenesis for three-carbon precursors. The first committed step in fatty acid synthesis is the carboxylation of acetyl CoA to form malonyl CoA. *Acetyl CoA carboxylase*, the enzyme that catalyzes this reaction, is yet another substrate for protein kinase A and is powerfully inhibited when phosphorylated. Inhibition of fatty acid synthesis not only preserves substrate for gluconeogenesis but also prevents oxidation of glucose by the hexose monophosphate shunt pathway (Figure 3). Nicotinamide adenine dinucleotide phosphate (NADP), which is required for shunt activity, is reduced in the initial reactions of this pathway and can be regenerated only by transferring protons to the elongating fatty acid chain.

Fatty acid synthesis and oxidation constitute another substrate cycle and another regulatory site for cyclic AMP action. The same reaction that inhibits fatty acid synthesis promotes fatty acid oxidation and consequently *ketogenesis* (ketone body formation) (Figure 7). Long-chain fatty acid molecules that reach the liver

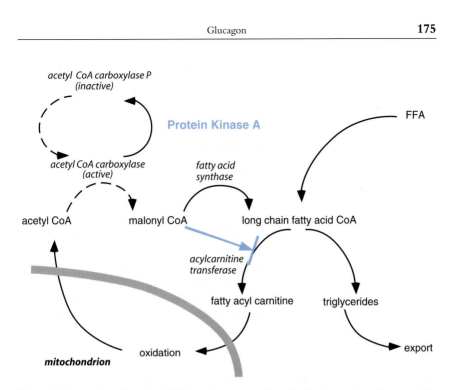

Figure 7 Protein kinase A (cyclic AMP-dependent protein kinase) indirectly stimulates ketogenesis by decreasing the formation of malonyl CoA, thus removing a restriction on accessibility of fatty acids to intramitochondrial oxidative enzymes.

can be either oxidized or esterified and exported to adipose tissue as the triglyceride component of low-density lipoproteins. To be esterified, fatty acids must remain in the cytosol, and to be oxidized they must enter the mitochondria. Long-chain fatty acids can cross the mitochondrial membrane only when linked to carnitine. *Carnitine acyltransferase*, the enzyme that catalyzes this linkage, is powerfully inhibited by malonyl CoA. Thus when malonyl CoA concentrations are high, coincident with fatty acid synthesis, fatty acid oxidation is inhibited. Conversely, when the formation of malonyl CoA is blocked, fatty acids readily enter mitochondria and are oxidized to acetyl CoA. Because long-chain fatty acids typically contain 16 and 18 carbons, each molecule that is oxidized yields eight or nine molecules of acetyl CoA. The ketone bodies β-hydroxybutyrate and acetoacetate are formed from condensation of two molecules of acetyl CoA and the subsequent removal of the CoA moiety.

By reducing the concentration of malonyl CoA, glucagon sets the stage for ketogenesis, but the actual amount of ketone production is determined by the amount of long-chain fatty acids available for oxidation. Most fatty acids oxidized in liver originate in adipose tissue, but glucagon, through cyclic AMP, may also activate a lipase in liver and thereby provide fatty acids from breakdown of hepatic triglycerides.

Ureogenesis

Whenever carbon chains of amino acids are used as substrate for gluconeogenesis, amino groups must be disposed of in the form of urea, which thus becomes a by-product of gluconeogenesis. By promoting gluconeogenesis, therefore, glucagon also increases the formation of urea (*ureogenesis*). Carbon skeletons of most amino acids can be converted to glucose, but because of peculiarities of peripheral metabolism alanine is quantitatively the most important glucogenic amino acid. By accelerating conversion of pyruvate to glucose (see above), glucagon indirectly accelerates transamination of alanine to pyruvate. Glucagon also accelerates ureogenesis by increasing transport of amino acids across hepatocyte plasma membranes by an action that requires synthesis of new RNA and protein. In addition, glucagon also promotes the synthesis of some urea cycle enzymes.

REGULATION OF GLUCAGON SECRETION

The concentration of glucose in blood is the most important determinant of glucagon secretion in normal individuals. When the plasma glucose concentration exceeds 200 mg/dl, glucagon secretion is maximally inhibited. Inhibitory effects of glucose are proportionately less at lower concentrations and disappear when glucose concentration falls below 50 mg/dl. Except immediately after a meal rich in carbohydrate, the blood glucose concentration remains constant at around 90 mg/dl. The set point for glucose concentration thus falls well within the range over which glucagon secretion is regulated, and alpha cells can respond to changes in blood glucose with either an increase or a decrease in glucagon output. The alpha cells appear to respond directly to changes in glucose concentration, but we do not yet understand how they monitor blood glucose concentration and translate that information to an appropriate rate of glucagon secretion. Little is understood of the intracellular molecular events that bring about an increase or decrease of glucagon secretion.

Low blood glucose (hypoglycemia) not only relieves inhibition of glucagon secretion, but this life-threatening circumstance stimulates the central nervous system to signal both parasympathetic and sympathetic nerve endings within the islet to release their neurotransmitters, acetylcholine and vasoactive intestinal peptide from parasympathetic endings, and norepinephrine and neuropeptide Y (NPY) from sympathetic endings. Alpha cells express receptors for these neurotransmitters, and they secrete glucagon in response to both parasympathetic and sympathetic stimulation. The sympathetic response to hypoglycemia also involves secretion of epinephrine and norepinephrine from the adrenal medulla (see Chapter 4). Adrenomedullary hormones further stimulate alpha cells to secrete glucagon.

Glucagon secretion is evoked by a meal rich in amino acids. Alpha cells respond directly to increased blood levels of certain amino acids, particularly arginine. In addition, digestion of protein-rich foods triggers the release of *cholecystokinin–pancreazymin* from cells in the duodenal mucosa. This gastrointestinal hormone is a secretagogue for islet hormones as well as pancreatic enzymes and may alert alpha cells to an impending influx of amino acids. Increased secretion of glucagon in response to a protein meal not only prepares the liver to dispose of excess amino acids by gluconeogenesis but also signals the liver to release glucose and thus counteracts the hypoglycemic effects of insulin, whose secretion is simultaneously increased by amino acids (see below).

Glucose is not the only physiological inhibitor of glucagon secretion. Insulin, somatostatin, GLP-1, glucose-dependent insulinotropic peptide, and free fatty acids (FFAs) also exert inhibitory influences on glucagon secretion (Figure 8). Insulin, which may reach alpha cells by either the endocrine or paracrine route, directly inhibits glucagon secretion and is required for expression of inhibitory effects of glucose. In fact, it has been suggested that glucose may inhibit glucagon

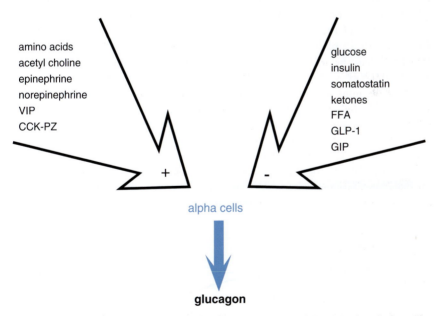

amino acids
acetyl choline
epinephrine
norepinephrine
VIP
CCK-PZ

glucose
insulin
somatostatin
ketones
FFA
GLP-1
GIP

+   -

alpha cells

glucagon

Figure 8 Stimulatory and inhibitory signals for glucagon secretion. VIP, Vasoactive intestinal peptide; CCK–PZ, cholecystokinin–pancreazymin; FFA, free fatty acids; GLP-1, glucagon-like peptide 1; GIP glucose-dependent insulinotropic peptide.

secretion indirectly through increased secretion of insulin. In persons suffering from insulin deficiency (see below), glucagon secretion is brisk despite high blood glucose concentrations.

INSULIN

BIOSYNTHESIS, SECRETION, AND METABOLISM

Insulin is composed of two unbranched peptide chains joined together by two disulfide bridges (Figure 9). The single gene that encodes the preproinsulin molecule consists of three exons and two introns and is located on chromosome 11. The two chains of insulin and their disulfide cross-bridges are derived from the single-chain proinsulin molecule, from which a 31-residue peptide, called the connecting peptide (C peptide), is excised by stepwise actions of two trypsinlike enzymes called *prohormone convertases*. Conversion of proinsulin to insulin takes place slowly within storage granules. The C peptide therefore accumulates within granules in equimolar amounts with insulin. When insulin is secreted, the entire contents of secretory vesicles are disgorged into extracellular fluid. Consequently, the C peptide and any remaining proinsulin and processing intermediates are released into the circulation. When secretion is rapid, proinsulin may comprise as much as 20% of the circulating peptides detected by insulin antibodies, but it contributes little biological activity. Although several biological actions of the C peptide have been described, no physiological role for the C peptide has yet been established.

The insulin storage granule contains a variety of proteins that are also released into the extracellular space whenever insulin is secreted. Most of these

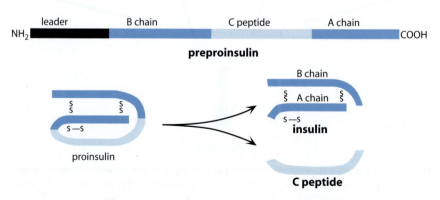

Figure 9 Posttranslational processing of preproinsulin.

proteins are thought to maintain optimal conditions for storage and processing of insulin, but some may also have biological activity. Their fate and actions, if any, are largely unknown. One such protein, however, called *amylin*, may contribute to the amyloid that accumulates in and around beta cells in states of insulin hypersecretion and may contribute to islet pathology. A wide variety of biological actions of amylin have been described, including antagonism to the actions of insulin in various tissues, suppression of appetite, and delaying of gastric emptying, but a physiological role for amylin remains to be established and is the subject of some controversy.

Insulin is cleared rapidly from the circulation with a half-life of 4–6 minutes and is destroyed by a specific enzyme, called *insulinase* (or insulin-degrading enzyme), that is present in liver, muscle, kidney, and other tissues. The first step in insulin degradation is receptor-mediated internalization through an endosomic mechanism. Degradation may take place within endosomes or after fusion of endosomes with lysosomes. The liver is the principal site of insulin degradation and inactivates about 30–70% of the insulin that reaches it in hepatic portal blood. Insulin degradation in the liver appears to be a regulated process governed by changes in availability of metabolic fuels and changing physiological circumstances. The liver may thus regulate the amount of insulin that enters the systemic circulation. The kidneys destroy about half of the insulin that reaches the general circulation following receptor-mediated uptake, both from the glomerular filtrate and from postglomerular blood plasma. Normally, little or no insulin is found in urine. Muscle and other insulin-sensitive tissues throughout the body apparently account for destruction of the remainder. Proinsulin has a half-life that is at least twice as long as that of insulin and is not converted to insulin outside the pancreas. The kidney is the principal site of degradation of proinsulin and the C peptide. Because little degradation of the C peptide occurs in the liver, its concentration in blood is useful for estimating the rate of insulin secretion and for evaluation of beta cell function in patients who are receiving injections of insulin.

PHYSIOLOGICAL ACTIONS OF INSULIN

Effects of Insulin Deficiency

In endocrinology, basic insights into the physiological role of a hormone can be gained from examining the consequences of its absence. Secondary or tertiary effects may overshadow the primary cellular lesion, but nevertheless, may ultimately broaden our understanding of cellular responses in the context of the whole organism. Insights into the physiology of insulin and the physiological processes it affects directly and indirectly were originally gained from clinical observation. Consideration of some of the classic signs of insulin-related disease therefore provides a good starting point for discussing the physiology of insulin.

1. Hyperglycemia. In the normal individual the concentration of glucose in blood is maintained at around 90 mg/dl of plasma. Blood glucose in diabetics may be 300–400 mg/dl and may even reach 1000 mg/dl on occasion. Diabetics have particular difficulty removing excess glucose from their blood. Normally, after ingestion of a meal rich in carbohydrate, there is only a small and transient increase in the concentration of blood glucose, and excess glucose disappears rapidly from plasma. The diabetic, however, is "intolerant" of glucose, and the ability to remove it from plasma is severely impaired.

Oral glucose tolerance tests, which assess the ability to dispose of a glucose load, are used diagnostically to evaluate existing or impending diabetic conditions. A standard load of glucose is given by mouth and the blood glucose concentration is measured periodically over the course of the subsequent 4 hours. In normal subjects, blood glucose concentrations return to base line values within 2 hours, and the peak value does not rise above 180 mg/dl. In the diabetic or "prediabetic," blood glucose values rise much higher and take a longer time to return to basal levels (Figure 10).

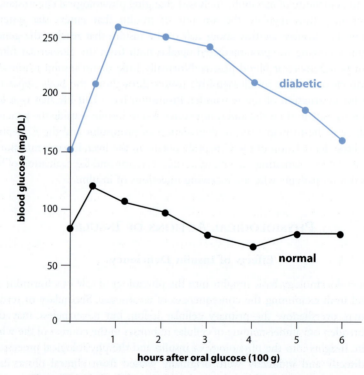

Figure 10   Idealized glucose tolerance tests in normal and diabetic subjects.

2. Glycosuria. Normally the renal tubule has adequate capacity to transport and reabsorb all the glucose filtered at the glomerulus, so that little or none escapes in the urine. Because of hyperglycemia, however, the concentration of glucose in the glomerular filtrate is so high that it exceeds the capacity for reabsorption and "spills" into the urine, causing glycosuria (excretion of glucose in urine).

3. Polyuria. Defined as excessive production of urine, polyuria occurs because more glucose is present in the glomerular filtrate than can be reabsorbed by proximal tubules. Glucose thus remains in the tubular lumen and exerts an osmotic hindrance to water and salt reabsorption in this portion of the nephron, which normally reabsorbs about two-thirds of the glomerular filtrate. The abnormally high volume of fluid that remains cannot be reabsorbed by more distal portions of the nephron, with the result that water excretion is increased (*osmotic diuresis*). Increased flow through the nephron increases urinary loss of sodium and potassium as well.

4. Polydipsia and Polyphagia. Dehydration results from the copious flow of urine and stimulates thirst, a condition called polydipsia, or excessive drinking. The untreated diabetic is characteristically thirsty and consumes large volumes of water to compensate for water lost in urine. Polydipsia is often the first symptom that is noticed by the patient or parents of a diabetic child. In what seems to be an effort to compensate for urinary loss of glucose, appetite is increased, leading to the condition called polyphagia (excessive food consumption). The mechanisms involved in this phenomenon are not yet understood.

5. Weight loss. Despite increased appetite and food intake, insulin deficiency reduces all anabolic processes and accelerates catabolic processes. Accelerated protein degradation, particularly in muscle, provides substrate for gluconeogenesis. Increased mobilization and utilization of stored fats indirectly leads to increased triglyceride concentration in plasma and often results in *lipemia* (high concentration of lipids in blood). Fatty acid oxidation by the liver results in increased production of the ketone bodies (*ketosis*), which are released into the blood and cause *ketonemia*. Because ketone bodies are small, readily filtrable molecules that are actively reabsorbed by a renal mechanism of limited capacity, high blood levels may result in loss of ketone bodies in the urine (*ketonuria*). Ketone bodies are organic acids and produce acidosis, which may be aggravated by excessive washout of sodium and potassium in the urine. Plasma pH may become so low that acidotic coma and death may follow unless insulin therapy is instituted.

The hyperglycemia that causes this whole sequence of events arises from an "underutilization" of glucose by muscle and adipose tissue and an "overproduction" of glucose by liver. Gluconeogenesis is increased at the expense of muscle protein, which is the chief source of the amino acid substrate. Consequently, there is marked wasting of muscle along with depletion of body fat stores. Devastating cardiovascular complications—atherosclerosis, coronary artery disease, and stroke—often result from high concentrations of blood lipids. Other less obviously related complications, including lesions in the microvasculature of the retina and kidneys and in peripheral nerves, result from prolonged hyperglycemia and complete the clinical picture.

The net effect of insulin lack is a severe reduction in the ability to store glycogen, fat, and protein. Conversely, the physiological role of insulin is to promote storage of metabolic fuel. Insulin has many effects on different cells. Even within a single cell it produces multiple effects that are both complementary and reinforcing. Insulin acts on adipose tissue, skeletal muscle, and liver to defend and expand reserves of triglyceride, glycogen, and protein. Within a few minutes after intravenous injection of insulin, there is a striking decrease in the plasma concentrations of glucose, amino acids, FFAs, ketone bodies, and potassium. If the dose of insulin is large enough, blood glucose may fall too low to meet the needs of the central nervous system, and *hypoglycemic coma* may occur. Insulin lowers blood glucose in two ways: (1) it increases uptake by muscle and adipose tissue and (2) it decreases output by liver. It also lowers the concentration of amino acids by stimulating their uptake by muscle and reducing their release. Insulin lowers the concentration of FFAs by blocking their release from adipocytes, and this action in turn lowers the blood ketone level. The decrease in potassium results from stimulation of the sodium/potassium ATPase (sodium pump) in the plasma membranes of muscle, liver, and fat cells. The physiological significance of this response to insulin is not understood.

Effects on Adipose Tissue

Storage of fat in adipose tissue depends on multiple insulin-sensitive reactions, including (1) synthesis of long-chain fatty acids from glucose, (2) synthesis of triglycerides from fatty acids and glycerol (*esterification*), (3) breakdown of triglycerides to release glycerol and long-chain fatty acids (*lipolysis*), and (4) uptake of fatty acids from the lipoproteins of blood. The relevant biochemical pathways are shown in Figure 11.

Lipolysis and esterification are central events in the physiology of the adipocyte. The rate of lipolysis depends on the activity of triglyceride lipase. Lipolysis proceeds at a basal rate in the absence of hormonal stimulation but increases dramatically when cyclic AMP is increased. Hormone-sensitive lipase catalyzes the breakdown of triglycerides into fatty acids and glycerol. Fatty acids can either escape from the adipocyte and become the FFA of blood or be reesterified to triglyceride. Fatty acid esterification requires a source of glycerol that is phosphorylated in its α-carbon; free glycerol cannot be used. Because adipose tissue lacks the enzyme α-glycerol kinase, all of the free glycerol that is produced by lipolysis escapes into the blood. The only source of α-glycerol phosphate available for esterification of fatty acids is derived from phosphorylated three-carbon intermediates formed from oxidation of glucose.

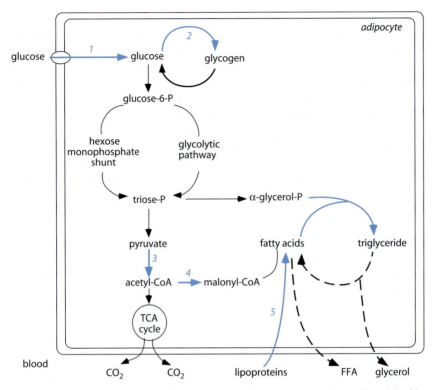

Figure 11   Carbohydrate and lipid metabolism in adipose tissue. Reactions enhanced by insulin (blue arrows) are as follows: (1) transport of glucose into adipose cell; (2) conversion of excess glucose to glycogen; (3) decarboxylation of pyruvate; (4) initiation of fatty acid synthesis; (5) uptake of fatty acids from circulating lipoproteins. Breakdown of triglycerides is inhibited by insulin (broken arrow). Esterification of fatty acids to triglycerides follows from availability of α-glycerol phosphate.

As its name implies, hormone-sensitive lipase is activated by lipolytic hormones, which stimulate the formation of cyclic AMP and thereby promote its phosphorylation by protein kinase A. Insulin accelerates the degradation of cyclic AMP by activating the enzyme cyclic AMP phosphodiesterase and thus interferes with activation of hormone-sensitive lipase. Simultaneously, insulin increases the rate of fatty acid esterification by increasing the availability of α-glycerol phosphate. The net result of these actions is preservation of triglyceride stores at the expense of plasma FFAs, which promptly decrease in concentration in blood plasma. Decreases in FFA concentrations are seen with doses of insulin that are too low to affect blood glucose, and appear to be the most sensitive response to insulin.

Because glucose does not readily diffuse across the plasma membrane, its entry into adipocytes and most other cells depends on carrier-mediated transport.

Insulin increases cellular uptake and metabolism of glucose by accelerating transmembrane transport of glucose and structurally related sugars. This action depends on the availability of glucose transporters in the plasma membrane. Glucose transporters (abbreviated GLUT) are large proteins that weave in and out of the membrane 12 times to form stereospecific channels through which glucose can diffuse down its concentration gradient. There are at least five isoforms of GLUT expressed in various cell types. In addition to GLUT 1, which is present in the plasma membrane of most cells, insulin-sensitive cells such as adipocytes contain pools of intracellular membranous vesicles that are rich in GLUT 4. Insulin increases the number of glucose transporters on the adipocyte surface by stimulating the translocation of GLUT 4-containing vesicles toward the cell surface and fusion of their membranes with the adipocyte plasma membrane (Figure 12).

Insulin may accelerate synthesis of fatty acids by increasing the uptake of glucose and by activating at least two enzymes that direct the flow of glucose carbons into fatty acids. Insulin increases conversion of pyruvate to acetyl CoA, which provides the building blocks for long-chain fatty acid synthesis, and it stimulates carboxylation of acetyl CoA to malonyl CoA, which is the initial and rate-determining reaction in fatty acid synthesis. In humans, adipose tissue is not an important site of fatty acid synthesis, particularly in Western cultures with fat-rich diets. Fat stored in adipose tissue is derived mainly from dietary fat and triglycerides synthesized in the liver. Fat destined for storage reaches adipose tissue in the form of low-density lipoproteins and chylomicrons. Uptake of fat from lipoproteins depends on cleavage of ester bonds in triglycerides by the enzyme *lipoprotein lipase*, to release fatty acids. Lipoprotein lipase is synthesized and secreted by adipocytes and adheres to the endothelium of adjacent capillaries. Insulin promotes synthesis of lipoprotein lipase and thus facilitates the transfer of fatty acids from lipoproteins to triglyceride storage droplets in adipocytes.

Effects on Muscle

Insulin increases uptake of glucose by muscle and directs its intracellular metabolism toward the formation of glycogen (Figure 13). Because muscle comprises nearly 50% of body mass, uptake by muscle accounts for the majority of the glucose that disappears from blood after injection of insulin. As in adipocytes, glucose utilization in muscle is limited by permeability of the plasma membrane. Insulin accelerates entry of glucose into muscle by mobilizing GLUT 4-containing vesicles by the same mechanism that is operative in adipocytes. Metabolism of glucose begins with conversion to glucose-6-phosphate catalyzed by either of the two isoforms of the enzyme *hexokinase* that are present in muscle. Insulin not only increases the synthesis of hexokinase II, but it also appears to enhance the efficiency of hexokinase II activity by promoting its association with the outer membrane of mitochondria, which optimizes access to ATP. In the basal state

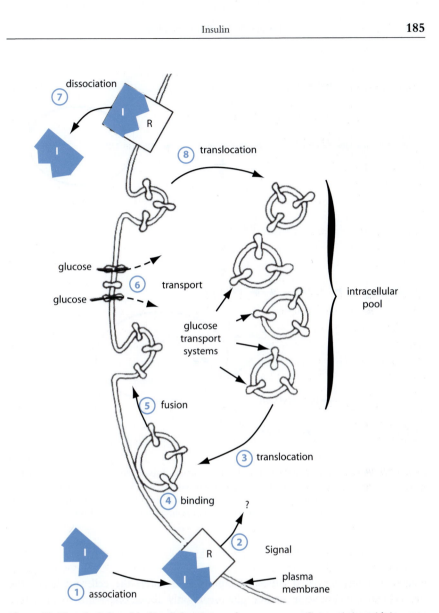

Figure 12 Hypothetical model of insulin's action on glucose transport. On associating with its receptors (R) in the cell membrane, insulin (I) signals the translocation of glucose transport systems to the plasma membrane. The stepwise sequence of events is indicated by the circled blue numbers. (From Karnieli, E., Zarnowski, M. J., Hissin, R. J., Salans, L. B., and Cushman, S. W., *J. Biol. Chem.* **256,** 4772–4777, 1981, with permission.)

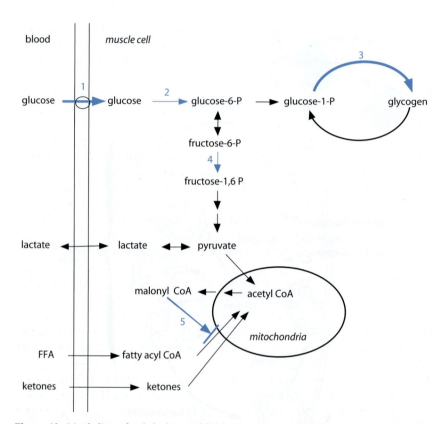

Figure 13 Metabolism of carbohydrate and lipid in muscle. Rate–limiting reactions accelerated by insulin (blue arrows) are as follows: (1) transport of glucose into muscle cells; (2) phosphorylation of glucose by hexokinase; (3) storage of glucose as glycogen; (4) addition of the second phosphate by phosphofructokinase; (5) inhibition of fatty acid entry into mitochondria by malonyl CoA.

glucose is phosphorylated almost as rapidly as it enters the cell, and hence the intracellular concentration of free glucose is only about one-tenth to one-third that of extracellular fluid.

Glucose-6-phosphate is an allosteric inhibitor of hexokinase and an allosteric activator of glycogen synthase. Stimulation of glycogen synthesis by insulin and glucose-6-phosphate protects hexokinase from the inhibitory effect of glucose-6-phosphate when entry of glucose into the muscle cell is rapid. Glycogen synthase activity is low when the enzyme is phosphorylated but increases when the enzyme is dephosphorylated. The degree of phosphorylation of glycogen synthase is determined by the balance of kinase and phosphatase activities. Insulin shifts the balance in favor of dephosphorylation in part by inhibiting the enzyme, *glycogen synthase kinase 3* (GSK-3), and in part by activating a phosphatase. Dephosphorylation of glycogen synthase not only increases its activity directly, but also increases its responsiveness to stimulation by its substrate, glucose-6-phosphate. Hence the powerful effects of insulin on muscle glycogen synthesis are achieved by the complementary effects of increased glucose transport, increased glucose phosphorylation, and increased glycogen synthase activity.

The alternative fate of glucose-6-phosphate, metabolism to pyruvate in the glycolytic pathway, is also increased by insulin. Access to the glycolytic pathway is guarded by phosphofructokinase, whose activity is precisely regulated by a combination of allosteric effectors, including ATP, ADP, and fructose-2,6-bisphosphate. This complex enzyme behaves differently in intact cells and in the broken cell preparations typically used by biochemists to study enzyme regulation. Because conflicting findings have been obtained under a variety of experimental circumstances, no general agreement has been reached on how insulin increases phosphofructokinase activity. In contrast to the liver, the isoform of the enzyme that forms fructose-2,6-bisphosphate in muscle is not regulated by cyclic AMP. The effects of insulin are likely to be indirect.

It should be noted that oxidation of fat profoundly affects the metabolism of glucose in muscle and that insulin also increases all aspects of glucose metabolism in muscle as an indirect consequence of its action on adipose tissue to decrease FFA production. When insulin concentrations are low, increased oxidation of fatty acids decreases oxidation of glucose by inhibiting the decarboxylation of pyruvate and the transport of glucose across the muscle cell membrane. In addition, products of fatty acid oxidation appear also to inhibit hexokinase, but recent studies have called into question the relevance of earlier findings that fatty acid oxidation may inhibit phosphofructokinase. Insulin not only limits the availability of fatty acids, but also inhibits their oxidation. Insulin increases the formation of malonyl CoA, which blocks entry of long-chain fatty acids into the mitochondria as described for liver (Figure 7). These effects are discussed in Chapter 9.

Protein synthesis and degradation are ongoing processes in all tissues, and in the nongrowing individual are completely balanced so that on average there is no net increase or decrease in body protein (Figure 14). In the absence of insulin there is net degradation of muscle protein and muscle becomes an exporter of amino acids, which serve as substrate for gluconeogenesis and ureogenesis in the liver. As with its effects on carbohydrate and fat metabolism, insulin intercedes in protein synthesis at several levels, and has both rapidly apparent and delayed effects. Insulin increases uptake of amino acids from blood by stimulating their transport across the plasma membrane. Insulin increases protein synthesis by promoting phosphorylation of the initiation factors (e.g., eIF-2, eukaryotic initiation factor-2) that govern translation of mRNA. Under the influence of insulin, attachment of mRNA to ribosomes is enhanced, as reflected by the higher content of polysomes compared

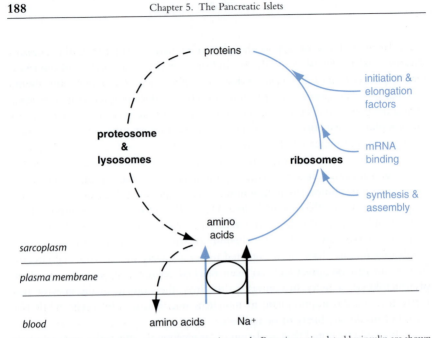

Figure 14 Effects of insulin on protein turnover in muscle. Reactions stimulated by insulin are shown in blue. The dashed arrows indicate inhibition.

to monosomes. This effect of insulin appears to be selective for mRNAs for specific proteins. On a longer time scale, insulin increases total RNA in muscle by increasing synthesis of ribosomal RNA and proteins. Understanding of how insulin decreases protein degradation is incomplete, but it appears that insulin decreases ATP-dependent protein degradation both by decreasing expression of various elements of the proteosomal protein degrading apparatus and by modulating the protease activity of its components.

Effects on Liver

Insulin reduces outflow of glucose from the liver and promotes storage of glycogen. It inhibits glycogenolysis, gluconeogenesis, ureogenesis, and ketogenesis, and it stimulates the synthesis of fatty acids and proteins. These effects are accomplished by a combination of actions that change the activity of some hepatic enzymes and rates of synthesis of other enzymes. Hence not all the effects of insulin occur on the same time scale. Although we use the terms "block" and "inhibit" to describe the actions of insulin, it is important to remember that these verbs are used in the relative and not the absolute sense. Rarely would inhibition of an enzymatic transformation be absolute. In addition, all of the hepatic effects of insulin are reinforced indirectly by actions of insulin on muscle and fat to reduce the influx of substrates for gluconeogenesis and ketogenesis. The actions of insulin on hepatic metabolism are always superimposed on a background of other regulatory influences exerted by metabolites, glucagon, and a variety of other regulatory agents. The magnitude of any change produced by insulin is thus determined not only by the concentration of insulin, but also by the strength of the opposing or cooperative actions of these other influences. Rates of secretion of both insulin and glucagon are dictated by physiological demand. Because of their antagonistic influences on hepatic function, however, it is the ratio, rather than the absolute concentrations, of these two hormones that determines the overall hepatic response.

Glucose Production

In general, liver takes up glucose when the circulating glucose level is high and releases it when the level is low. Glucose transport into or out of hepatocytes depends on a high-capacity insulin-insensitive isoform of the glucose transporter, GLUT 2. Because the movement of glucose is passive, net uptake or release depends on whether the concentration of free glucose is higher in extracellular or intracellular fluid. The intracellular concentration of free glucose depends on the balance between phosphorylation and dephosphorylation of glucose (Figure 2, cycle II). The two enzymes that catalyze phosphorylation are hexokinase, which has a high affinity for glucose and other six-carbon sugars, and glucokinase, which is specific for glucose. The kinetic properties of glucokinase are such that phosphorylation increases proportionately with glucose concentration over the entire physiological range. In addition, glucokinase activity is regulated by glucose. When glucose concentrations are low, much of the glucokinase is bound to an inhibitory protein that sequesters it within the nucleus. An increase in glucose concentration releases glucokinase from its inhibitor and allows it to move into the cytosol, where glucose phosphorylation can take place.

Phosphorylated glucose cannot pass across the hepatocyte membrane. Dephosphorylation of glucose requires the activity of glucose-6-phosphatase. Insulin suppresses synthesis of glucose-6-phosphatase and increases synthesis of glucokinase, thereby decreasing net output of glucose while promoting net uptake. This response to insulin is relatively sluggish and contributes to long-term adaptation rather than to minute-to-minute regulation. The rapid effects of insulin to suppress glucose release are exerted indirectly through decreasing the availability of glucose-6-phosphate, hence starving the phosphatase of substrate. The process of uptake and phosphorylation by glucokinase is only one source of glucose-6-phosphate. Glucose-6-phosphate is also produced by gluconeogenesis and glycogenolysis. Insulin not only inhibits these processes, but it also drives them in the opposite direction.

Most of the hepatic actions of insulin are opposite to those of glucagon, as discussed earlier, and can be traced to inhibition of cyclic AMP accumulation. Rapid actions of insulin largely depend on changes in the phosphorylation state of enzymes already present in hepatocytes. Insulin decreases hepatic concentrations of cyclic AMP by accelerating its degradation by cyclic AMP phosphodiesterase, and may also interfere with cAMP formation and, perhaps, activation of protein kinase A. The immediate consequences can be seen in Figure 15 and are in sharp contrast to the changes in glucose metabolism produced by glucagon shown in Figure 2. Insulin promotes glycogen synthesis and inhibits glycogen breakdown. These effects are accomplished by the combination of interference with cyclic AMP-dependent processes that drive these reactions in the opposite direction (see Figure 3), inhibition of glycogen synthase kinase (which, like protein kinase A, inactivates glycogen synthase), and by activation of the phosphatase that dephosphorylates both glycogen synthase and phosphorylase. The net effect is that glucose-6-phosphate is incorporated into glycogen.

By lowering cAMP concentrations, insulin decreases the breakdown and increases the formation of fructose-2,6-phosphate, which potently stimulates phosphofructokinase and promotes the conversion of glucose to pyruvate. Insulin affects several enzymes in the PEP substrate cycle (Figure 2, cycle IV) and in so doing directs substrate flow away from gluconeogenesis and toward lipogenesis (Figure 16). With relief of inhibition of pyruvate kinase, PEP can be converted to pyruvate, which then enters mitochondria. Insulin activates the mitochondrial enzyme that catalyzes decarboxylation of pyruvate to acetyl CoA and indirectly accelerates this reaction by decreasing the inhibition imposed by fatty acid oxidation. Decarboxylation of pyruvate to acetyl CoA irreversibly removes these carbons from the gluconeogenic pathway and makes them available for fatty acid synthesis. The roundabout process that transfers acetyl carbons across the mitochondrial membrane to the cytoplasm, where lipogenesis occurs, requires condensation with oxaloacetate to form citrate. Citrate is transported to the cytosol and cleaved to release acetyl CoA and oxaloacetate. It might be recalled from earlier discussion that oxaloacetate is a crucial intermediate in gluconeogenesis and is converted to PEP by PEP carboxykinase. Insulin bars the flow of this lipogenic substrate into the gluconeogenic pool by inhibiting synthesis of PEP carboxykinase. The only fate left to cytosolic oxaloacetate is decarboxylation to pyruvate.

Finally, insulin increases the activity of acetyl CoA carboxylase, which catalyzes the rate-determining reaction in fatty acid synthesis. Activation is accomplished in part by relieving cyclic AMP-dependent inhibition and in part by promoting the polymerization of inactive subunits of the enzyme into an active complex. The resulting malonyl CoA not only condenses to form long-chain fatty acids but also prevents oxidation of newly formed fatty acids by blocking their entry into mitochondria (Figure 7). On a longer time scale, insulin increases the synthesis of acetyl coA carboxylase.

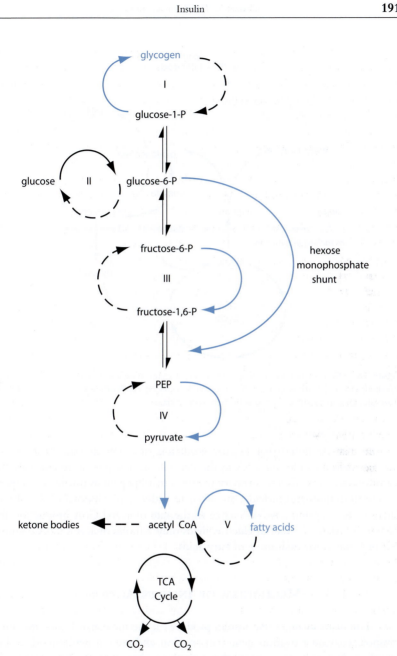

Figure 15 Effects of insulin on glucose metabolism in hepatocytes. Blue arrows indicate reactions that are increased, and broken arrows indicate reactions that are decreased.

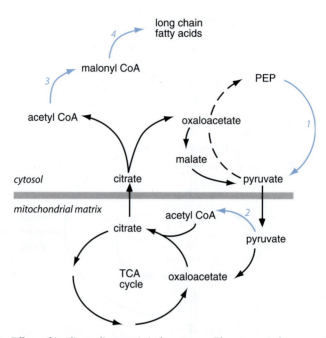

Figure 16 Effects of insulin on lipogenesis in hepatocytes. Blue arrows indicate reactions that are increased, and broken arrow indicates reaction that is decreased. (1) Pyruvate kinase; (2) pyruvate dehydrogenase; (3) acetyl CoA carboxylase; (4) fatty acid synthase.

It may be noted that hepatic oxidation of either glucose or fatty acids increases delivery of acetyl CoA to the cytosol, but ketogenesis results only from oxidation of fatty acids. The primary reason is that lipogenesis usually accompanies glucose utilization and provides an alternate pathway for disposal of acetyl CoA. There is also a quantitative difference in the rate of acetyl CoA production from the two substrates: 1 mole of glucose yields only 2 moles of acetyl CoA compared to 8 or 9 moles for each mole of fatty acids.

MECHANISM OF INSULIN ACTION

The many changes that insulin produces at the molecular level—membrane transport, enzyme activation, gene transcription, and protein synthesis—have been described. The molecular events that link these changes with the interaction of insulin and its receptor are still incompletely understood but are the subjects of intense investigation. Many of the intermediate steps in the action of insulin have been uncovered, but others remain to be identified. It is clear that transduction of

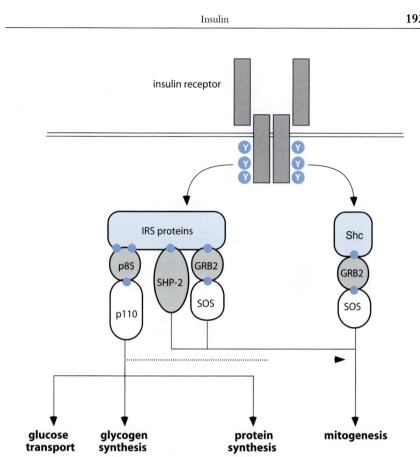

glucose   glycogen   protein   mitogenesis
transport   synthesis   synthesis

Figure 17 Current model of the insulin receptor signaling. Phosphorylated tyrosine residues (Y) on the insulin receptor serve as anchoring sites for cytosolic proteins (IRS proteins and Shc), which in turn are phosphorylated on tyrosines (dark blue circles) and dock with other proteins. IRS, Insulin receptor substrate; Shc, Src homology-containing protein; SHP-2, protein tyrosine phosphatase-2; GRB2, growth factor receptor binding protein 2; SOS, son of sevenless (a GTPase-activating protein); p85 and p110 are subunits of phosphoinositol-3-kinase. (From Virkamäki, A., Ueki, K., and Kahn, C. R., *J. Clin. Invest.* **103**, 931, 1999, with permission.)

the insulin signal is not accomplished by a linear series of biochemical changes, but rather that multiple intracellular signaling pathways are activated simultaneously and may intersect at one or more points before the final result is expressed (Figure 17).

The insulin receptor is a tetramer composed of two alpha and two beta glycoprotein subunits that are held together by disulfide bonds that link the alpha subunits to the beta subunits and the alpha subunits to each other (Figure 18). The alpha and beta subunits of insulin are encoded in a single gene that contains

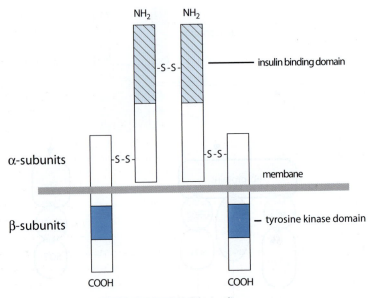

Figure 18 Model of the insulin receptor.

22 exons. The alpha subunits are completely extracellular and contain the insulin-binding domain. The beta subunits span the plasma membrane and contain tyrosine kinase activity in the cytosolic domain. Binding to insulin is thought to produce a conformational change that relieves the beta subunit from the inhibitory effects of the alpha subunit, allowing it to phosphorylate itself and other proteins at tyrosine residues. Autophosphorylation of the kinase domain is required for full activation. Tyrosine phosphorylation of the receptor also provides docking sites for other proteins that participate in transducing the hormonal signal. Docking on the phosphorylated receptor may position proteins optimally for phosphorylation by the receptor kinase.

Among the proteins that are phosphorylated on tyrosine residues by the insulin receptor kinase are four cytosolic proteins called *insulin receptor substrates* (IRS-1, IRS-2, IRS-3, and IRS-4). These relatively large proteins contain multiple tyrosine phosphorylation sites and act as scaffolds, on which other proteins are assembled to form large signaling complexes. IRS-1 and IRS-2 appear to be present in all insulin target cells, whereas IRS-3 and IRS-4 have more limited distribution. Despite their names the IRS proteins are not functionally limited to transduction of the insulin signal, but are also important for expression of effects of other hormones and growth-promoting factors. Moreover, they are not the only substrates for the insulin receptor kinase. A variety of other proteins that are tyrosine phosphorylated by the insulin receptor kinase have also been identified. Proteins recruited to the insulin receptor and IRS proteins may have enzymatic activity or they may in turn recruit other proteins by providing sites for protein:protein interactions. The assemblage of proteins initiates signaling cascades that ultimate express the various actions of insulin described above. One of the most important of the proteins that is activated is phosphatidylinositol-3 (PI-3) kinase. PI-3 kinase plays a critical role in activating many downstream effector molecules, including protein kinase B, which is thought to mediate the effects of insulin on glycogen synthesis and GLUT 4 translocation. PI-3 kinase, however, is also activated by a variety of other hormones, cytokines, and growth factors whose actions do not necessarily mimic those of insulin. The uniqueness of the response to insulin probably reflects the unique combination of biochemical consequences produced by the simultaneous activity of multiple signaling pathways and the particular set of effector molecules expressed in insulin target cells. Although insulin is known to regulate expression of more than 150 genes, few of the nuclear regulatory proteins that are activated by insulin are known, and precisely how the insulin receptor communicates with these regulatory proteins is unknown. A more detailed discussion of the complex molecular events that govern insulin action can be found in the suggested readings listed at the end of this chapter.

REGULATION OF INSULIN SECRETION

As might be expected of a hormone whose physiological role is promotion of fuel storage, insulin secretion is greatest immediately after eating and decreases during between-meal periods (Figure 19). Coordination of insulin secretion with nutritional state as well as with fluctuating demands for energy production is achieved through stimulation of beta cells by metabolites, hormones, and neural signals. Because insulin plays the primary role in regulating storage and mobilization of metabolic fuels, the beta cells must be constantly apprised of bodily needs, not only with regard to feeding and fasting, but also to the changing demands of the environment. Energy needs differ widely when an individual is at peace with the surroundings and when fighting for survival. Maintaining constancy of the internal environment is achieved through direct monitoring of circulating metabolites by beta cells. This input can be overridden or enhanced by hormonal or neural signals that prepare an individual for rapid storage of an influx of food or for massive mobilization of fuel reserves to permit a suitable response to environmental demands.

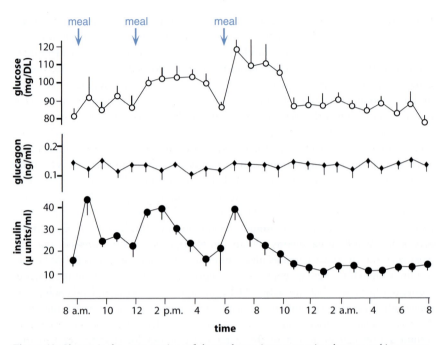

Figure 19 Changes in the concentrations of plasma glucose, immunoreactive glucagon and immunore-active insulin throughout the day. Values are the mean ± SEM (*n* = 4). (From Tasaka, Y., Sekine, M., Wakatsuki, M., Ohgawara, H., and Shizume, K., *Horm. Metab. Res.* **7**, 205–206, 1975, with permission.)

Metabolite Control

Glucose

Glucose is the most important regulator of insulin secretion. In the normal individual its concentration in blood is maintained within the narrow range of about 70 or 80 mg/dl after an overnight fast to about 150 mg/dl immediately after a glucose-rich meal. When blood glucose increases above a threshold value of about 100 mg/dl, insulin secretion increases proportionately. At lower concentrations adjustments in insulin secretion are largely governed by other stimuli (see below) that act as amplifiers or inhibitors of the effects of glucose. The effectiveness of these agents therefore decreases as glucose concentration decreases.

Other Circulating Metabolites

Amino acids are important stimuli for insulin secretion. The transient increase in plasma amino acids after a protein-rich meal is accompanied by increased secretion of insulin. Arginine, lysine, and leucine are the most potent amino acid stimulators of insulin secretion. Insulin secreted at this time may facilitate storage of dietary amino acids as protein and prevents their diversion to gluconeogenesis. Amino acids are effective signals for insulin release only when blood glucose concentrations are adequate. Failure to increase insulin secretion when glucose is in short supply prevents hypoglycemia that might otherwise occur after a protein meal containing little carbohydrate. Fatty acids and ketone bodies may also increase insulin secretion, but only when they are present at rather high concentrations. Because fatty acid mobilization and ketogenesis are inhibited by insulin, their ability to stimulate insulin secretion provides a feedback mechanism to protect against excessive mobilization of fatty acids and ketosis.

Hormonal and Neural Control

In response to carbohydrate in the lumen, the intestinal mucosa secretes one or more factors, called *incretins*, that reach the pancreas through the general circulation and stimulate the beta cells to release insulin, even though the increase in blood glucose is still quite small. Incretins are thought to act by amplifying the stimulatory effects of glucose. This anticipatory secretion of insulin prepares tissues to cope with the coming influx of glucose and dampens what might otherwise be a large increase in blood sugar. Various gastrointestinal hormones, including gastrin, secretin, cholecystokinin–pancreozymin, glucagon-like peptide, and glucose-dependent insulinotropic peptide (GIP), can evoke insulin secretion when tested experimentally, but of these hormones, only GLP-1 and GIP appear to be physiologically important incretins.

Secretion of insulin in response to food intake is also mediated by a neural pathway. The taste or smell of food or the expectation of eating may increase insulin secretion during this so-called *cephalic phase* of feeding. Parasympathetic fibers in the vagus nerve stimulate beta cells by releasing acetylcholine or the neuropeptide VIP. Activation of this pathway is initiated by integrative centers in the brain and involves input from sensory endings in the mouth, stomach, small intestine, and portal vein. An increase in the concentration of glucose in portal blood is detected by glucose sensors in the wall of the portal vein and the information is relayed to the brain via vagal afferent nerves. In response, vagal efferent nerves stimulate the pancreas to secrete insulin and the liver to take up glucose.

Insulin secretion by the human pancreas is virtually shut off by epinephrine or norepinephrine delivered to beta cells, by either the circulation or sympathetic neurons. This inhibitory effect is seen not only as a response to low blood glucose levels, but may occur even when the blood glucose level is high. It is mediated through α_2-adrenergic receptors on the surface of beta cells. Physiological circumstances that activate the sympathetic nervous system thus can shut down insulin secretion and thereby remove the major restraint on mobilization of metabolic fuels needed to cope with an emergency.

Secretory activity of beta cells is also enhanced by growth hormone and cortisol by mechanisms that are not yet understood. Although they do not directly evoke a secretory response, basal insulin secretion is increased when these hormones are present in excess, and beta cells become hyperresponsive to signals for insulin secretion. Conversely, insulin secretion is reduced when either is deficient. Excessive growth hormone or cortisol decreases tissue sensitivity to insulin and can produce diabetes (see Chapter 9). The factors that regulate insulin secretion are shown in Figure 20.

Cellular Events

Beta cells increase their rates of insulin secretion within 30 seconds of exposure to increased concentrations of glucose and can shut down secretion as rapidly. The question of how the concentration of glucose is monitored and translated into a rate of insulin secretion has not been answered completely, but many of the important steps are known. The beta cell has specific receptors for glucagon, acteylcholine, GLP-1, and other compounds that increase insulin secretion by promoting the formation of cyclic AMP or IP_3 and DAG (see Chapter 1), but it does not appear to have specific receptors for glucose. To affect insulin secretion, glucose must be metabolized by the beta cell, indicating that some consequence of glucose oxidation, rather than simply the glucose, is the critical determinant. Beta cell membranes contain the glucose transporter GLUT 2, which has a high capacity

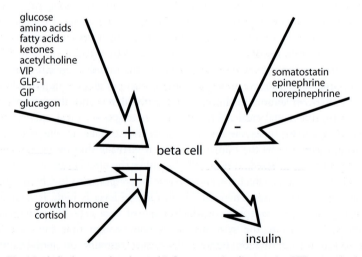

Figure 20   Metabolic, hormonal, and neural influences on insulin secretion. VIP, vasoactive intestinal peptide; GLP-1, glucagon-like peptide-1; GIP, glucose-dependent insulinotropic peptide.

but relatively low affinity for glucose. Consequently, as glucose concentrations increase above about 100 mg/dl, glucose enters the beta cell at a rate that is limited by its concentration and not by availability of transporters. It is likely that glucokinase, which is specific for glucose and catalyzes the rate-determining reaction for glucose metabolism in beta cells, has the requisite kinetic characteristics to behave as a glucose sensor. Mutations that affect the function of this enzyme result in decreased insulin secretion in response to glucose that may be severe enough to cause a form of diabetes.

Secretion of insulin, like other peptide hormones, requires increased cytosolic calcium. Perhaps through the agency of a calmodulin-activated protein kinase, calcium promotes movement of secretory granules to the periphery of the beta cell, fusion of the granular membrane with the plasma membrane, and the consequent extrusion of granular contents into the extracellular space. To increase insulin secretion, increased metabolism of glucose must somehow bring about an increase in intracellular calcium. Linkage between glucose metabolism and intracellular calcium concentration appears to be achieved by their mutual relationship to cellular concentrations of ATP and ADP.

In resting pancreatic beta cells, efflux of potassium through open potassium channels maintains the membrane potential at about −70 mV. Some potassium channels in these cells are sensitive to ATP, which inhibits (closes) them, and to ADP, which activates (opens) them. When blood glucose concentrations are low, the effects of ADP predominate even though its concentration in beta cell cytoplasm is about 1000 times lower than that of ATP. Because glucose transport is not rate limiting in beta cells, increased concentrations in blood accelerate glucose oxidation and promote ATP formation at the expense of ADP. As a result, ADP levels become insufficient to counter the inhibitory effects of ATP, and potassium channels close. The consequent buildup of positive charge within the beta cell causes the membrane to depolarize, which activates voltage-sensitive calcium channels. When the depolarizing membrane potential reaches about −50 mV, calcium channels open. Influx of positively charged calcium reverses the membrane potential. Electrical recording of these events produces a pattern of voltage changes that resembles an action potential. The frequency and duration of electrical discharges in beta cells increase as glucose concentrations increase. In addition to triggering insulin secretion, elevated intracellular calcium inhibits voltage-sensitive calcium channels and activates calcium-sensitive potassium channels, allowing potassium to exit and the cell to repolarize (Figure 21).

Although entry of calcium triggers insulin secretion, it appears that glucose and the various hormonal modulators may stimulate secretion by acting at additional regulatory sites downstream from calcium. Hormones and neurotransmitters that increase insulin secretion act through either the cyclic AMP or DAG/IP$_3$ second messenger pathways to enhance the stimulatory effect of glucose. Some evidence suggests that the voltage-sensitive calcium channels may be substrates for

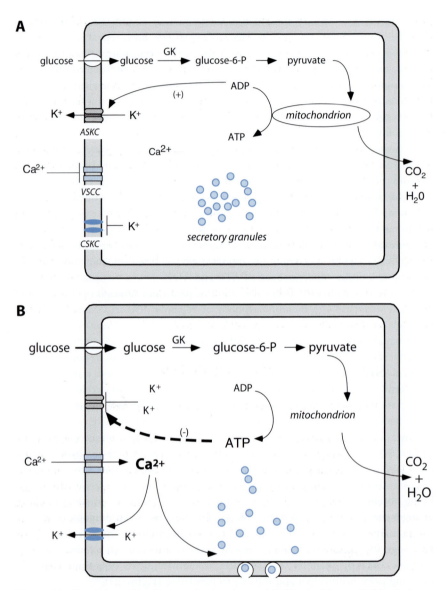

Figure 21 Regulation of insulin secretion by glucose. (A) "Resting" beta cell (blood glucose <100 mg/dl). ADP/ATP ratio is high enough so that ATP-sensitive potassium channels (ASKC) are open, and the membrane potential is about −70 mV. Voltage-sensitive calcium channels (VSCC) and calcium-sensitive potassium channels (CSKC) are closed. (B) Beta cell response to increased blood glucose. In response to increased glucose entry and metabolism, the ratio of ADP/ATP decreases, and ATP-sensitive potassium channels close. Voltage-sensitive calcium channels are activated; calcium enters and stimulates insulin secretion. Increased cytosolic calcium inhibits voltage-sensitive calcium channels

protein kinase A, and that phosphorylation may lower their threshold for activation. Additional actions of protein kinase A appear to enhance later steps in the secretory pathway and further increase insulin secretion under conditions when the rate of calcium influx is maximal. This might explain how glucagon and other hormones that activate adenylyl cyclase increase insulin secretion. Agents such as acetylcholine increase IP_3 and may thus stimulate release of calcium from intracellular storage sites. In addition, activation of protein kinase C enhances aspects of the secretory process that are independent of calcium. Norepinephrine and somatostatin block insulin secretion by way of the inhibitory guanine nucleotide binding protein (G_i; see Chapter 1), which may directly inhibit voltage-sensitive calcium channels as well as adenylyl cyclase.

In addition to serving as the principal signal for insulin secretion, glucose appears to be the most important stimulator of insulin synthesis. Both glucose and cyclic AMP increase transcription of the insulin gene. The mRNA template for insulin turns over slowly and has a half-life of about 30 hours. Glucose also appears to regulate its stability. Hyperglycemia prolongs its half-life more than twofold whereas hypoglycemia accelerates its degradation. In addition, glucose increases translation of the preproinsulin mRNA by stimulating both the initiation and elongation reactions. Concurrently, glucose also up-regulates production of the enzymes needed to process preproinsulin to insulin.

SOMATOSTATIN

BIOSYNTHESIS, SECRETION, AND METABOLISM

Somatostatin was originally isolated from hypothalamic extracts that inhibited the secretion of growth hormone. Somatostatin is widely distributed in many neural tissues, where it presumably functions as a neurotransmitter. It is found in many secretory cells (delta cells) outside of the pancreatic islets, particularly in the lining of the gastrointestinal tract. Somatostatin is stored in membrane-bound vesicles and secreted by exocytosis. Measurable increases in the somatostatin concentration can be found in peripheral blood after ingestion of a meal rich in fat or protein, with the vast majority secreted by intestinal cells rather than islet cells. It is cleared rapidly from the blood and has a half-life of only about 3 minutes.

---

and activates calcium-sensitive potassium channels, thereby allowing the cell membrane to repolarize and calcium channels to close. Persistence of high glucose results in repeated spiking of electrical discharges and oscillation of intracellular calcium concentrations.

PHYSIOLOGICAL ACTIONS OF SOMATOSTATIN

The physiological importance of pancreatic somatostatin is not understood. Because it can inhibit secretion of both insulin and glucagon it has been suggested that somatostatin, by acting in a paracrine fashion, may contribute to the regulation of glucagon and insulin secretion. However, anatomical relationships and the direction of flow in the microcirculation in the islets are inconsistent with such a role. Somatostatin also inhibits secretion of various gastrointestinal hormones and decreases acid secretion by the gastric mucosa and enzyme secretion by the acinar portion of the pancreas. In addition, somatostatin decreases intestinal motility and may slow the rate of absorption of nutrients from the digestive tract. Increased fecal excretion of fat is a prominent feature in patients suffering from somatostatin-secreting tumors. At the cellular level the inhibitory effects of somatostatin are mediated by surface receptors that act through the inhibitory guanine nucleotide binding protein to inhibit adenylyl cyclase, and by beta/gamma subunits that activate potassium channels and hyperpolarize cell membranes (see Chapter 1).

REGULATION OF SOMATOSTATIN SECRETION

Increased concentrations of glucose or amino acids in blood stimulate somatostatin secretion by intestinal delta cells. In addition, glucose or fat in the gastrointestinal tract elicits a secretory response by pancreatic delta cells, mediated perhaps by glucagon or gastrointestinal hormones. Somatostatin secretion is also increased by norepinephrine and inhibited by acetylcholine.

SUGGESTED READING

Becker, A. B., and Roth, R. A. (1990). Insulin receptor structure and function in normal and pathological conditions. *Annu. Rev. Med.* **41**, 99–116.

Burant, C. F., Sivitz, W. I., Fukumoto, H., Kayano, T., Nagamatsu, S., Seino, S., Pessin, J. E., and Bell, G. I. (1991). Mammalian glucose transporters: Structure and molecular regulation. *Recent Prog. Horm. Res.* **47**, 349–387.

Cheatham, B., and Kahn, C. R. (1995). Insulin action and the insulin signaling network. *Endocr. Rev.* **16**, 117–142.

Jefferson, L. S., and Cherrington, A. D. (2001). "The Endocrine Pancreas and Regulation of Metabolism, Handbook of Physiology, Section 7, Volume II." Oxford University Press, New York. (This volume covers a wide range of topics relevant to items discussed in this chapter.)

Kieffer, T. J., and Habener, J. F. (1999). The glucagon-like peptides. *Endocr. Rev.* **20**, 876–913.

Kimball, S. R., Vary, T. C., and Jefferson, L. S. (1994). Regulation of protein synthesis by insulin. *Annu. Rev. Physiol.* **56**, 321–348.

Miller, R. E. (1981). Pancreatic neuroendocrinology: Peripheral neural mechanisms in the regulation of the islets of Langerhans. *Endocr. Rev.* **2**, 471–494.

Pilkis, S. J., and Granner, D. K. (1992). Molecular physiology of the regulation of hepatic gluconeo-genesis and glycolysis. *Annu. Rev. Physiol.* **54**, 885–909.

Rajan, A. S., Aguilar-Bryan, L., Nelson, D. A., Yaney, G. C., Hsu, W. H., Kunze, D. L., and Boyd, A. E. III. (1990). Ion channels and insulin secretion. *Diabetes Care* **13**, 340–363.

Taylor, S. I., Cama, A., Accili, D., Barbetti, F., Quon, M. J., de la Luz Sierra, M., Suzuki, Y., Koller, E., Levy-Toledano, R., Wertheimer, E., Moncada, V. Y., Kadowaki, H., and Kadowaki, T. (1992). Mutations in the insulin receptor gene. *Endocr. Rev.* **13**, 566–595.

Unger, R. H., and Orci, L. (1976). Physiology and pathophysiology of glucagon. *Physiol. Rev.* **56**, 778–838.

CHAPTER 6

Principles of Hormonal Integration

Redundancy
Reinforcement
Push–Pull Mechanisms
Modulation of Responding Systems
  The Concepts of Sensitivity and Response Capacity
  Modulation of Sensitivity
    Effects of Receptor Abundance
    Postreceptor Effects
    Abundance of Competent Target Cells
    Permissive Actions

Until now, we have considered individual endocrine glands, and some basic information about their physiological functions. Although it is helpful for the student first to understand one hormone at a time or one gland at a time, it must be recognized that life is considerably more complex, and that endocrinological solutions to physiological problems require integration of a large variety of simultaneous events. In this chapter we will consider some of the general principles of endocrine integration. In the ensuing chapters, we will see the application of these principles to solution of physiological problems.

REDUNDANCY

Survival in a hostile environment has been made possible by the evolution of fail-safe mechanisms to govern crucial functions. Just as each organ system has built in excess capacity giving it the potential to function at levels beyond the usual day-to-day demands, so too, is there excess regulatory capacity provided in the form of seemingly duplicative or overlapping controls. Simply put, the body has more than one way to achieve a given end. For example, as we have seen, conversion of liver glycogen to blood glucose can be signaled by at least two hormones, glucagon from the alpha cells of the pancreas, and epinephrine from the adrenal medulla (Figure 1). Both of these hormones increase cyclic AMP production in the

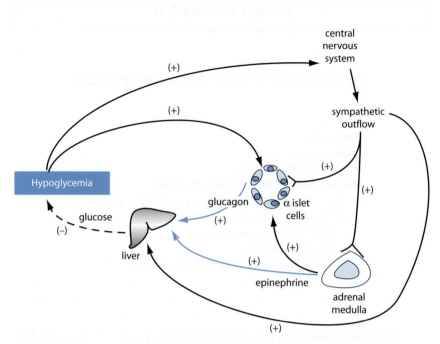

Figure 1   Redundant mechanisms to stimulate hepatic glucose production.

liver, and thereby activate the enzyme, phosphorylase, which catalyzes glycogenolysis. Two hormones secreted from two different tissues, sometimes in response to different conditions, thus produce the same end result.

Further redundancy can also be seen at the molecular level. Using the same example of conversion of liver glycogen to blood glucose, there are even two ways that epinephrine can activate phosphorylase. By stimulating β-adrenergic receptors, epinephrine increases cyclic AMP formation, as already mentioned. By stimulating α_1-adrenergic receptors epinephrine also activates phosphorylase, but these receptors operate through the agency of increased intracellular calcium concentrations produced by the release of IP_3 (Figure 2).

Redundant mechanisms not only assure that a critical process will take place, but they also offer opportunity for flexibility and subtle fine tuning of a process. Though redundant in the respect that two different hormones may have some overlapping effects, the actions of the two hormones are usually not identical in all respects. Within the physiological range of its concentrations in blood, glucagon's action is restricted to the liver; epinephrine produces a variety of other responses in many extrahepatic tissues while increasing glycogenolysis in the liver. Variations in the relative input from both hormones allow for a wide spectrum of changes in blood glucose concentrations relative to other effects of epinephrine, such as increased heart rate.

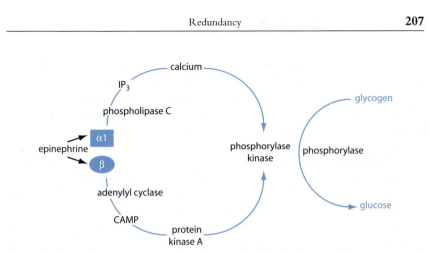

Figure 2   Redundant mechanisms to activate glycogen phosphorylase by a single hormone, epineph-rine, acting through both alpha and beta receptors.

Two hormones that produce common effects may differ not only in their range of actions, but also in their time constants (Figure 3). One may have a more rapid onset and short duration of action, while another may have a longer duration of action, but a slower onset. For example, epinephrine increases blood concentrations of free fatty acids (FFA) within seconds or minutes and this effect dissipates as rapidly when epinephrine secretion is stopped. Growth hormone similarly increases blood concentrations of FFAs, but its effects are seen only after a lag period of 2–3 hours and persist for many hours. A hormone such as epinephrine may therefore be used to meet short-term needs, and another, such as growth hormone, may satisfy sustained needs.

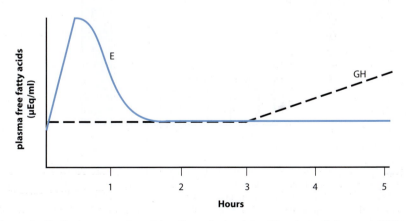

Figure 3   Idealized representation of the effects of epinephrine (E) and growth hormone (GH) on plasma concentrations of free fatty acids.

Redundancy also pertains to processes in which the same end may be achieved by more than one physiological means. For example, blood concentrations of calcium may be increased by an action of the parathyroid hormone to mobilize calcium stored in bone crystals (Chapter 8), or by an action of vitamin D to promote calcium absorption from the gut. These processes, as might be expected, have different time constants as well.

Finally, redundancy may also lead to the phenomenon of *synergism*, or *potentiation*. Two or more hormones are said to act synergistically when the response to their simultaneous administration is greater than the sum of the responses when each is given alone. For example, both growth hormone and cortisol modestly increase lipolysis in adipocytes. When given simultaneously, however, glycerol production is nearly twice as great as the sum of the effects of each (Figure 4).

One of the implications of redundancy for understanding both normal physiology and endocrine disease is that partial, or perhaps even complete, failure of one mechanism can be compensated by increased reliance on another mechanism. This point is particularly relevant to interpretation of results of gene knockout studies. In some cases elimination of what was thought to be an essential protein resulted in no apparent functional changes, perhaps because of redundancy.

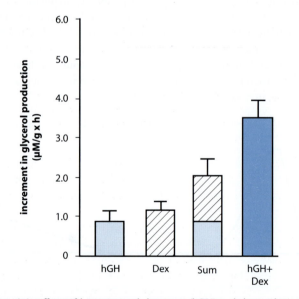

Figure 4 Synergistic effects of human growth hormone (hGH) and the synthetic glucocorticoid dexamethasone (Dex) on lipolysis as measured by the increase in glycerol release from rat adipocytes. Both hGH and Dex were effective when added individually, but when added together their effect was greater than the sum of their individual effects (From Gorin *et al.*, *Endocrinology* **126**, 2973,1990, by permission of the Endocrine Society).

Thus functional deficiencies may be evident only in subtle ways and may not show up readily as overt disease. Some deficiencies may become apparent only after appropriate provocation or perturbation of the system. Conversely, strategies for therapeutic interventions designed to increase or decrease the rate of a process must take into account the redundant inputs that regulate that process. Merely accelerating or blocking one regulatory input may not produce the desired effect because independent adjustments in redundant pathways may completely compensate for the intervention.

REINFORCEMENT

It is an oversimplification to think of any hormone as simply having a single unique effect. In accomplishing any end, most hormones act at several locales either within a single cell, or in different tissues or organs to produce separate but mutually reinforcing responses. In some cases, a hormone may produce radically different responses at different locales, which, nevertheless reinforce each other from the perspective of the whole organism. Let us consider, for example, just some of the ways insulin acts on the fat cell to promote storage of triglycerides:

1. Increased uptake of glucose, which serves as substrate for fatty acid synthesis, and for the α-glycerol phosphate needed to trap any free fatty acids formed by spontaneous lipolysis of triglyceride stores.
2. Activation of several enzymes critical for fatty acid synthesis, e.g., pyruvate dehydrogenase, pyruvate carboxylase, and acetyl CoA carboxylase.
3. Inhibition of breakdown of already formed triglycerides.
4. Induced synthesis of the extracellular enzyme lipoprotein lipase, needed to take up lipids from the circulation.

Any one of these effects might accomplish the end of increasing fat storage, but collectively, these different effects make possible an enormously broader range of response in a shorter time frame. These effects of insulin will be considered further in Chapter 9.

Reinforcement can also take the form of a single hormone acting in different ways in different tissues to produce complementary effects. A good example of this is the action of glucocorticoid hormones to promote gluconeogenesis. As we have seen (Chapter 4, Figure 12), glucocorticoids promote protein breakdown in muscle and lymphoid tissues, and the consequent release of amino acids into the blood. In adipose tissue glucocorticoids promote triglyceride lipolysis and the release of glycerol. In the liver, glucocorticoids induce the formation of the enzymes necessary to convert amino acids, glycerol, and other substrates into glucose. Either the extrahepatic action to provide substrate or the hepatic action to increase capacity to utilize that substrate would increase gluconeogenesis. Together, these complementary actions increase the overall magnitude and speed of the response.

PUSH–PULL MECHANISMS

As discussed in Chapter 1, many critical processes are under dual control by agents that act antagonistically either to stimulate or to inhibit. Such dual control allows for more precise regulation through negative feedback than would be possible with a single control system. The example cited was hepatic production of glucose, which is increased by glucagon and inhibited by insulin. In emergency situations or during exercise, epinephrine and norepinephrine released from the adrenal medulla and sympathetic nerve endings override both negative feedback systems by inhibiting insulin secretion and stimulating glucagon secretion (Figure 5). The effect of adding a stimulatory influence while simultaneously removing an inhibitory influence is a rapid and large response, more rapid and larger than could be achieved by simply affecting either hormone alone, or than could be accomplished by the direct glycogenolytic effect of epinephrine or norepinephrine.

Another type of push–pull mechanism can be seen at the molecular level. Net synthesis of glycogen from glucose depends on the activities of two enzymes, glycogen synthase, which catalyzes the formation of glycogen from glucose, and glycogen phosphorylase, which catalyzes glycogen breakdown (Figure 6). The net reaction rate is determined by the balance of the activity of the two enzymes. The activity of both enzymes is subject to regulation by phosphorylation, but in opposite directions: addition of a phosphate group activates phosphorylase, but inactivates synthase. In this case, a single agent, cyclic AMP, which activates protein kinase A, increases the activity of phosphorylase and simultaneously inhibits synthase.

MODULATION OF RESPONDING SYSTEMS

THE CONCEPTS OF SENSITIVITY AND RESPONSE CAPACITY

In discussing the responses of tissues to stimulation by hormones we have spoken as though any particular amount of hormone secreted always produces the same magnitude of response. This is an oversimplification. In actuality the relationship between the amount of hormone available and the magnitude of the response is subject to regulation by many factors, including the actions of other hormones (Figure 7). Clearly two of the most important determinants of the magnitude and duration of responses are the concentration of hormone present in

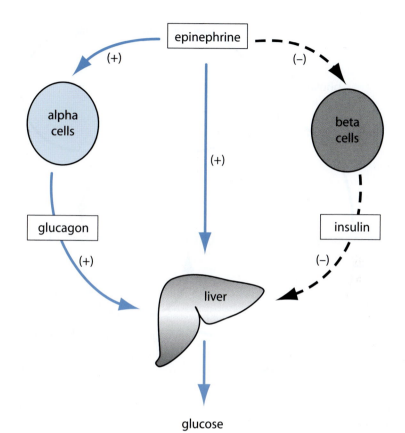

Figure 5 Push–pull mechanism. Epinephrine inhibits insulin secretion while promoting glucagon secretion. This combination of effects on the liver stimulates glucose production while simultaneously relieving an inhibitory influence. (+), stimulates; (–), inhibits.

the extracellular fluid surrounding target cells and the length of time that that concentration is maintained. For hormones that are secreted in a pulsatile fashion, both the amount secreted in each pulse and the frequency of secretory pulses are important determinants. In addition to its plasma concentration, the rate of delivery of a hormone to the extracellular fluid bathing its target cells also depends on the rate of blood flow, capillary permeability, and the fraction of hormone that is bound to plasma proteins, all of which are sensitive to changing physiological circumstances and the actions of hormones. Finally, the rate of hormone degradation either in the target cell or in blood and nontarget tissues may also change with changing circumstances.

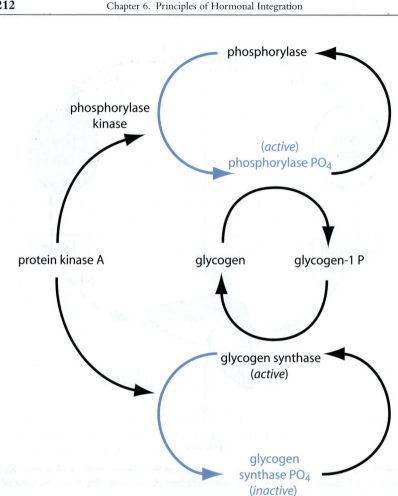

Figure 6 Push–pull mechanism to activate glycogen phosphorylase while simultaneously inhibiting glycogen synthase.

Neither the sensitivity of target tissues to hormonal stimulation nor their capacity to respond is constant. Sensitivity to stimulation and capacity to respond are two separate though related aspects of hormonal responses. Sensitivity describes the acuity of the ability a cell or organ to recognize and respond to a signal in proportion to the intensity of that signal. The capacity to respond, or the maximum response that a tissue or organ is capable of giving, depends on the number of target cells and their competence. Hormones regulate both the sensitivity and the capacity of target tissues to respond, either to themselves or to other hormones.

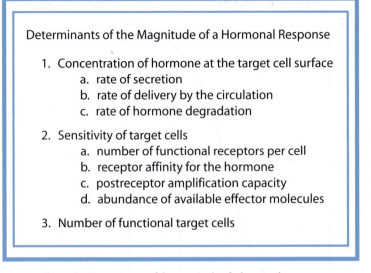

Figure 7   Determinants of the magnitude of a hormonal response.

MODULATION OF SENSITIVITY

The relationship between the magnitude of a hormonal response and the concentration of hormone that produces the response can be described by a sigmoidal curve (Figure 8). Sensitivity is often described in terms of the concentration of hormone needed to produce a half-maximal response. An increase in sensitivity lowers the concentration of hormone needed to elicit a half-maximal response, and a decrease in sensitivity increases the concentration of hormone needed to evoke the same response. In the example shown in Figure 8 we may assume that curve B represents the basal sensitivity that may be increased (curve A) or decreased (curve C) in different physiological or pathological conditions. Changes in the capacity to respond are illustrated in Figure 9. In this case the maximum response may be increased (curve A) or decreased (curve C), but the sensitivity (i.e., the concentration of hormone needed to produce the half-maximal response) remains unchanged at 1 ng/ml.

Effects of Receptor Abundance

One mechanism by which hormones or locally acting paracrine or autocrine factors may adjust the sensitivity of target tissues is by regulation of the availability of hormone receptors. It should be recalled that the initial event in a

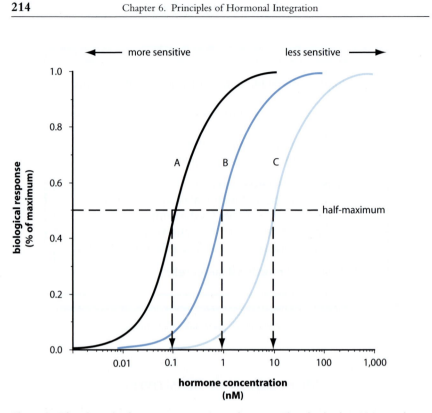

Figure 8 The relationship between concentration and response. Three levels of sensitivity are shown. Arrows indicate the concentration of hormone that produces a half-maximal response for each level of sensitivity.

hormonal response is the binding of the hormone to its receptor (Chapter 1). The higher the concentration of hormone, the more likely it is to interact with its receptors. If there are no hormone receptors, there can be no response, and the more receptors that are available to interact with any particular amount of hormone, the greater is the likelihood of a response. In other words, the probability that a molecule of hormone will encounter a molecule of receptor is related to the abundance of both the hormone and the receptor. The effects of changing receptor abundance on hormone sensitivity are shown in Figure 10. It may be noted that the relationship is not linear, and that increasing the receptor number by 40% decreases the hormone concentration needed to produce a half-maximal response by about a factor of 2, while decreasing the number of receptors by 40% increases the needed concentration of hormone by a factor of about 5.

Many hormones decrease the number of their own receptors in target tissues. This so-called down-regulation was originally recognized as a real phenomenon

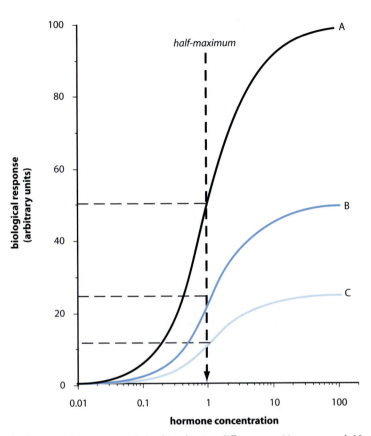

Figure 9   Concentration response relationships showing different capacities to respond. Note that the concentration needed to produce the half-maximal response is identical for all three response capacities.

in modern endocrinology when it was shown that the decreased sensitivity of some cells to insulin in hyperinsulinemic states resulted from a decrease in the number of receptors on the cell surface. However, a similar phenomenon was observed many years earlier and described as the "supersensitivity of denervated tissues." The original discovery of this phenomenon concerned the hypersensitivity of the denervated heart to circulating epinephrine and norepinephrine. The generality of this phenomenon for both endocrine and neural control systems is further indicated by the increase in acetylcholine receptors that occurs after a muscle is denervated and the restoration to normal after reinnervation. The phenomenon of *tachyphylaxis*, or loss of responsiveness to a pharmacological agent on repeated or constant exposure, may be another example of

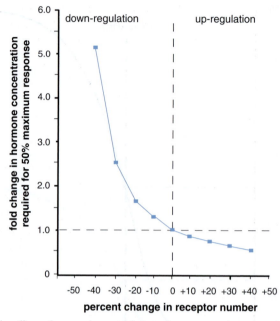

Figure 10   The effects of up- or down-regulation of receptor number on sensitivity to hormonal stimulation.

down-regulation of receptors. Down-regulation may result from inactivation of the receptors at the cell surface, from an increased rate of sequestration of internalized membrane receptors, or a change in the rates of receptor degradation or synthesis.

Down-regulation is not limited to the effects of a hormone on its own receptor, or to the surface receptors for the water-soluble hormones. One hormone can down-regulate receptors for another hormone. This appears to be the mechanism by which T3 decreases the sensitivity of the thyrotropes of the pituitary to TRH (see Chapter 3). Similarly, progesterone may down-regulate both its own receptor and that of estrogen as well. Up-regulation, or the increase in available receptors, also occurs. Prolactin and growth hormone may up-regulate their own receptors in responsive cells. Estrogen up-regulates both its own receptors and those of LH in ovarian cells during the menstrual cycle (see Chapter 12).

The other determinant of sensitivity is the affinity of the receptor for the hormone. Affinity reflects the "tightness" of binding or the likelihood that an encounter between a hormone and its receptor will result in binding. Affinity is usually defined in terms of the concentration of hormone needed to occupy half of the available receptors. Although the affinity of the receptor for its hormone may be adjusted by covalent modifications such as phosphorylation or dephosphorylation, in general, it appears the number of receptors is modulated, rather than their affinity.

Postreceptor Effects

Biological responses do not necessarily parallel hormone binding, and therefore are not limited by the affinity of the receptor for the hormone. Because they depend on many postreceptor events, responses to some hormones may be at a maximum at concentrations of hormone that do not saturate all of the receptors (Figure 11). When fewer than 100% of the receptors need to be occupied to obtain a maximum response, cells are said to express "spare receptors." For example, glucose uptake by the fat cell is stimulated in a dose-dependent manner by insulin, but the response reaches a maximum when only a small percentage of the receptors is occupied by insulin. Consequently, the sensitivity of the cells to insulin is considerably greater than the affinity of the receptor for insulin. Recall that sensitivity is measured in terms of a biological response, which is the physiologically meaningful parameter, whereas affinity is independent of the postreceptor events that produce the biological response. The magnitude of a cellular response to a hormone is determined by summation of the signals generated by each of the occupied receptors, and therefore is related to the number of receptors that is activated rather than the fraction of the total receptor pool that is bound to hormone. However, because the fraction of available receptors that binds to hormone is determined by the hormone concentration, the number of activated receptors needed to a produce a half-maximal response will be equivalent to smaller and smaller fraction as the total number of receptors increases. In the example shown in Figure 11 expression of five times more receptors than needed for a maximum response increases the sensitivity sevenfold.

Another consequence of spare membrane receptors for peptide hormones such as insulin relates to the rapidity with which hormone can be cleared from the blood. It may be recalled that degradation of insulin depends on receptor-mediated internalization of the hormone and hence access to proteolytic enzymes (Chapter 5). It appears that cells have a greater capacity to degrade hormone than to generate a hormone response. Receptors that may be spare with respect to producing a hormonal response may nevertheless play a physiologically important role in degrading its ligand. In fact, some membrane receptors, such as the so-called clearance receptors for the atrial natriuretic hormone (see Chapter 7), lack the biochemical components needed for signal transduction, and function only in hormone degradation. Spare receptors thus may blunt potentially harmful over-responses to rapid changes in hormone concentrations and facilitate clearing hormone from blood.

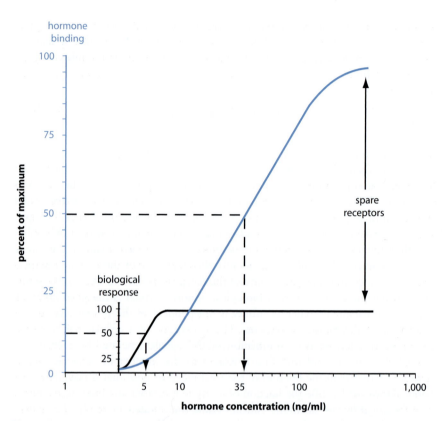

Figure 11 Spare receptors. Note that the concentration of hormone needed to produce a half-maximal response is considerably lower than that needed to occupy half of the receptors.

Sensitivity to hormonal stimulation can also be modulated in ways that do not involve changes in receptor number or affinity. Postreceptor modulation may affect any of the steps in the biological pathway through which hormonal effects are produced. Up- or down–regulation of effector molecules such as enzymes, ion channels, and contractile proteins may amplify or dampen responses and hence change the relationship between receptor occupancy and magnitude of response. For example, the activity of cAMP phosphodiesterase increases in adipocytes in the absence of pituitary hormones. It may be recalled (see Chapter 1) that this enzyme catalyzes the degradation of cAMP, and when its activity is increased, less cAMP can accumulate after stimulation of adenylyl cyclase by a hormone such as epinephrine. Therefore, if all other things were equal, a higher concentration of epinephrine would be needed to produce a given amount of lipolysis than might be necessary in the presence of normal amounts of pituitary hormones, and hence sensitivity to epinephrine would be reduced. Increased activity of phospho-diesterase is only one of several factors contributing to decreased responses to epinephrine in adipocytes of hypopituitary animals. These tissues also express less hormone-sensitive lipase and changes in G-proteins. Consequently, even when all of the receptors are occupied, the maximum response that these cells can make is below normal. In this example, both the sensitivity and the capacity to respond are decreased.

Abundance of Competent Target Cells

At the tissue, organ, or whole body level, the response to a hormone is the aggregate of the contributions of all of the stimulated cells, so that the magnitude of the response is determined both by the number of responsive cells and by their competence. For example, as we have seen, ACTH produces a dose-related increase in blood cortisol concentration in normal individuals. However, immediately after removal of one adrenal gland, changes in the concentration of cortisol in response to ACTH administration would be only half as large as is seen when both glands are present. Therefore, a much higher dose of ACTH will be needed to achieve the same change as was produced preoperatively. With time, however, as the adrenal cortical cells up-regulate ACTH receptors and increase their capacity for steroidogenesis, the concentration of ACTH needed to achieve a particular rate of cortisol secretion will decline. Another example of how changes in cell number and competence are reflected at the whole body level is the gradual increase in estradiol secretion seen in the early part of the menstrual cycle, despite an unchanging or even decreasing concentration of FSH (Chapter 12). This change in sensitivity to FSH reflects the activity of an increasing number of estradiol-producing cells.

Permissive Actions

Another aspect of hormone modulation that is related to the above examples has been called *permissive action*. This phenomenon has already been mentioned in Chapter 4. A hormone acts permissively when its presence is necessary for, or permits, a biological response to occur, even though the hormone does not initiate the response. Permissive effects were originally described for the adrenal cortical hormones, but appear to occur for other hormones as well. Permissive actions are not limited to responses to hormones, but pertain to any cellular response to any signal.

Numerous examples of the foregoing principles of hormonal integration can be seen in the succeeding chapters, which focus on the interactions of multiple hormones in addressing physiological problems.

CHAPTER 7

Regulation of Sodium and Water Balance

221

OVERVIEW

Sodium and water balance are precisely regulated by the endocrine system. Osmolality of the extracellular fluid is monitored and adjusted by regulating water excretion by the kidney in response to antidiuretic hormone (ADH), which is secreted by the posterior lobe of the pituitary gland. Constancy of blood osmolality ensures constancy of cellular volume, but if it were attained only by adjusting water retention, the vascular volume would fluctuate widely. Therefore blood volume must also be precisely regulated so that it is sufficient to ensure perfusion of body tissues but not so expanded that blood pressure is increased. Maintenance of vascular volume depends on maintenance of sodium balance. Renal mechanisms that govern retention or loss of sodium are regulated by the rennin–angiotensin–aldosterone system and the atrial natriuretic hormone. ADH also contributes directly to volume regulation, and when demands for constancy of osmolality are in conflict with demands for constancy of volume, the latter prevail. These hormonal mechanisms operate largely by regulating renal function, but they also regulate salt and water intake.

GENERAL CONSIDERATIONS

All cells of the body have an uninterrupted requirement for nutrients and oxygen and produce waste materials. In addition, many cells must receive signals from other parts of the body to perform their specialized tasks, which in some cases includes production of some product that must be transported to other specialized cells. The cardiovascular system accommodates these nutritive, excretory, and communicative needs. Blood readily equilibrates with the fluid that bathes each cell and thereby preserves the integrity of the extracellular environment, delivers chemical messages, and transports excretory and secretory products. The integrity of the blood is restored as it flows through organs such as kidneys, lungs, and liver whose specialized functions renew the blood's composition and maintain its constancy.

Perfusion of tissues is ensured by maintaining both a sufficient volume of arterial blood and adequate pressure to drive it through the capillaries. Blood flow to a region is matched to changing requirements of cells by locally initiated adjustments in arteriolar tone. Circularly oriented smooth muscle cells in arterioles relax in response to products of cellular metabolism and thus decrease resistance to flow. Increased flow washes away accumulated products, and with removal of the signal for vasodilation, arterioles regain their former tone. The circulatory system can thus be viewed as a central reservoir of pressurized fluid that can be tapped on demand at any locale to provide needed renewal of the cellular environment.

Several factors go into maintenance of the central reservoir of pressure: (1) the beating of the heart, which provides energy; (2) a high degree of arteriolar tone, which slows dissipation of the energy imparted by each beat of the heart; (3) low compliance of the arterial tree, which allows pressure to build up; and (4) sufficient volume of blood, which fills the system. Central control exerted through the autonomic nervous system provides the minute-to-minute adjustments to cardiac function and arteriolar constriction that maintain blood pressure relatively constant. Volume is regulated largely by the endocrine system, but volume and pressure are closely interrelated. Changes in volume can offset changes in arteriolar tone and vice versa to maintain constancy or at least adequacy of the central pressure reservoir. It is not surprising, therefore, that hormones that play decisive roles in regulating blood volume also constrict or dilate arterioles.

Permeability of capillaries to small molecules allows blood in the vascular compartment to equilibrate quickly with interstitial fluid. Blood and interstitial fluid together comprise the extracellular compartment, which contains about one-third of total body water (Figure 1). Water distributes freely between the vascular compartment and the interstitial compartment, usually in a ratio of 1:3. In some pathological states, however, the interstitial compartment becomes disproportionally enlarged, and edema may be considerable. Major determinants of this

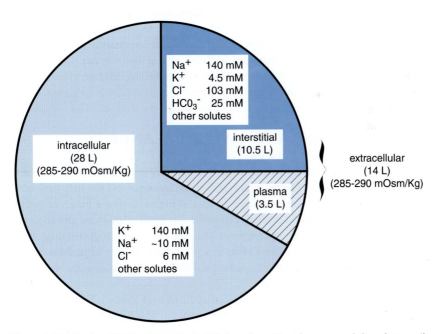

Figure 1 Distribution of body water and principal electrolytes. Note that water and electrolytes equilibrate freely between plasma and interstitial fluid, but only water equilibrates between the intracellular and extracellular compartments.

distribution are the protein content of plasma, principally albumin, and blood pressure within the capillaries. It appears that the volume of the interstitial compartment is not directly monitored or regulated. Rather, control of interstitial volume is achieved indirectly by controlling pressure, composition, and volume of the vascular compartment.

The volumes of fluid in the intracellular and extracellular compartments are also determined by their solute contents. With a few important exceptions in the kidney, biological membranes are freely permeable to water. Net movement of water into or out of cells is determined by the osmotic gradient. Osmotic flow of water is independent of the identity of solutes and responds simply to the discrepancy in number of solute particles (osmolytes) on either side of the cell membrane. Addition or depletion of water in one compartment therefore is followed by compensatory changes in the other. Concentrations of particular solutes on the extracellular or intracellular sides of the plasma membrane are different, however, and are determined by the properties of the membrane.

The major intracellular cation is potassium, which exists in cellular water at a concentration nearly 35 times higher than that in extracellular water. The major extracellular cation is sodium, which is present at about 10 times its intracellular concentration. Blood plasma is in osmotic equilibrium with interstitial and intracellular fluids; regulation of plasma osmolality therefore regulates total body osmolality. Because sodium is the major contributor to osmolality of blood, and because it is largely excluded from the intracellular compartment, changes in sodium balance can change both the distribution of body water and its total volume. Thus homeostatic regulation of blood volume depends on regulation of intake and excretion of sodium as well as water.

SALT AND WATER BALANCE

Salt and water balance are maintained remarkably constant despite wide variations in intake and loss of both sodium and water. Intake of sodium may vary from almost none in saltpoor environments to several grams during a binge of potato chips and pretzels. Output is primarily in urine, but smaller losses are also incurred in sweat and feces. Large losses can result from excessive sweating, vomiting, diarrhea, burns, or hemorrhage. The kidney is a powerful regulator of sodium output and can preserve sodium balance even when daily intake varies over the 4000-fold range between 50 mg and 200 g.

Under basal conditions the typical adult turns over about 1.75 liters of water each day. Most of it originates in the diet in the form of solid and liquid foods, and the remainder is formed metabolically from the oxidation of carbohydrate and fat. Unavoidable losses occur by evaporation from the lungs and skin, as well as by elimination of wastes in the urine and feces. Environment, climate, daily activities, and personal habits impose additional needs for either intake or excretion that must be perfectly offset to maintain physiological balance. So long as intake exceeds obligatory losses, balance can be achieved by controlling excretion. Intake, however, is a voluntary act and varies widely. Thirst and salt appetite are increased when intake falls below the amount needed to maintain balance.

Blood volume is monitored indirectly, primarily as a function of pressure. The concentration of sodium, which is the principal osmolyte of plasma and the primary determinant of blood volume, is monitored only indirectly as a function of osmolality. The kidney is the primary effector of regulation, and at least four hormones—ADH, aldosterone, angiotensin II, and atrial natriuretic factor (also called atrial natriuretic peptide, ANP)—are used to signal regulatory adjustments. To understand how these hormones regulate water and electrolyte balance, we first consider briefly some aspects of renal function.

Each human kidney is contains about 1 million *nephrons* (Figure 2), which are the functional units that adjust the composition of the urine, and hence the blood, by selective reabsorption or secretion of solutes from the ultrafiltrate of plasma formed at each *glomerulus*. Each glomerulus contains a specialized tuft of capillaries situated within the swollen proximal end of the nephron. Glomerular capillaries lie between two resistance vessels, an *afferent arteriole*, which brings blood to the capillaries, and an *efferent arteriole*, which carries blood away. Efferent arterioles give rise to a second capillary network, the peritubular capillaries in the cortex or the vasa rectae in the medulla. Because blood pressure in the glomerular capillaries is relatively high and the capillary endothelium is specialized for filtration, a large volume of nearly protein-free fluid (~180 litres/day) filters through the endothelium to become the precursor of the urine. The increased concentration of proteins in postglomerular blood provides a strong colloid osmotic force for absorption of interstitial fluid into peritubular capillaries and the vasa rectae. About 99% of the glomerular filtrate is reabsorbed, driven through the tubular epithelium by the osmotic forces created by active transport of sodium. Because so large a volume of fluid is processed each day, changes in tubular transport mechanisms that affect reabsorption of only a small percentage of the filtered sodium or water are sufficient to maintain homeostasis. Likewise, small changes in intrarenal blood pressure or flow can influence both filtration and reabsorption and provide an additional means of regulating renal function.

Initial processing of the glomerular filtrate takes place in the proximal convoluted tubule, where about two-thirds of the sodium and water is reabsorbed isosmotically, driven by sodium-coupled transport of glucose, amino acids, phosphate, bicarbonate, and other solutes. The electrochemical gradient for sodium across the luminal membrane of the tubular cells provides the driving force for sodium-coupled uptake of these compounds. Energy that maintains the sodium gradient is provided by the sodiumpotassium ATPase located in the basolateral membranes of these cells. This enzyme "pumps" out three ions of sodium in exchange for two ions of potassium. Passive movements of potassium and other ions maintain electrical neutrality, and the passive movement of water created by

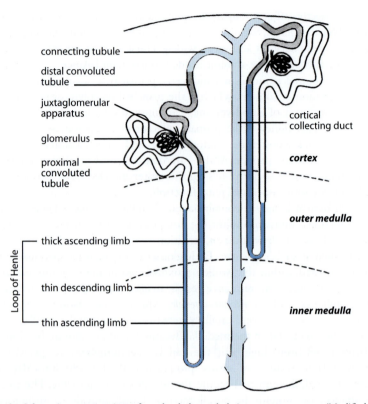

connecting tubule

distal convoluted
tubule

juxtaglomerular
apparatus

glomerulus

proximal
convoluted
tubule

cortical
collecting duct

cortex

outer medulla

thick ascending limb

Loop of Henle

thin descending limb

thin ascending limb

inner medulla

Figure 2 Schematic representation of renal tubules and their component parts. (Modified from Kriz, W., *Am. J. Physiol.* **254**, F1–F8, 1988, with permission.)

accumulation of solutes in the interstitium drives the resorption of water. Although multiple different cotransporters, exchangers, and ion channels are present in different portions of the nephron, the general pattern of passive uptake of sodium at the luminal surface coupled by active sodium/potassium exchange at the basolateral surfaces accounts for virtually all of the renal tubular reabsorptive activity.

The volume of the glomerular filtrate is further reduced by about 20–25% in the *loop of Henle*, which is shaped like a hairpin and doubles back on itself in the renal medulla. The loop of Henle consists of three functionally distinct segments: the *thin descending limb*, which is freely permeable to water, the *thin ascending limb*, which is permeable to sodium but impermeable to water, and the *thick ascending limb*, which is also impermeable to water and which actively transports sodium chloride from the tubular lumen to the interstitium. Because sodium chloride is transported while water is held back, the solute concentration of the interstitium is increased, while the fluid in the thick ascending limb is diluted. Water leaves the thin descending limb and equilibrates rapidly with the sodium chloride-enriched interstitial fluid and hence tubular fluid becomes increasingly concentrated as it flows by the thick ascending limb. Diffusion of sodium, but not water, out of the thin ascending limb increases the sodium concentration in the inner medullary interstitium. The combination of the geometry of Henle's loop, selective permeability to water, active transport of sodium chloride in the thick ascending limb, and countercurrent flow of fluid through the tubule sets up a countercurrent multiplier effect that produces a gradient of increasing osmolarity so that the interstitial sodium concentration in the inner medulla is more than twice that of plasma (Figure 3).

Further reabsorption of solutes in the distal convoluted tubule, the connecting tubule, and the cortical portion of the collecting duct is partially under hormonal control and provides fine tuning of the ionic composition of the urine, and hence the extracellular fluid. Final steps in the processing of urine take place in the collecting ducts, which extend from the cortex through the deepest reaches of the medulla and empty into the renal pelvis. Fluid in the collecting ducts passes through the interstitial osmotic gradient, which provides the driving force for water reabsorption and the production of a concentrated urine. Hormonally regulated permeability of the collecting ducts to water and urea determines the final composition of the urine.

ANTIDIURETIC HORMONE

Antidiuretic hormone (ADH), another name for arginine vasopressin, is discussed in Chapters 2 and 4. ADH is synthesized in magnocellular hypothalamic neurons in the supraoptic and paraventricular nuclei. Axons of these cells pass down the pituitary stalk and terminate in the posterior lobe of the pituitary gland, from whence the hormone is secreted. As we have seen, the same peptide is also produced by other hypothalamic neurons and may act as a hypophysiotropic hormone (Chapter 4) or as a neurotransmitter. ADH is the hormone that signals the kidney to conserve water when plasma osmolarity is increased or when plasma volume is decreased. The hormone is "antidiuretic" because there is a prompt decrease in urine volume following hormone administration. ADH is also called arginine vasopressin, in reference to the acute increase in blood pressure seen as a result of arteriolar constriction produced when the hormone is administered in sufficient dosage. Antidiuretic effects are seen at lower hormone concentrations, as compared to vasoconstrictor effects. The two actions of the hormone are produced by two different G-protein-coupled receptors. The V1 receptor is coupled through $G\alpha_q\beta\gamma$ to phospholipase B and therefore signals through the diacylglycerol/inositol trisphosphate second messenger system (see Chapter 1) and produces its vasopressor effect by constricting vascular smooth muscle. The antidiuretic effect is produced by the V2 receptor, which signals through $G\alpha_s\beta\gamma$ to activate adenylyl cyclase and cyclic AMP production in renal tubular cells.

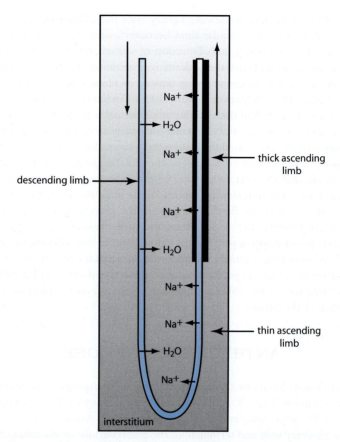

Figure 3 The countercurrent multiplier in the loop of Henle. Selective permeability of the tubular epithelium and active transport of sodium by the thick ascending limb create osmotic gradients. Tubular to interstitial flow of water concentrates sodium (shown in blue) in the descending limb. Sodium movement across the water-impermeable ascending limb creates the osmotic gradient in the interstitium. Vertical arrows indicate the direction of flow. Note that active sodium transport in the thick ascending limb creates the gradient in the interstitium and makes the tubular fluid hypoosmotic by the time it emerges from Henle's loop.

ANTIDIURETIC EFFECT

Osmolarity of blood plasma can be increased or decreased by adjusting the proportion of water relative to solute that is excreted in the urine, thus producing reciprocal changes in the osmolality of the urine. Water that is reabsorbed or excreted in excess of solute is referred to as *free water*. Conservation of free water lowers, and excretion of free water raises, plasma osmolality. As already mentioned, the geometry of the nephron and the permeability characteristics of some segments provide a mechanism for varying water reabsorption without an accompanying change in sodium reabsorption. Antidiuretic hormone conserves free water by increasing collecting duct permeability to water. As a result of sodium chloride reabsorption by the water-impermeable thick ascending limb and distal tubule, fluid that enters the collecting ducts is hypoosmolar. As it flows through the cortical and medullary collecting ducts toward the renal pelvis it passes through regions of increasing osmolarity, but in the absence of ADH, the tubules are impermeable to water and urea so that the low solute concentration is preserved, and a relatively large volume of dilute urine is excreted. In the presence of ADH the collecting ducts in the cortex and outer medulla become permeable to water but remain impermeable to sodium and urea. As a consequence of water reabsorption, the concentration of urea in the terminal regions of the collecting ducts may become as high as 700 mM compared to about 15 mM in plasma. In this region ADH increases the permeability of the principal cells to urea as well as to water. Urea can then diffuse out the collecting ducts and into the interstitium, where it provides about half of the total osmolytes at the tip of the medullary papilla. In rodents, which can produce a much more highly concentrated urine than humans, ADH also stimulates sodium chloride reabsorption in the thick ascending limb and in more distal segments of the nephron. It is possible that similar effects also occur in human kidneys.

CELLULAR MECHANISMS

Plasma membranes of most cells are permeable to water because of the presence of specialized proteins, called *aquaporins* (AQPs), that form water channels. At least four of these proteins (AQP-1, -2, -3 and -4) are expressed in mammalian kidneys. In the proximal convoluted tubule, water and solute are reabsorbed proportionally because AQP-1 is abundantly expressed in the both apical and basolateral membranes of tubular cells and because the intercellular junctions are leaky. Segments of the nephron that are impermeable to water, such as the ascending limb of the loop of Henle, do not express AQPs on their luminal surfaces. Aquaporin-3 and AQP-4 are expressed by the principal cells of the collecting ducts but only in the basolateral membranes. These cells also express AQP-2, which may be found in their luminal membranes or in submembranous vesicles (Figure 4). ADH stimulates an exocytosis-like process in which the AQP-2-bearing vesicles fuse with the luminal membrane and thus insert AQP-2 in much the same manner that insulin increases the abundance of GLUT 4 in adipocyte membranes (see Chapter 5). In the absence of ADH, AQP-2 is removed from the luminal membrane by endocytosis and stored in vesicular membranes. The presence of AQP-3

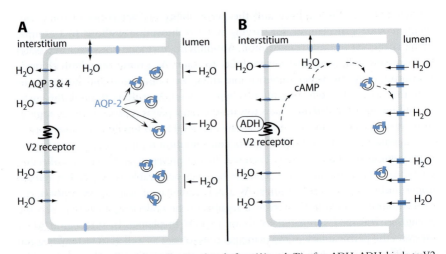

Figure 4 Principal cells of the collecting duct before (A) and (B) after ADH. ADH binds to V2 receptors to induce formation of cAMP, which promotes insertion of aquaporin-2 (AQP-2) into the luminal membrane, making it permeable to water. In the presence of ADH, water can pass through the principal cell from lumen to interstitium, driven by the osmotic gradient. Deep in the medulla, urea transporters are also inserted in the luminal membrane in addition to AQP-2. Expression of aquaporins-3 and -4 in the basolateral membranes allows osmotic equilibration between intercellular and interstitial water.

and AQP-4 in the basolateral surfaces of the principal cells allows intracellular fluid to equilibrate rapidly with interstitial fluid.

Water channels formed by AQP-2 do not allow passage of urea. Permeability to urea depends on the presence of specialized urea transport proteins. ADH promotes the insertion of urea transporters into the luminal membranes of the principal cells of the terminal portions of the inner medullary collecting ducts in much the same way that it regulates AQP-2. Similarly, ADH may also stimulate sodium reabsorption in the thick ascending limb of the loop of Henle and in more distal parts of the nephron by promoting the insertion of transporters or channels into the luminal membranes.

All of the effects of ADH described above depend on the formation of cyclic AMP. V2 receptors are found in the basolateral membranes of the principal cells of the collecting ducts and the cells of the thick portion of Henle's loop. Substrates for cyclic AMP-dependent protein kinase have not been identified, but are thought to include proteins that regulate vesicle trafficking. Stimulation of water and urea permeability is seen in less than 10 minutes after addition of ADH. On a longer time scale, ADH increases transcription of the genes that encode AQP-2, AQP-3, and the urea transporter.

EFFECTS ON BLOOD PRESSURE

Vasopressin may be the most potent naturally occurring constrictor of vascular smooth muscle. On a molar basis, it is at least 10 times more active than norepinephrine or angiotensin II in stimulating contraction of isolated strips of artery. Increases in total peripheral resistance are observed at concentrations of ADH that fall within the upper part of the range that promotes water reabsorption. Small increases in peripheral resistance produced by ADH are not accompanied by increased systemic blood pressure, however, because the baroreceptor reflexes mediate compensatory decreases in heart rate and cardiac output and because ADH may decrease cardiac contractility secondary to coronary arteriolar constriction. Because not all arterioles are equally sensitive to it, ADH does not increase resistance uniformly in all vascular beds. Consequently, there is a redistribution of blood flow, which decreases most profoundly in skin and skeletal muscle. Redistribution can compensate for decreased blood volume by making a disproportionate share of the cardiac output available to essential tissues such as the brain. Arteriolar constriction also changes the distribution of fluid between the vascular and interstitial compartments. It may be recalled that filtration and reabsorption of fluid in the capillaries are balanced largely between the outward hydrostatic force and the inward colloid osmotic force. By constricting arterioles, ADH lowers downstream capillary and venular blood pressure, and thereby promotes net reabsorption of interstitial fluid in skin and skeletal muscle. Intravascular volume increases, therefore, at the expense of the interstitial compartment.

REGULATION OF ADH SECRETION

Plasma Osmolality

The most important stimulus for ADH secretion is an increase in blood osmolality. Little ADH is secreted so long as the osmolality of plasma remains at or below a threshold value of about 280 mOsm/liter. Osmoreceptors are exquisitely sensitive and elicit increased secretion of ADH when the osmolality increases by as little as 1–2%. Above the osmolal threshold, the concentration of ADH in plasma changes in direct proportion to the increase in plasma osmolality (Figure 5) and decreases urine volume significantly.

The finding that injection of a small volume of hyperosmotic fluid into the internal carotid artery elicits ADH secretion, even though peripheral osmolality remains unchanged, indicated that osmolality is detected in the region of the hypothalamus. In rats, intracarotid injection of concentrated saline increased the electrical activity of nerve cells in the supraoptic and paraventricular nuclei. Osmoreceptive cells are thought to reside in the circumventricular organs, particularly the organum

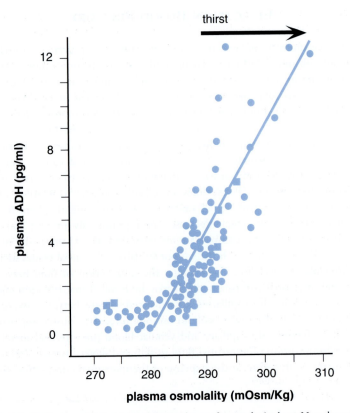

Figure 5 Relation between ADH and osmolality in plasma of unanesthetized rats. Note the appearance of thirst (increased drinking behavior) when plasma osmolality exceeds 290 mOsm/kg. (From Robertson, G. L., and Berl, T., *In* "The Kidney," 5th Ed., p. 881. Saunders, Philadelphia, 1996, with permission.)

vasculosum and subfornical organ located in the vicinity of the third ventricle. This region lies outside the blood–brain barrier and derives its blood supply from the internal carotid artery. Axons project from this region to make synaptic connections with ADH-producing cells in the supraoptic and paraventricular nuclei and stimulate them to release hormone from nerve terminals in the posterior pituitary gland. Some evidence also indicates that the ADH-producing neurons may be directly activated by increased osmolality. In addition, peripheral input from cells in the area drained by the mesenteric and portal veins detects changes in osmolality associated with intestinal absorption and stimulate or inhibit ADH secretion through a neural pathway.

Osmolality is detected as a change in volume of the whole cell or specialized vesicular components. Cells bounded by membranes that allow water to pass freely but restrict movement of most solutes swell in a hypoosmotic environment and shrink when the extracellular fluid is hyperosmotic. Hyperosmolarity activates nonselective cation channels and causes the membrane to depolarize and generate an action potential. Conversely closure of these channels when the cell swells results in hyperpolarization. Sodium chloride, which is largely excluded from cells, is perhaps the most potent osmolyte, as judged by its ability to increase ADH secretion; a more permeant molecule such as urea has only a small effect.

Blood Volume

Changes in blood volume are sensed by receptors in both arterial (high pressure) and venous (low pressure) sides of the circulation. Volume is monitored indirectly via the tension exerted on stretch receptors located (1) on the arterial side, in the carotid sinuses and aortic arch, and (2) on the venous side, in the atria and perhaps the thoracic veins. Low-pressure receptors monitor central venous pressure, which can vary widely with redistribution of blood, as might occur with changes in posture, physical activity, and ambient temperature. Central venous pressure can fall by as much as 10–15% when an individual simply rises from a recumbent to an upright posture. Thus there is a wide range over which deviations in venous pressure are not reliable indicators of true variations in volume. Because of the extensive buffering capacity of the baroreceptor reflexes, changes in arterial pressure are seen only after large decreases in blood volume. Thus changes in volume of the order a few per cent are difficult to detect and do not elicit acute compensatory adjustments in fluid balance.

At normal osmolality, ADH secretion is minimal so long as blood volume remains at or above its physiological threshold or set point. Because volume receptors are not equipped to detect small changes, stimulation of ADH secretion is not initiated until a relatively large depletion has occurred. The minimal change needed for low-pressure receptors to signal ADH secretion is a 10–15% reduction in volume. Similarly, a loss of 10–15% of the blood volume can occur before the threshold for high-pressure volume receptors is reached. Secretion of ADH in response to a decrease in volume is thus an emergency response and not a fine tuner of blood volume. Retention of free water alone is not an effective means of defending the plasma volume. Because it distributes in all compartments, only about 1 ml of every 12 ml of free water retained remains in the vascular compartment. However, if volume falls below a critical threshold value, a potentially life-threatening event is perceived and vigorous secretion of ADH increases blood levels exponentially (Figure 6).

Because the ADH-secreting cells of the supraoptic and paraventricular nuclei are stimulated by two inputs—increased osmolality and decreased volume—they must be able to integrate these signals and respond appropriately. Decreases in volume or pressure heighten the sensitivity of the osmoreceptors and lower the

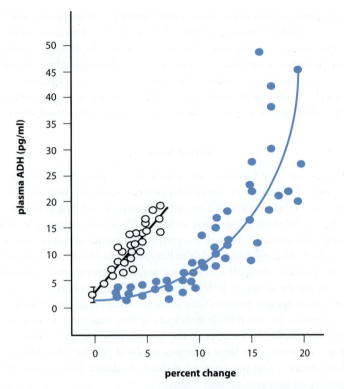

Figure 6 Relation between changes in osmolality (o) or volume (●) and ADH in plasma of unanesthetized rats. (From Dunn *et al.*, *J. Clin. Invest.* **52**, 3212, 1973, with permission.)

threshold for ADH secretion in response to increased osmolality (Figure 7). Volume depletion and increased osmolality, as might result from dehydration, for example, stimulate ADH secretion and reinforce each other. Situations can arise, however, when inputs from osmotic and volume receptors are conflicting. As a general rule, osmolality is preferentially guarded when depletion of volume is small. When volume loss is large, however, osmolality is sacrificed in order to maintain the integrity of the circulation.

Dysfunctional States

The disease state associated with a deficiency of ADH, diabetes insipidus, is characterized by copious production of dilute urine. With this condition more than 20 liters of water may be excreted per day. If not balanced by water intake, osmolality of body fluids and the concentration of plasma sodium increase dramatically

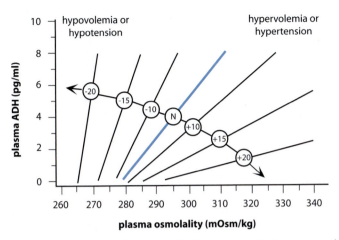

Figure 7 The effects of increases or decreases in blood volume or blood pressure on the relation between ADH concentrations and osmolality in the plasma of unanesthetized rats. Circled numbers indicate percentage change from normal (N). (Modified from Robertson, G. L., and Berl, T., *In* "The Kidney," 5th Ed., p. 881, Saunders, Philadelphia, 1996, with permission.)

and catastrophically. The term *insipidus*, meaning 'tasteless," was adopted to distinguish the consequences of ADH deficiency from those of insulin deficiency (diabetes mellitus) in which there is copious production of glucose-laden urine (see Chapter 5). Nephrogenic diabetes insipidus is the disease that results from failure of the kidney to respond to ADH and may result from defects in the V2 receptor, aquaporin-2, or any of the regulatory proteins that govern cellular responses to ADH. In the syndrome of inappropriate secretion of ADH, death may result from profound dilution of plasma electrolytes because of an inability to excrete free water.

THE RENIN–ANGIOTENSIN–ALDOSTERONE SYSTEM

As already described in Chapter 4, aldosterone is an adrenal steroid that plays a pivotal role in maintaining salt and water balance. Aldosterone is secreted by cells of the zona glomerulosa and acts primarily on the principal cells in the cortical collecting ducts to promote reabsorption of sodium and excretion of potassium. It may be recalled that aldosterone does not stimulate a simple one-for-one exchange of sodium for potassium in the nephron. Sodium reabsorption exceeds potassium excretion by the principal cells. However, because sodium and potassium are also regulated at other renal sites, the net effects of administered aldosterone on sodium and potassium excretion in the urine differ in different physiological states.

Retention of sodium obligates simultaneous reabsorption of water by the nephron and expands the interstitial and vascular volume accordingly. The effects of aldosterone to promote sodium retention by the kidney are augmented by similar effects on sweat and salivary glands and by a poorly understood effect on the brain that increases the appetite for sodium chloride.

Aldosterone secretion is controlled by angiotensin II, whose complementary actions on a variety of target tissues play a critical role in maintaining the central pressure:volume reservoir. Angiotensin II is an octapeptide formed in blood by proteolytic cleavage of a circulating precursor, angiotensinogen (Figure 8). Angiotensinogen is a glycoprotein with a mass of about 60,000–65,000 Da, depending on its degree of glycosylation, and belongs to the serine protease inhibitor (SERPIN) superfamily of plasma proteins. It is present in blood at a concentration of about 1 μM and is constitutively secreted by the liver, which is the major, though not exclusive, source of angiotensinogen in blood. Hepatic production of angiotensinogen varies in different physiological conditions, but although its rate of cleavage to angiotensin is sensitive to changes in its concentration, it is normally present in adequate amounts to satisfy demands for angiotensin production.

The initial cleavage of angiotensinogen, catalyzed by the enzyme renin, releases the amino terminal decapeptide that is called angiotensin I. Angiotensin I is biologically inactive and is rapidly converted to angiotensin II by the angiotensin-converting enzyme (ACE), which removes two amino acids from the carboxyl terminus to produce the biologically active octapeptide, angiotensin II. Angiotensin-converting enzyme is an ectopeptidase that is anchored to the plasma membranes of endothelial cells by a short carboxyl-terminal tail. It is widely distributed in vascular epithelium and may also be secreted into the blood as a soluble enzyme. Angiotensin I is converted to angiotensin II mainly during passage through the pulmonary circulation, but some angiotensin II is also produced throughout the circulation, including the glomerular capillaries. The reaction appears to be limited only by the concentration of angiotensin I. The rate of angiotensin II formation is therefore governed by the rate of release of angiotensin I from angiotensinogen, which, in turn, is primarily regulated by secretion of renin by the kidneys. Angiotensin II has a very short half-life and may be further metabolized to form angiotensin III, and to angiotensin IV by successive removal of the N-terminal and C-terminal amino acids. Some data indicate that these compounds may have biological activity, but their physiological importance has not been established.

Figure 8 Formation of angiotensin II. ACE, Angiotensin-converting enzyme.

Renin is an aspartyl protease that is synthesized and secreted by the juxtaglomerular cells, which are modified smooth muscle cells in the walls of the afferent glomerular arterioles. These cells and cells of the macula densa, which are located in the wall of the distal convoluted tubule of the nephron where it loops back to come in contact with its own glomerulus, make up the juxtaglomerular apparatus (Figure 9). Prorenin is encoded by a single gene located on chromosome 1 and is converted to its enzymatically active form by removal of a 43-amino acid peptide at the N terminus during maturation of its storage granules. Renin is secreted along with some prorenin by a exocytotic process that is activated in response to a decrease in blood volume that is sensed as a corresponding decrease in pressure. At the cellular level, this secretory process is stimulated by cyclic AMP and, contrary to most secretory processes, is inhibited by increased intracellular calcium.

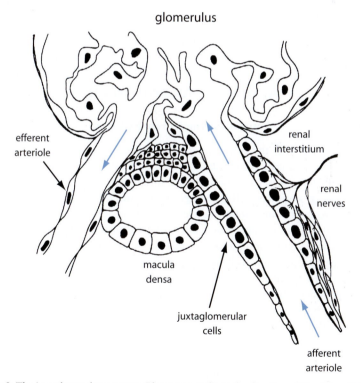

glomerulus

efferent arteriole

renal interstitium

renal nerves

macula densa

juxtaglomerular cells

afferent arteriole

Figure 9 The juxtaglomerular apparatus. Blue arrows indicate the direction of blood flow. (Modified from Davis, J. O., *In* "Handbook of Physiology, Section 7: Endocrinology, Volume IV: Adrenal Gland." American Physiological Society, Washington, D. C., 1975, with permission.)

Three different but related inputs signal increased secretion of renin.

1. The juxtaglomerular cells are richly innervated by sympathetic nerve fibers. These fibers are activated reflexly by a decrease in arterial pressure that is sensed by baroreceptors in the carotid sinuses, aortic arches, and perhaps the great veins. Release of norepinephrine from sympathetic nerve terminals stimulates cyclic AMP production in the juxtaglomerular cells by activating β- adrenergic receptors and adenylyl cyclase.

2. Blood pressure (volume) is also sensed as tension exerted on the smooth muscle cells of the afferent glomerular arterioles. Stretch-activated ion channels in the membranes of juxtaglomerular smooth muscle cells produce partial membrane depolarization, activation of voltage-sensitive calcium channels, and increased intracellular calcium concentrations. Conversely, a decrease in pressure lowers intracellular calcium and relieves inhibition of renin secretion.

3. Decreased pressure in the afferent glomerular arterioles also results in decreased glomerular filtration, which in turn decreases the rate of sodium chloride delivery to the distal convoluted tubules. Cells in the macula densa sense the decrease in sodium chloride by mechanisms that are not fully understood, and in response release adenosine, which activates adenosine II receptors in afferent arteriolar cells and increases cyclic AMP.

ACTIONS OF ANGIOTENSIN II

Actions on the Adrenal Cortex

Angiotensin II is the primary signal for increased aldosterone secretion by adrenal glomerulosa cells. Administration of angiotensin II to normal or sodium-deficient humans increases aldosterone concentrations in blood plasma. Conversely, drugs that block angiotensin II receptors or that lower angiotensin II concentration by blocking the angiotensin-converting enzyme (ACE inhibitors) decrease plasma concentrations of aldosterone. On a longer time scale, angiotensin II causes the volume of the zona glomerulosa to increase by stimulating an increase in both cell size (hypertrophy) and cell number (hyperplasia). Such an effect is seen in individuals who maintain high plasma levels of angiotensin II as a result of a sodium-poor diet. These individuals show an increased sensitivity of aldosterone secretion in response to angiotensin II in part because of up-regulation of angiotensin II receptors and in part because of the increase in the number of responsive cells and the increased capacity of their biosynthetic machinery.

Actions on the Kidney

In addition to its indirect effects to promote salt and water reabsorption through stimulation of aldosterone secretion, angiotensin II also defends the vascular volume directly through actions exerted on both vascular and tubular elements of the kidney. By constricting renovascular smooth muscles, angiotensin II increases vascular resistance in the kidney and hence decreases renal blood flow and glomerular filtration. Decreased glomerular filtration may also be augmented by constriction of the glomerular mesangial cells, which may alter the efficiency of filtration by regulating blood flow in individual glomerular capillaries. Because reabsorptive mechanisms are not 100% efficient, a small fraction of the glomerular filtrate is inevitably lost in the urine. Decreased glomerular filtration, therefore, ultimately results in decreased sodium and water excretion. Angiotensin II also directly increases sodium bicarbonate reabsorption by stimulating sodium–proton exchange in the luminal membranes of proximal tubular cells and activating the sodium bicarbonate cotransporter in the basolateral membrane of these cells (Figure 10).

Cardiovascular Effects

Angiotensin II produces profound long- and short-term effects on the cardiovascular system. Stimulation of angiotensin II receptors in vascular smooth

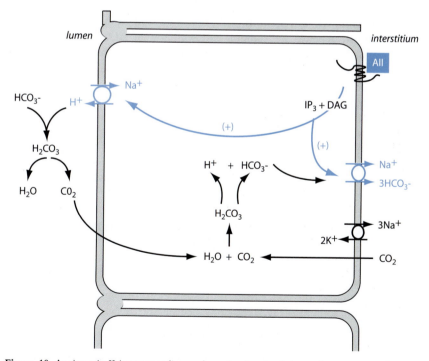

Figure 10 Angiotensin II increases sodium reabsorption by stimulating sodium–proton exchange in the luminal brush border and sodium–bicarbonate cotransport in the basolateral membrane. Hydrogen ions and bicarbonate are regenerated in the cell cytosol from CO_2 and water.

muscle activates the diacylglycerol/inositol trisphosphate second messenger system (Chapter 1) and results in increased intracellular calcium concentrations and sustained vasoconstriction. These direct effects on smooth muscle tone are reinforced by activation of vasomotor centers in the brain to increase sympathetic outflow to vascular smooth muscle and decrease vagal inhibitory input to the heart. Angiotensin II also acts directly on cardiac myocytes to increase calcium influx and therefore cardiac contractility. The combination of these effects and the expansion of vascular volume markedly increase blood pressure and make angiotensin II the most potent pressor agent known. Vasoconstrictor effects are not uniformly expressed in all vascular beds, however, probably because of differences in receptor abundance. In addition to increasing volume and pressure, angiotensin II also redistributes blood flow to brain, heart, and skeletal muscle at the expense of skin and visceral organs. However, at high concentrations it may also constrict the coronary arteries and compromise cardiac output. Chronically high concentrations of angiotensin can lead to remodeling of cardiac and vascular muscle because angiotensin II may act as a growth factor.

Central Nervous System Effects

Angiotensin II, acting both as a hormone and as a neurotransmitter, stimulates thirst, appetite for sodium, and secretion of ADH through actions exerted on the hypothalamus and perhaps other regions of the brain. Blood-borne angiotensin II can interact with receptors present on hypothalamic cells in the subfornical organ and the organum vasculosum of the stria terminalis, which lie outside the blood–brain barrier and project to the supraoptic and paraventricular nuclei and other hypothalamic sites, including vasomotor regulatory centers. In addition, ADH-producing cells in the paraventricular nuclei express receptors for angiotensin II and release ADH when angiotensin II is presented to them experimentally by intraventricular injection or when released from impinging axons. These diverse actions of angiotensin II are summarized in Figure 11.

REGULATION OF THE RENIN–ANGIOTENSIN–ALDOSTERONE SYSTEM

The rennin–angiotensin–aldosterone system is regulated by negative feedback, but neither the concentration of aldosterone, or angiotensin II, nor the concentration of sodium per se, is the controlled variable. Although preservation of body sodium is the central theme of aldosterone action, the concentration of sodium in blood does not appear to be monitored directly, and fluctuations in plasma concentrations have little direct effect on the secretion of renin. Reabsorption of sodium results in reabsorption of a proportionate volume of

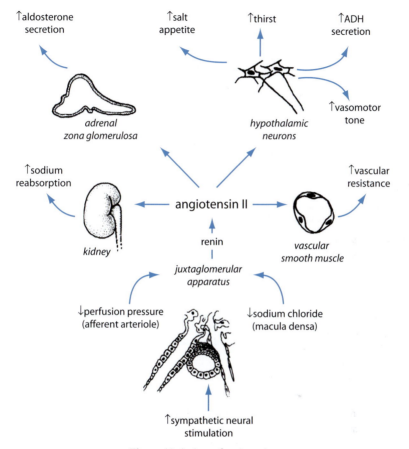

Figure 11 Actions of angiotensin.

water. Increased blood volume, which is the ultimate result of sodium retention, provides the negative feedback signal for regulation of renin and aldosterone secretion (Figure 12). It is noteworthy that even though angiotensin II directly increases sodium reabsorption and exerts a variety of complementary actions that contribute to maintenance of the central pressure–volume reservoir, it cannot sustain an adequate vascular volume to ensure survival in the absence of aldosterone. Despite apparent redundancies in their actions, both aldosterone and angiotensin II are critical for maintaining salt and water balance.

The kidney is the primary regulator of the angiotensin II concentration in blood, but angiotensin II is also produced locally in a variety of other tissues, including walls of blood vessels, adipose tissue, and brain, where it functions as a

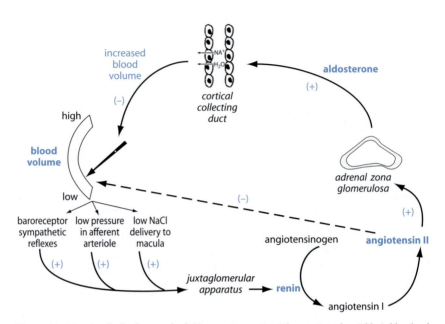

Figure 12 Negative feedback control of aldosterone secretion. The monitored variable is blood volume; (+), stimulates; (−), inhibits. Note that the angiotensin II also contributes directly to maintenance of blood volume, but its influence in this regard (indicated by the dashed arrow) is inadequate in the absence of mineralocorticoid.

neurotransmitter. These extrarenal tissues synthesize angiotensinogen as well as renin and ACE and may form angiotensin II intracellularly. Locally produced angiotensin II may serve a paracrine function to stimulate prostaglandin production and in some instances may act as a local growth factor. The extent to which such localized production of angiotensin II contributes to the regulation of sodium and water balance is unclear.

ATRIAL NATRIURETIC FACTOR

Atrial natriuretic factor (ANF), as its name implies, promotes the excretion of sodium (*natrium* in Latin) in the urine. It is synthesized, stored in membrane-bound granules, and secreted by exocytosis from cardiac atrial myocytes. ANF is a 28-amino-acid peptide that corresponds to the carboxyl terminus of a 126-amino-acid prohormone, which is the principal storage form. Secretion of ANF is stimulated by increased vascular volume, which is sensed as increased stretch of the atrial wall. A second natriuretic peptide originally isolated from pig brain, and therefore called brain natriuretic peptide (BNP), is also produced in the atria and ventricles of the human heart. ANF and BNP are products of separate genes, but have similar structures and actions, although BNP is considerably less potent than ANF. A third related gene encodes CNP, which is expressed principally, but not exclusively, in the central nervous system and lacks natriuretic activity. ANF produces its biological effects by stimulating the formation of cyclic guanosine monophosphate (cyclic GMP), which may modify cellular functions by activating cyclic GMP-dependent protein kinase, activating a cyclic nucleotide phosphodiesterase that degrades cyclic AMP, interacting directly with membrane ion channels, and regulating gene expression. BNP binds to the same receptors as ANF, but with 10-fold lower affinity. Receptors that mediate the natriuretic effects of ANF and ANP and the closely related CNP receptor consist of an extracellular hormone binding domain, a single membrane-spanning domain, and an intracellular domain that catalyzes formation of cyclic GMP from GTP. Other ANF receptors, the so-called clearance receptors, bind all three peptides with similar affinity and contain the hormone-binding and membrane-spanning domains, but lack the guanylyl cyclase domain. These abundant receptors remove ANF, BNP, and CNP from blood and extracellular fluid and deliver them to the lysosomes for degradation. ANF disappears from plasma with a half-life of about 3 minutes, due in part to the action of the clearance receptors and in part to proteolytic cleavage at the brush border of renal proximal tubular cells.

PHYSIOLOGICAL ACTIONS

The physiological role of ANF is to protect against volume overload. Through its combined effects on the cardiovascular system, the kidneys, and the adrenal glands it lowers mean arterial blood pressure and decreases the effective blood volume. Its physiological effects are essentially opposite to those of angiotensin II (Figure 13).

Cardiovascular Actions

Increased concentrations of ANF in blood produce a prompt decrease in mean arterial blood pressure. Initial responses include relaxation of resistance vessels and stimulation of cardiac afferent nerves that project to central vasomotor centers to suppress sympathetic reflexes. Some evidence indicates that ANF also decreases norepinephrine release from sympathetic nerve endings and the adrenal medullae. An overall decrease in sympathetic input to vascular smooth muscle attenuates the pressor responses that might otherwise counteract vasodilatory effects of ANF. In addition, decreased sympathetic stimulation of the juxtaglomerular cells combined with direct inhibitory effects of ANF on renin secretion

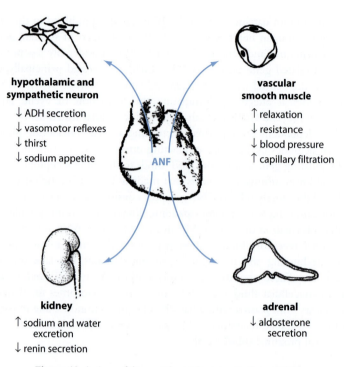

hypothalamic and sympathetic neuron

↓ ADH secretion
↓ vasomotor reflexes
↓ thirst
↓ sodium appetite

vascular smooth muscle

↑ relaxation
↓ resistance
↓ blood pressure
↑ capillary filtration

ANF

kidney

↑ sodium and water excretion
↓ renin secretion

adrenal

↓ aldosterone secretion

Figure 13 Actions of the peptide atrial natriuretic factor (ANF).

lowers circulating levels of angiotensin II. Together, these effects enable the decrease in blood pressure to be sustained. Cardiac rate and contractility are reduced both as a consequence of decreased sympathetic stimulation of the heart and by direct actions of ANF on cardiac muscle. The decrease in arteriolar tone results in increased capillary pressure and favors net filtration of fluid from the vascular to the interstitial compartment and thus decreases vascular volume.

Renal Actions

Vascular volume is further decreased by actions on the kidney that promote excretion of water and sodium (Figure 14). ANF relaxes the afferent glomerular arterioles and the glomerular mesangial cells while constricting efferent arterioles. The resulting increase in capillary hydrostatic pressure and surface area produces an increase in the glomerular filtration rate that accounts in large measure for increased urinary loss of salt and water. ANF also decreases sodium reabsorption in the proximal tubule by inhibiting the effects of angiotensin II on sodium bicarbonate reabsorption, and perhaps by directly inhibiting the sodium–proton

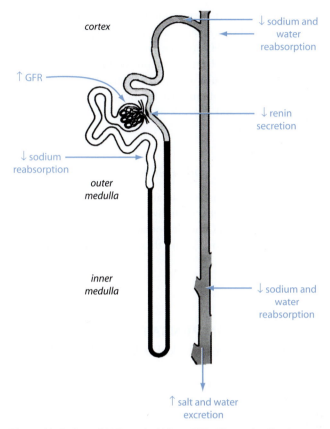

cortex

↑ GFR

↓ sodium
reabsorption

outer
medulla

inner
medulla

↓ sodium and
water
reabsorption

↓ renin
secretion

↓ sodium and
water
reabsorption

↑ salt and water
excretion

Figure 14 Actions of ANF on the kidney. GFR, Glomerular filtration rate.

antiporter. As a consequence of these actions, increased amounts of salt and water reach the loop of Henle and partially "wash out" the osmotic gradient that provides the osmotic driving force for water reabsorption in the collecting ducts. In addition, ANF acts directly on the collecting ducts to decrease salt and water reabsorption. The net result is increased sodium excretion in a large volume of dilute urine and a reduction in blood volume and total body sodium.

Effects on Aldosterone Secretion

As already discussed in Chapter 4, ANF receptors are present in adrenal glomerulosa cells, where ANF directly inhibits aldosterone synthesis and secretion. ANF blocks the stimulatory effects of angiotensin II and high potassium on aldosterone secretion and also decreases aldosterone secretion indirectly by inhibiting renin secretion and thereby decreasing the availability of angiotensin II. In addition, ANF decreases ACTH secretion and thus deprives glomerulosa cells of the supportive effects of ACTH on the steroid synthetic apparatus. Although the consequences of the cardiovascular and renal actions of ANF are apparent without delay, the decrease in blood volume that follows from inhibition of aldosterone secretion is slower in onset and depends on the rapidity of cellular degradation of aldosterone-induced proteins.

Other Effects

Acting through mechanisms that are not yet understood, ANF inhibits the secretion of ADH. This effect on hypothalamic neurons is reinforced by the decrease in angiotensin II. Other hypothalamic effects include suppression of thirst and salt-seeking behavior. All of these effects are opposite to those of angiotensin II.

INTEGRATED COMPENSATORY RESPONSES TO CHANGES IN SALT AND WATER BALANCE

The three hormones, ADH, angiotensin II, and aldosterone, collaborate to maintain or increase the effective volume of the blood plasma. Their properties and characteristics are summarized in Table 1. In addition to reinforcing each other's effects, each of these hormones acts at multiple sites to reinforce its own effects. Physiological responses to these hormones are countered by ANF and brain natriuretic peptide (BNP), which act at many of the same target sites. To some extent all of these hormones are present in the circulation simultaneously, though in different relative amounts, and target cells such as vascular smooth muscle cells and the principal cells of the cortical collecting ducts must integrate these and other conflicting and reinforcing signals. The following discussion focuses on the endocrine adjustments that play decisive roles in maintaining salt and water balance. Students should be aware, however, that the sympathetic nervous system and a variety of locally produced paracrine factors, some of which are listed in Table 2, may also contribute, particularly by adjusting arteriolar tone. To gain some understanding of how the various endocrine pathways interact, we consider several examples of perturbations in salt and water balance and the hormonal mechanisms that restore homeostasis. Volume changes can take several forms and may or may not be accompanied by changes in osmolality (sodium balance), as shown in Table 3.

Table 1

Properties of the Principal Hormones that Regulate Water and Sodium Balance

Hormone	Chemistry	Mode of action	Major target cells	Major cellular actions	Physiological responses
ADH	peptide	cAMP (V2 receptors); DAG and IP_3 (V1 receptors)	principal cells of collecting ducts; vascular smooth muscle	↑ water and urea permeability vasoconstriction	Water conservation ↑ blood pressure
Aldosterone	steroid	Gene transcription (primarily)	principal cells of cortical collecting ducts	↑ Na^+ reabsorption, ↑ K^+ excretion	Expand vascular volume
Angiotensin II	peptide	DAG and IP_3 (AT1 receptors)	adrenal glomerulosa vascular smooth muscle afferent arterioles proximal tubule cells	↑ aldosterone secretion vasoconstriction vasoconstriction ↑ Na^+/H^+ exchange, ↑ $NaHCO_3$ reabsorption	(see above); ↑ blood pressure, ↓ GFR; Na^+ retention
			hypothalamic neurons	↑ ADH secretion ↑ thirst and salt appetite	(see above) salt and water ingestion
ANF	peptide	cGMP	vascular smooth muscle glomerular mesangial cells proximal tubule cells adrenal glomerulosa hypothalamic neurons	vasodilation relaxation ↓ Na^+ reabsorption ↓ aldosterone secretion ↓ vasomotor reflexes, ↓ ADH secretion	↓ blood pressure ↓ blood volume ↑ GFR natriuresis diuresis

Table 2

Some Locally Produced Hormone-like Agents That Affect Cardiovascular Functions Related to Salt and Water Balance

Agent	Chemistry	Source	Relevant actions
Nitric oxide	Gas	Vascular endothelium	Relax vascular smooth muscle
Adrenomedullin	Peptide	Ubiquitous	Relax vascular smooth muscle
Endothelin	Peptide	Ubiquitous	Constrict vascular smooth muscle
Bradykinin	Peptide	Plasma, many tissues	Relax vascular smooth muscle; natriuresis and diuresis
Prostaglandins	Arachidonic acid derivatives	Ubiquitous	Constrict or relax vascular smooth muscle, increase capillary permeability

Table 3

Examples of Changes in Fluid Volume

Condition	Expansion	Contraction
Isosmolal	↑Salt and water ingestion Hyperaldosteronism Heart failure	Hemorrhage Hypoalbuminemia
Hypoosmolal	Excessive water intake Syndrome of inappropriate ADH secretion	Excessive sweating followed by water intake
Hyperosmolal	Excessive salt intake	Dehydration

Hemorrhage

With hemorrhage, the vascular volume is decreased without a change in osmolality. To cope with blood loss, especially if it is large, a three-part strategy is usually followed: prevention of further fluid loss, redistribution of remaining fluid to maximize its usefulness, and replacement of the water and sodium losses. The sympathetic nervous system is indispensable for survival during the initial moments after hemorrhage. Hormonal contributions may augment the initial sympathetic reactions and are largely responsible for mediating the later aspects of recovery.

The immediate response to hemorrhage is massive vasoconstriction driven by the sympathetic nervous system. This response sustains arterial pressure and redistributes the cardiac output to ensure adequate blood flow to essential tissues. Renal blood flow and glomerular filtration are markedly reduced. Although slower in onset, hormonal responses nevertheless may contribute to maintenance of arterial blood pressure through vasoconstrictor actions of angiotensin II, ADH, and

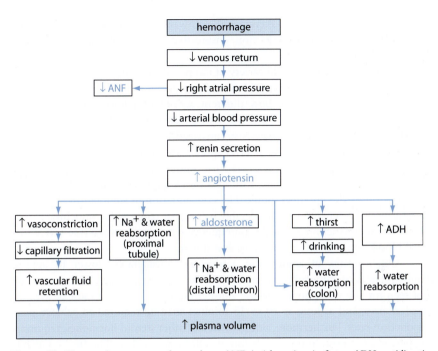

Figure 15   Hormonal responses to hemorrhage. ANF, Atrial natriuretic factor; ADH, antidiuretic hormone.

adrenomedullary hormones. The sum of these responses transfers extracellular water to the vascular compartment by promoting net fluid absorption by capillaries and venules. Figure 15 shows the pathways that eventually lead to restoration of blood volume. Initially, hemorrhage, especially if it is severe (30% of blood volume), decreases venous return and thereby reduces right atrial pressure. Cardiac output is thus decreased, which at least transiently decreases arterial blood pressure and triggers the sympathetic response.

RESPONSE OF THE RENIN–ANGIOTENSIN SYSTEM

Decreased arterial pressure is one of the signals for renin secretion and is directly sensed by the juxtaglomerular cells in the afferent arterioles. This input is nullified if arterial pressure is fully restored by the increase in total peripheral resistance, but direct sympathetic stimulation of the juxtaglomerular cells also increases renin secretion. In addition, decreased renal blood flow and glomerular filtration decrease sodium chloride flux through the distal tubule, which acts as yet another stimulus for renin secretion. Redundant pathways for evoking renin secretion, and therefore angiotensin production, ensure that this crucial system is activated by hemorrhage.

Angiotensin II has a wide variety of temporally and spatially separate actions that summate to restore plasma volume and compensate for hemorrhage. This situation is a good example of how different actions of a hormone expressed in different target cells reinforce each other to produce a cumulative response. This was discussed above and therefore is summarized here only in terms of the temporal sequence. Constriction of arteries and arterioles occurs within seconds, but the consequent mobilization of extravascular fluid requires several minutes. Also on the order of minutes may be direct stimulation by angiotensin II of salt and water reabsorption by the proximal tubule. The concentration of angiotensin II required to activate this mechanism is considerably below that required for vasoconstriction, but obviously this action is of little consequence when hemorrhage is so severe that renal blood flow is nearly completely shut down.

Other results of angiotensin action are considerably slower to appear. Consequences of stimulating adrenal glomerulosa cells to secrete aldosterone are not seen for almost an hour. Because aldosterone is not stored, it must be synthesized *de novo*, and as long as 10 to 15 minutes may be required to achieve peak production rates. Furthermore, aldosterone, like other steroid hormones, requires a lag period of at least 30 minutes before its effects are evident. The final contribution of angiotensin II is stimulation of salt appetite, thirst, and fluid absorption by the colon. Depending on the severity of the blood loss and the availability of water and salt, many hours or even days may pass before the renin–angiotensin system can restore the plasma volume to prehemorrhage levels.

Response of the ADH System

Even though osmolality is unchanged, decreased pressure sensed by receptors in the atria, aorta, and carotid sinuses stimulates ADH secretion. In addition, ADH secretion is also increased by angiotensin II. Here we have another case of redundancy, because ADH and angiotensin II have overlapping actions on arteriolar smooth muscle. These hormones also reinforce each other's actions at the level of the renal tubule, because they increase water reabsorption at different sites and by different mechanisms. Although ADH is secreted almost instantaneously in response to hemorrhage, its physiological importance for the early responses is questionable because vascular smooth muscle may already be maximally constricted by sympathetic stimulation, which is even faster. In addition, when renal shutdown is severe, little urine reaches the collecting ducts and hence even maximal antidiuresis can conserve little water. ADH, however, is an indispensable component of the recovery phase. Thirst and salt-conserving mechanisms would be of little benefit without ADH to promote renal retention of water.

RESPONSE OF ALDOSTERONE

Like ADH, aldosterone is of little consequence for the immediate reactions to hemorrhage. It acts too slowly. Furthermore, decreased glomerular filtration is far more important quantitatively in conserving sodium. Increased secretion of aldosterone, which is initiated promptly by the renin–angiotensin system and reinforced by increased ACTH secretion, can be regarded as an anticipatory response to ensure sodium conservation when renal blood flow is restored. Aldosterone is indispensable for replenishing blood volume by conserving sodium ingested during recovery and probably by stimulating sodium intake.

RESPONSE OF ANF

It almost goes without saying that depleted vascular volume reduces or eliminates signals for the secretion of ANP; this situation is the converse of that depicted in Figure 9. We thus have a push–pull mechanism wherein secretion of an inhibitory influence on the actions of angiotensin II and ADH is shut off by the same events that increase secretion of these hormones.

DEHYDRATION

Dehydration (water deficit) is a commonly encountered derangement of homeostasis and may result from severe sweating, diarrhea, vomiting, fever, excessive alcohol ingestion, or simply insufficient fluid intake. Because dehydration usually involves a greater deficit of water than solute, the osmolality of both the intracellular and extracellular compartments increases. Consequently, the ADH pathway is the principal means for correcting this derangement in water homeostasis. As osmolality increases above its threshold value, ADH secretion promptly increases, and water is reabsorbed in excess of solute until osmolality is restored. This action prevents further loss of water in urine, but cannot restore the volume deficit that usually accompanies dehydration. Decreased volume stimulates the renin–angiotensin–aldosterone system to facilitate vascular adjustments, stimulate thirst, and prepare for replenishment from increased intake of salt and water. Decreased volume also reinforces osmotic stimulation of ADH secretion. Again, ANF secretion is not activated and the actions of ADH and angiotensin II are unopposed. Figure 16 illustrates the endocrine responses to dehydration and the series of events that restore osmolality and volume to normal.

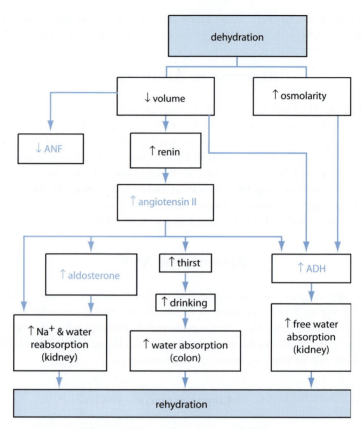

Figure 16 Hormonal responses to dehydration.

SALT LOADING AND DEPLETION

Although sodium chloride is scarce in many regions of the world, it is in oversupply in most Western diets. The endocrine system plays a pivotal role in maintaining homeostasis and normal blood pressure in the face of salt loading or depletion. Figure 17 shows the responses of normal human subjects who volunteered to consume diets that contained high, low, or standard amounts of sodium. Salt-loaded subjects excreted more than 10 times as much sodium in their urine as did salt-deprived subjects, but the concentration of sodium in plasma and systolic blood pressure were nearly identical in all three groups. Although extracellular fluid volume was expanded on the high-salt diet and contracted on the low salt diet, osmolality and sodium concentration of body fluids remained remarkably constant.

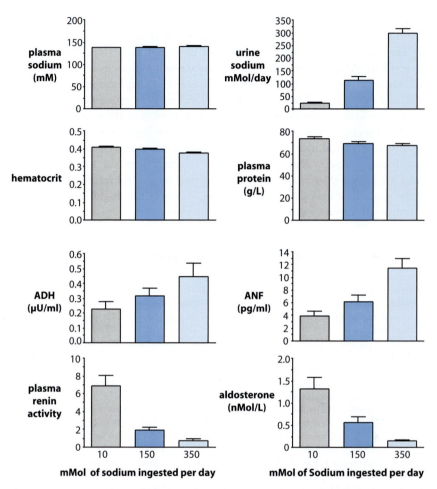

Figure 17 Responses of normal subjects to low or high intake of sodium chloride for 5 days. Plasma sodium concentrations were maintained within less than 2% by compensating rates of sodium excretion. The small increases or decreases in hematocrit and plasma protein concentrations indicate contraction of the plasma volume in the low-sodium diet group and expansion in the high-sodium diet group. The changes in hormone concentrations are in response to the small changes in osmolality and volume. Plasma renin activity is a measure of renin concentration expressed as nanograms of angiotensin I formed per milliliter of plasma in 1 hour. (Drawn from the data of Sagnella, G. A., Markandu, N. D., Shore, A. C., Forsling, M. L., and MacGregor, G. A., *Clin. Sci.* **72**, 25–30, 1987.)

Changes in blood concentrations of angiotensin II (as reflected in renin levels), aldosterone, ADH, and ANF elicited by different amounts of sodium intake are shown in Figure 17. Plasma renin activity and aldosterone secretion decreased as sodium intake increased. The high rate of sodium loss by subjects on the high-sodium diet may be explained by decreased reabsorption of sodium in the proximal tubule as a result of decreased angiotensin II and increased ANP, and in the cortical collecting ducts as a result of decreased aldosterone. As might be expected, the ANF concentration was increased when sodium intake was increased and was decreased when sodium intake was low. The reciprocal relation between ANF and angiotensin II acts as a push–pull mechanism to promote sodium loss in the sodium-loaded individual and sodium conservation in the salt-deprived subject. ADH secretion was also increased in subjects on high salt intake, probably in response to the small increase in plasma osmolality. The plasma sodium concentration in these subjects was 2% higher than in subjects on the low-sodium diet and 1.4% higher than in subjects on the normal diet. ADH secreted in response to the osmotic stimulus prevented the loss of water that might otherwise have accompanied increased amounts of sodium in urine.

SUGGESTED READING

Andreoli, T. E., Reeves, W. B., and Bichet, D. G. (2000). Endocrine control of water balance. *In* "Endocrine Regulation of Water and Electrolyte Balance, Volume III, Handbook of Physiology, Section 7, The Endocrine System" (J. C. S. Fray, ed.), pp. 530–569. Oxford University Press, New York.

Ballerman, B. J., and Oniugbo, M. A. C. (2000). Angiotensins. *In* "Endocrine Regulation of Water and Electrolyte Balance, Volume III, Handbook of Physiology, Section 7, The Endocrine System" (J. C. S. Fray, ed.), pp. 104–155. Oxford University Press, New York.

Brenner, B. M., Ballermann, B. J., Gunning, M. E., and Zeidel, M. L. (1990). Diverse biological actions of atrial natriuretic peptide. *Physiol. Rev.* **70**, 665–699.

de Bold, A. J. (1985). Atrial natriuretic factor: A hormone produced by the heart. Science, **230**, 767–769.

Fray, J. (2000). Endocrine control of sodium balance. *In* "Endocrine Regulation of Water and Electrolyte Balance, Volume. III, Handbook of Physiology Section 7, The Endocrine System" (J. C. S. Fray, ed.), pp. 250–305. Oxford University Press, New York.

Gibbons, G. H., Dzau, V. J., Farhi, E. R., and Barger, A. C. (1984). Interaction of signals influencing renin release. *Annu. Rev. Physiol.* **46**, 291–308.

Hackenthal, E., Paul, M., Ganten D., and Taugner, R. (1990). Morphology, physiology, and molecular biology of renin secretion. *Physiol. Rev.* **70**, 1067–1098.

Laragh, J. H. (1985). Atrial natriuretic hormone, the renin–aldosterone axis, and blood pressure–electrolyte homeostasis. *N. Engl. J. Med.* **313**, 1330–1340.

Reid, I. A., and Schwartz, J. (1984). Role of vasopressin in the control of blood pressure. *Front. Neuroendocrinol.* **8**, 177–197.

Wade, J. B. (1986). Role of membrane fusion in hormonal regulation of epithelial transport. *Annu. Rev. Physiol.* **48**, 213–224.

CHAPTER 8

Hormonal Regulation of Calcium Metabolism

OVERVIEW

Adequate amounts of calcium in its ionized form, Ca^{2+}, are needed for normal function of all cells. Calcium ion regulates a wide range of biological processes and is one of the principal constituents of bone. In terrestrial vertebrates, including humans, maintenance of adequate concentrations calcium ion[1] in the extracellular fluid requires the activity of two hormones, parathyroid hormone (PTH) and a derivative of vitamin D called $1\alpha,25$-dihydroxycholecalciferol [$1,25(OH)_2D_3$] (also called calcitriol). In more primitive vertebrates living in a marine environment, guarding against excessively high concentrations of calcium requires another hormone, calcitonin, which appears to have only vestigial activity in humans.

Body calcium ultimately is derived from the diet, and daily intake is usually offset by urinary loss. The skeleton acts as a major reservoir of calcium and can buffer the concentration of calcium in extracellular fluid by taking up or releasing calcium phosphate. PTH promotes the transfer of calcium from bone, the glomerular filtrate, and intestinal contents into the extracellular fluid. It acts on bone cells to promote calcium mobilization and on renal tubules to promote reabsorbtion calcium and excretion of phosphate. It promotes intestinal transport of calcium and phosphate indirectly by increasing the formation of $1,25(OH)_2D_3$, required for calcium uptake by intestinal cells. This vitamin D metabolite also promotes calcium mobilization from bone and reinforces the actions of PTH on this process. In addition, $1,25(OH)_2D_3$ promotes reabsorption of calcium and phosphate by renal tubules. The rate of PTH secretion is inversely related to the concentration of blood calcium, which directly inhibits secretion by the chief cells of the parathyroid glands. Calcitonin inhibits the activity of bone-resorbing cells, and thus blocks inflow of calcium to the extracellular fluid compartment. Its secretion is stimulated by high concentrations of blood calcium.

GENERAL FEATURES OF CALCIUM BALANCE

Calcium enters into a wide range of cellular and molecular processes. Changes of its concentration within cells regulate enzymatic activities and fundamental

[1]Calcium is present in several forms within the body, but only the ionized form, Ca^{2+}, is monitored and regulated. In this discussion, calcium refers to the ionized form except when otherwise specified.

cellular events such as muscular contraction, secretion, and cell division. As already discussed (see Chapter 1), calcium and calmodulin also act as intracellular mediators of hormone action. In the extracellular compartment, calcium is vital for blood clotting and maintenance of normal membrane function. Calcium is the basic mineral of bones and teeth and thus plays a structural as well as a regulatory role. Not surprisingly, its concentration in extracellular fluid must be maintained within narrow limits. Deviations in either direction are not readily tolerated and, if severe, may be life-threatening.

Electrical excitability of cell membranes increases when the extracellular concentration of calcium is low, and the threshold for triggering action potentials may be lowered almost to the resting potential, which results in spontaneous, asynchronous, and involuntary skeletal muscle contractions called *tetany*. A typical attack of tetany involves muscular spasms in the face and characteristic contortions of the arms and hands. Laryngeal spasm and contraction of respiratory muscles may compromise breathing. Pronounced *hypocalcemia* (low blood calcium) may produce more generalized muscular contractions and convulsions.

Increased concentrations of calcium in blood (*hypercalcemia*) may cause calcium salts to precipitate out of solution because of their low solubility at physiological pH. "Stones" form, especially in the kidney, where they may produce severe painful damage (renal colic), which may lead to renal failure and hypertension.

DISTRIBUTION OF CALCIUM IN THE BODY

The adult human body contains approximately 1000 g of calcium, about 99% of which is sequestered in bone, primarily in the form of hydroxyapatite crystals $[Ca_{10}(PO_4)_6(OH)_2]$. In addition to providing structural support, bone serves as an enormous reservoir for calcium salts. Each day about 600 mg of calcium is exchanged between bone mineral and the extracellular fluid. Much of this exchange reflects resorption and reformation of bone as the skeleton undergoes constant remodeling, but some also occurs by exchange with a labile calcium pool in bone.

Most of the calcium that is not in bone crystals is found in cells of soft tissues bound to proteins within the sarcoplasmic reticulum, mitochondria, and other organelles. Energy-dependent transport of calcium by these organelles and the cell membrane maintains the resting concentration of free calcium in cytosol at low levels of about 0.1 μM. Cytosolic calcium can increase 10-fold or more, however, with just a brief change in membrane permeability or affinity of intracellular binding proteins. The rapidity and magnitude of changes in cytosolic calcium are consistent with its role as a biological signal.

The concentration of calcium in interstitial fluid is about 1.5 mM. Interstitial calcium, consists mainly of free, ionized calcium, but about 10% is complexed with anions such as citrate, lactate, or phosphate. Ionized and complexed calcium passes freely through capillary membranes and equilibrates with calcium in blood plasma. The total calcium concentration in blood is nearly twice that of interstitial fluid because calcium is avidly bound by albumin and other proteins. Total calcium in blood plasma is normally about 10 mg/dl (5 mEq/liter or 2.5 mM), but only the ionized component appears to be monitored and regulated. Because so large a fraction of blood calcium is protein bound, diseases that produce substantial changes in albumin concentrations may produce striking abnormalities in total plasma calcium content, even though the concentration of ionized calcium may be normal.

CALCIUM BALANCE

Normally, adults are in calcium balance; that is, on average, daily intake equals daily loss in urine and feces. Except for lactation and pregnancy, deviations from balance reflect changes in the metabolism of bone. Immobilization of a limb, bed rest, weightlessness, and malignant disease are examples of circumstances that produce negative calcium balance, whereas growth of the skeleton produces positive calcium balance. Dietary intake of calcium in the United States typically varies between 500 and 1500 mg per day, primarily in the form of dairy products. For example, an 8-ounce glass of milk contains about 290 mg of calcium. Calcium absorbed from the gut exchanges with the various body pools and ultimately is lost in the urine so that there is no net gain or loss of calcium in the extracellular pool in young adults. These relations are illustrated in Figure 1. It is noteworthy that the entire extracellular calcium pool turns over many times in the course of a day. Hence even small changes in any of these calcium fluxes can have profound effects.

Intestinal Absorption

Calcium is taken up along the entire length of the small intestine, but uptake is greatest in the ileum and jejunum. Secretions of the gastrointestinal tract are rich in calcium and add to the minimum load that must be absorbed to maintain balance. Net uptake is usually in the range of 100–200 mg per day. Absorption of calcium requires metabolic energy and the activity of specific carrier molecules in the luminal membrane (brush border) of intestinal cells. Although detailed understanding is not yet at hand, it appears that carrier-mediated transport across the brush border determines the overall rate. Calcium is carried down its concentration gradient into the cytosol of intestinal epithelial cells and is extruded from the basolateral surfaces in exchange for sodium, which must then be pumped out at metabolic expense. Overall transfer of calcium from the intestinal lumen to interstitial fluid proceeds against a concentration gradient and is largely dependent on $1,25(OH)_2D_3$ (see below). Although some calcium is taken up passively, simple

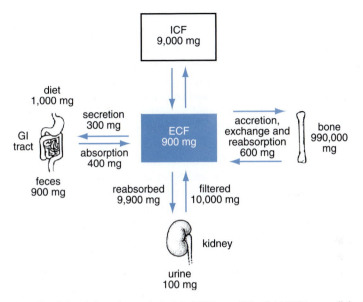

Figure 1 Daily calcium balance in a typical adult. ICF, Intracellular fluid; ECF, extracellular fluid.

diffusion is not adequate to meet body needs even when the concentration of calcium in the intestinal lumen is high.

Bone

Understanding regulation of calcium balance requires at least a rudimentary understanding of the physiology of bone. Metabolic activity in bone must satisfy two needs. The skeleton must attend to its own structural integrity through continuous remodeling and renewal, and it must respond to systemic needs for adequate amounts of calcium in the extracellular fluid. By and large, maintenance of adequate concentrations of calcium in blood takes precedence over maintenance of structural integrity of bone. However, the student must recognize that these two homeostatic functions, though driven by different forces, are not completely independent. Diseases of bone that disrupt skeletal homeostasis may have consequences for overall calcium balance, and, conversely, inadequacies of calcium balance lead to inadequate mineralization of bone.

The *extracellular matrix* is the predominant component of bone. One-third of the bony matrix is organic, and two-thirds is composed of highly ordered mineral crystals. The organic component, called *osteoid*, is composed primarily of collagen and provides the framework on which bone mineral is deposited. Collagen molecules in osteoid aggregate and cross-link to form fibrils of precise structure. Spaces between the ends of collagen molecules within fibrils provide initiation sites for crystal formation. Most calcium phosphate crystals are found within collagen fibrils and have their long axes oriented in parallel with the fibrils.

The most prevalent form of bone is *cortical* (*compact*) bone, it is found in the shafts of long bones and on the surfaces of the pelvis, skull, and other flat bones. The basic unit of cortical bone is called an *osteon* and consists of concentric layers, or lamellae, of bone arranged around a central channel (*haversian canal*), which contains the capillary blood supply. Osteons are usually 200–300 mm in diameter and several hundred millimeters long. They are arranged with their long axes oriented in parallel with the shaft of bone. Other canals, which run roughly perpendicularly, penetrate the osteons and form an anastomosing array of channels through which blood vessels in haversian canals connect with vessels in the *periosteum*. Tightly packed osteons are surrounded on both inner and outer aspects by several lamellae that extend circumferentially around the shaft. The entire bone is surrounded on its outer surface by the periosteum and is separated from the marrow by the *endosteum*.

Cancellous (*trabecular*) bone is found at the ends of the long bones, in the vertebrae, and in the internal portions of the pelvis, skull, and other flat bones. It is also called spongy bone, a term that well describes its appearance in section (Figure 2). Although only about 20% of the skeleton is composed of trabecular bone, its sponge like organization provides at least five times as much surface area for metabolic exchange, compared to compact bone. The trabeculae of spongy bone are not penetrated by blood vessels, but the spaces between them are filled with blood sinusoids or highly vascular marrow. The trabeculae are completely surrounded by endosteum.

Distributed throughout the lamellae of both forms of bone are tiny chambers, or lacunae, each of which houses an *osteocyte*. The lacunae are interconnected by an extensive network of canaliculi, which extend to the endosteal and periosteal surfaces. Osteocytes receive nourishment and biological signals by way of cytoplasmic processes that extend through the canaliculi to form gap junctions with each other and with cells of the endosteum or periosteum (Figure 3). The space in the lacunae and canaliculi that lies between the osteocytes and the bone matrix is filled with fluid. The surface area of bone matrix that is in contact with this pool of bone extracellular fluid has been estimated to be between 1000 and 5000 m^2 and therefore represents a major site for mineral exchange between soluble and crystallized calcium phosphate.

It is important to recognize that the mineralized matrix in both forms of bone and the bone extracellular fluid are separated from the extracellular compartment of the rest of the body by a continuous layer of cells, sometimes called the *bone membrane*. This layer of cells is composed of the endosteum, periosteum, osteocytes, and cells that line the haversian canals. Crystallization or solubilization of bone mineral is determined by physicochemical equilibria related to the concentrations of calcium, phosphate, hydrogen, and other constituents in bone

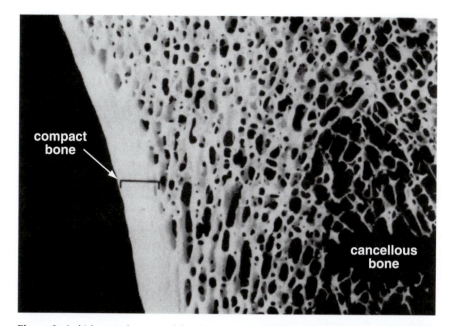

Figure 2 A thick ground section of the tibia, illustrating cortical compact bone and the lattice of trabeculae of cancellous bone. (From Fawcett, D.W., "A Textbook of Histology," 11th Ed., p. 201. Saunders, Philadelphia, 1986, with permission.)

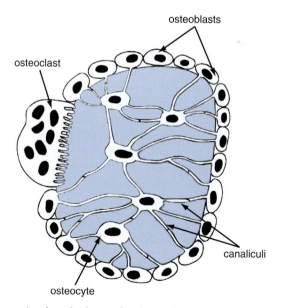

Figure 3 Cross section through a bony trabecula. The shaded area indicates mineralized matrix.

water. The fluxes of calcium and phosphate into or out of the bone extracellular fluid involve active participation of the cells of the bone membrane. Hormones regulate the calcium concentration in the extracellular fluid compartment and the mineralization of bone by regulating the activities of these cells.

Osteoblasts are the cells responsible for formation of bone. They arise from progenitors in connective tissue and marrow stroma and form a continuous sheet on the surface of newly forming bone. When actively laying down bone, osteoblasts are cuboidal or low columnar in shape. They have a dense rough endoplasmic reticulum, consistent with synthesis and secretion of collagen and other proteins of bone matrix. Osteoblasts probably also promote mineralization, but their role in this regard and details of the mineralization process are somewhat controversial. Under physiological conditions, calcium and phosphate are in metastable solution. That is, their concentrations in extracellular fluid would be sufficiently high for them to precipitate out of solution were it not for other constituents, particularly pyrophosphate, which stabilize the solution. During mineralization osteoblasts secrete alkaline phosphatase, which cleaves pyrophosphate and thus removes a stabilizing influence and at the same time increases local concentrations of phosphate, which promotes crystallization. In addition, during bone growth and perhaps during remodeling of mature bone, osteoblasts secrete calcium-rich vesicles into the calcifying osteoid.

During growth or remodeling of bone, some osteoblasts become entrapped in matrix and differentiate into osteocytes. Osteocytes are the most abundant cells in bone and are about 10 times more abundant than osteoblasts in human bone. On completion of growth or remodeling, surface osteoblasts dedifferentiate to become the flattened, spindle-shaped cells of the endosteum that lines most of the surface of bone. These cells may be reactivated in response to stimuli for bone formation. Thus osteoblasts, osteocytes, and quiescent lining cells represent three stages of the same cellular lineage and together comprise most, or perhaps all, of the bone membrane. In subsequent discussion these cells, along with their stromal and periosteal precursors, are referred to as *osteoblastic* cells.

Osteoclasts are responsible for bone resorption. They are large cells that arise by fusion of mononucleated hematopoietic cells; they may have as many as 20–40 nuclei. Precursors of osteoclasts originate in bone marrow and migrate through the circulation from thymus and other reticuloendothelial tissues to sites of bone destined for resorption. Differentiation and activation of osteoclasts require direct physical contact with osteoblastic cells that govern these processes by producing at least two indispensable cytokines. Osteoclasts arise from the cellular lineage that also gives rise to macrophages, which express receptors for macrophage colony-stimulating factor (M-CSF) on their surface membranes. Osteoblastic cells secrete M-CSF. Osteoclasts and their precursors also express receptor activators of NF-κB (RANK) on their surfaces. NF-κB is a transcription factor that translocates from the cytosol to the nucleus on activation (Chapter 4). These receptors belong to the tumor necrosis factor α (TNFα) family of cytokine receptors. Osteoblastic cells also express a membrane-bound cytokine called RANK ligand (RANKL). This cytokine is a member of the TNFα family of cytokines. It binds to and activates RANK on the surface of osteoclasts or their precursors that come in contact with osteoblastic cells. Members of this cytokine family are transmembrane proteins, some of which are sometimes cleaved at the cell surface by an exopeptidase to release soluble forms. When produced under experimental conditions the soluble form of RANKL can cause osteoclast differentiation and activation, but under physiological circumstances it appears that RANKL remains membrane bound. Steps in the differentiation of osteoclasts are illustrated in Figure 4. Another member of the TNF receptor family exerts negative control over osteoclast formation and activity. This compound, called osteoprotegerin (OPG), binds RANKL with high affinity and blocks its activation of RANK. Unlike typical members of the TNFα receptor family, OPG lacks a transmembrane domain, and is secreted as a soluble protein. Osteoprotegerin, which is produced by many different cells, competes with RANK for RANKL on the surface of osteoblastic cells and thereby limits osteoclast formation and activity. An unusual and confusing aspect of osteoclast regulation is the seemingly inverted participation of a membrane-bound ligand (RANKL) and a soluble receptor (OPG). Although the factors that regulate expression of RANKL and M-CSF are known (see below), the factors that regulate expression of OPG have not been identified.

In histological sections, osteoclasts are usually found on the bone surface in pits created by erosive osteoclast action. Integrins on the osteoclast surface form tight bonds with osteocalcin, osteopontin, and other proteins of the bone matrix and create a sealed-off region of extracellular space between the osteoclasts and the surface of the bony matrix. The specialized part of the osteoclast that faces the bony surface bone is thrown into many folds, called the ruffled border. The ruffled

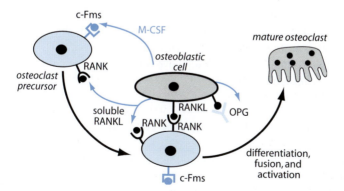

Figure 4 Differentiation and activation of osteoclasts. c-FMS, Receptor for macrophage colony-stimulating factor; M-CSF, macrophage colony-stimulating factor; RANK, receptor activators of NF-κB; RANKL, RANK ligand; OPG, osteoprotegerin. (Modified from Khosla, S., *Endocrinology* **124**, 5050–5055, 2001, by permission of The Endocrine Society.)

border sweeps over the surface of bone, continuously changing its configuration as it releases acids and hydrolytic enzymes that dissolve the bone mineral crystals and the protein matrix . Small bone crystals are often seen in deep folds of phagocytic vesicles. On completion of resorption, osteoclasts are inactivated and lose some of their nuclei. Complete inactivation involves fission of the giant, polynucleated cells back to mononuclear cells, which may undergo apoptosis, but some multinuclear cells remain quiescent on the bone surface interspersed among the lining cells.

Resorption of bone is precisely coupled with bone formation. The pattern of events in bone remodeling typically begins with differentiation and activation of osteoclasts followed sequentially by bone resorption, osteoblast activation and migration to the site of bone resorption, and finally bone formation. Details of the signaling mechanisms that couple bone formation with bone resorption are not yet known, but it appears that osteoblasts secrete a variety of autocrine and paracrine growth factors that are trapped and stored in the bone matrix during osteogenesis. Resorption of the matrix by osteoclastic activity appears to release these factors, which in turn may activate quiescent osteoblasts. In addition, these growth factors promote differentiation of new osteoblasts from progenitor cells that are attracted to the site by peptide fragments of partially degraded osteoid.

Kidney

Ionized and complexed calcium can pass freely through glomerular membranes. Normally 98–99% of the 10,000 mg of calcium filtered by the glomeruli each day is reabsorbed by the renal tubules. About two-thirds of the reabsorption occurs in the proximal tubule, tightly coupled to sodium reabsorption and, for the most part, dragged passively along with water. Much of the remaining calcium is resorbed in the loop of Henle and is also tightly coupled to sodium reabsorption. Normally only about 10% of the filtered calcium reaches the distal nephron. Reabsorption of calcium in the vicinity of the junction of the distal convoluted tubules and the collecting ducts is governed by an active, saturable process that is independent of sodium reabsorption. Active transport of calcium in this region is hormonally regulated (see below).

PHOSPHORUS BALANCE

Because of their intimate relationship, the fate of calcium cannot be discussed without also considering phosphorus. Calcium is usually absorbed in the intestines, accompanied by phosphorus, and deposition and mobilization of calcium in bones always occur in conjunction with phosphorus. Phosphorus is as ubiquitous in its distribution and physiological role as is calcium. The high-energy phosphate bond of ATP and other metabolites is the coinage of biological energetics. Phosphorus is indispensable for biological information transfer. It is a component of nucleic acids and second messengers such as cyclic AMP and IP$_3$, and is the addend that increases or decreases enzymatic activities or guides protein:protein interactions.

About 90% of the 500–800 g of phosphorus in the adult human is deposited in the skeleton. Much of the remainder is incorporated into organic phosphates distributed throughout soft tissues in the form of phospholipids, nucleic acids, and soluble metabolites. Daily intake of phosphorus is in the range of 1000 to 1500 mg, mainly in dairy products. Organic phosphorus is digested to inorganic phosphate before it is absorbed in the small intestine by both active and passive processes. Net absorption is linearly related to intake and appears not to saturate. The concentration of inorganic phosphate in blood is about 3.5 mg/dl. About 55% is present as free ions, about 35% is complexed with calcium or other cations, and 10% is protein bound. Phosphate concentrations are not tightly controlled and may vary widely under such influences as diet, age, and sex. Ionized and complexed phosphate can pass freely across glomerular and other capillary membranes. Phosphate in the glomerular filtrate is actively reabsorbed by a sodium-coupled cotransport process in the proximal tubule. These relations in daily phosphorus balance are shown in Figure 5.

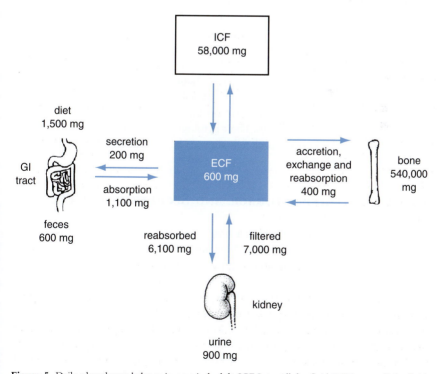

Figure 5 Daily phosphorus balance in a typical adult. ICF, Intracellular fluid; ECF, extracellular fluid.

PARATHYROID GLANDS AND PARATHYROID HORMONE

MORPHOLOGY

The parathyroid glands arose relatively recently in vertebrate evolution, coincident with the emergence of ancestral forms onto dry land. They are not found in fish and are seen in amphibians such as the salamander only after metamorphosis to the land-dwelling form. The importance of the parathyroids in normal calcium economy was established during the latter part of the nineteenth century when it was found that parathyroidectomy resulted in lethal tetany. Diseases resulting from overproduction or underproduction of parathyroid hormone (PTH) are relatively uncommon.

Human beings typically have four parathyroid glands, but as few as two and as many as eight have been observed. Each gland is a flattened ellipsoid measuring about 6 mm in its longest diameter. The aggregate mass of the adult parathyroid glands is about 120 mg in men and about 140 mg in women. These glands adhere to the posterior surface of the thyroid gland or occasionally are embedded within thyroid tissue. They are well vascularized and derive their blood supply mainly from the inferior thyroid arteries. Parathyroid glands are composed of two cell types (Figure 6). *Chief cells* predominate and are arranged in clusters or cords. They are the source of PTH and have all of the cytological characteristics of cells that produce protein hormones: rough endoplasmic reticulum, prominent Golgi apparatus, and some membrane-bound storage granules. *Oxyphil cells*, which appear singly or in small groups, are larger than chief cells and contain a remarkable number of mitochondria. Oxyphil cells have no known function and are thought by some to be degenerated chief cells. Their cytological properties are not characteristic of secretory cells. Few oxyphil cells are seen before puberty, but their number increases thereafter with age.

BIOSYNTHESIS, STORAGE, AND SECRETION OF PTH

The secreted form of PTH is a simple straight-chain peptide of 84 amino acids. There are no disulfide bridges. As many as 50 amino acids can be removed from the carboxyl terminus without compromising biological potency, but removal of just the serine at the amino terminus virtually inactivates the hormone. All of the known biological effects of PTH can be reproduced with a peptide corresponding to amino acids 1–34. PTH is expressed as a larger "preprohormone" and is the product of a single-copy gene located on chromosome 11. Sequential cleavage forms first, a 90-amino-acid prohormone, and then the mature hormone. The larger, transient forms have little or no biological activity and are not released

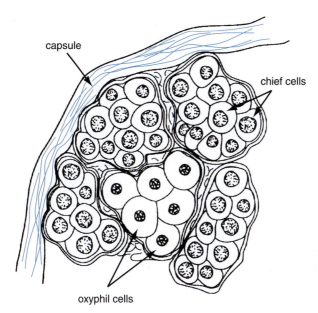

Figure 6   Section through a human parathyroid gland showing small chief cells and larger oxyphil cells. The cells are arranged in cords surrounded by loose connective tissue. (From M. Borysenko, and T. Beringer, "Functional Histology," 2nd Ed., p. 316. Little, Brown, Boston, 1984, by permission of Lippincott Williams & Wilkins.)

into the circulation even at times of intense secretory activity. Synthesis of PTH is regulated at both transcriptional and posttranscriptional sites. Cyclic AMP, acting through the cyclic AMP response element binding protein, up-regulates PTH gene transcription, whereas vitamin D down-regulates transcription. Low plasma calcium concentrations prolong the survival time of the mRNA transcript, and low concentrations of plasma phosphate accelerate message degradation. How extracellular ion concentrations produce these effects is not understood. PTH is synthesized continuously, but the glands store little hormone; only enough to sustain maximal secretion rates for about 90 minutes.

Parathyroid cells are unusual in the respect that hormone degradation and synthesis are adjusted according to physiological demand for secretion. As much as 90% of the hormone synthesized may be destroyed within the chief cells, which degrade PTH at an accelerated rate when plasma calcium concentrations are high. Proteolytic enzymes incorporated into secretory granules cleave the PTH molecules so that fragments representing middle and C-terminal portions are co-secreted with the intact 84-amino-acid PTH. Apparently no amino-terminal fragments are released into the circulation. Similar fragments are produced by degradation of PTH in liver and kidney and also enter the bloodstream.

PTH fragments are cleared from blood by filtration at the glomeruli, but remain in the blood hours longer than intact hormone, which has a half-life of only 2–4 minutes. The intact 84-amino-acid peptide is the only biologically active form of PTH in the blood. Standard radioimmunoassays overestimate active PTH concentrations in blood because most antisera cannot distinguish fragments from intact hormone. Use of two-site immunometric, or "sandwich," assays (see Chapter 1) has overcome this difficulty. These assays use two antibodies that recognize epitopes in carboxyl- and amino-terminal segments and therefore detect only the intact hormone present in the plasma sample.

A substance closely related to PTH, parathyroid hormone-related peptide (PTHrP), is found in the plasma of patients suffering from certain malignancies and accounts for the accompanying hypercalcemia. The gene for PTHrP was isolated from some tumors and was found to encode a peptide whose first 13 amino acid residues are remarkably similar to the first 13 amino acids of PTH; thereafter the structures of the two molecules diverge. The similarities of the N-terminal primary sequence and presumably the secondary structure of subsequent segments allow PTHrP to bind with high affinity to the PTH receptor, therefore producing the same biological effects as PTH. PTH and PTHrP are immunologically distinct and do not cross-react in immunoassays. PTHrP is synthesized in a wide range of tissues and acts locally as a paracrine factor to regulate a variety of processes that are unrelated to regulation of calcium concentrations in extracellular fluid. Little or no PTHrP is found in blood plasma of normal individuals.

MECHANISMS OF PARATHYROID HORMONE ACTIONS

Binding of PTH to G-protein-coupled receptors on the surfaces of target cells increases the formation of cyclic AMP and of IP_3 and diacylglycerol (see Chapter 1). The PTH receptor is coupled to adenylyl cyclase through a stimulatory G protein (G_s) and to phospholipase C through G_q. Consequently, protein kinases A and C are also activated and intracellular calcium is increased. Rapid responses result from protein phosphorylation whereas delayed responses result from altered expression of genes regulated by cyclic AMP response element binding protein. It is likely that the two second messenger pathways activated by PTH are redundant and reinforce each other. The importance of cyclic AMP for the action of PTH is underscored by the occurrence of a rare disease called *pseudohypoparathyroidism*, in which patients are unresponsive to PTH. About one-half of the reported cases of unresponsiveness to PTH are attributable to a genetic defect in the GTP-binding protein (G_s) that couples the hormone receptor with adenylyl cyclase. These patients also have decreased responses to some other cyclic AMP-dependent signals, but because there are four distinct genes for $G\alpha_s$, each expressed with particular receptors, not all cyclic AMP-dependent responses are affected.

PHYSIOLOGICAL ACTIONS OF PTH

Parathyroid hormone is the principal regulator of the extracellular calcium pool. It increases the calcium concentration and decreases the phosphate concentration in blood by various direct and indirect actions on bone, kidney, and intestine. In its absence, the concentration of calcium in blood, and hence interstitial fluid, decreases dramatically over a period of several hours while the concentration of phosphate increases. Hypoparathyroidism may result from insufficient production of active hormone or defects in the responses of target tissues; acutely, all the symptoms of hypocalcemia are seen, including tetany and convulsions. Chronically, neurological, ocular, and cardiac deficiencies may also be seen. Hyperparathyroidism results in kidney stones and excessive demineralisation, leading to weakening of bone.

Actions on Bone

Increases in PTH concentration in blood result in mobilization of calcium phosphate from the bone matrix, due primarily, and perhaps exclusively, to increased osteoclastic activity. The initial phase is seen within 1–2 hours and results from activation of preformed osteoclasts already present on the bone surface. A later and more pronounced phase becomes evident after about 12 hours and is characterized by widespread resorption of both mineral and organic components of bone matrix, particularly in trabecular bone. Evidence of osteoclastic activity is reflected not only by calcium phosphate mobilization, but also by increased urinary excretion of hydroxyproline and other products of collagen breakdown. Although activity of all bone cell types is affected by PTH, it appears that only cells of osteoblastic lineage express receptors for PTH. Osteoclasts are thus not direct targets for PTH. Activation, differentiation, and recruitment of osteoclasts in response to PTH result from increased expression of at least two cytokines by cells of osteoblastic lineage. PTH induces osteoblastic cells to synthesize and secrete M-CSF and to express RANKL on their surfaces (see above, section on osteoclasts). These cytokines increase the differentiation and activity of osteoclasts and protect them from apoptosis. In addition, PTH may decrease osteoblastic expression of OPG, the competitive inhibitor of RANK for binding RANKL. PTH also induces retraction of osteoblastic lining cells and thus exposes bony surfaces to which osteoclasts can bind.

Synthesis and secretion of collagen and other matrix proteins by osteoblasts are inhibited in the early phases of PTH action, but are reactivated subsequently as a result of biological coupling (discussed above), and new bone is laid down. At this stage PTH stimulates osteoblasts to synthesize and secrete growth factors, including insulin-like growth factor I (IGF-I) and transforming growth factor β (TGFβ), which are sequestered in the bone matrix and also act in an autocrine or paracrine manner to stimulate osteoblast progenitor cells to divide and differentiate. At least part of the stimulus for osteoblastic activity that follows osteoclastic activity may come from liberation of growth factors that were sequestered in the bony matrix when the bone was laid down. With prolonged continuous exposure to high concentrations of PTH, as seen with hyperparathyroidism, osteoclastic activity is greater than osteoblastic activity, and bone resorption predominates. However, a therapeutic regimen of intermittent stimulation with PTH leads to net formation of bone.

In addition to its critical role in maintaining blood calcium concentrations, PTH is also important for skeletal homeostasis. As already mentioned, bone remodeling continues throughout life. Remodeling of bone not only ensures renewal and maintenance of strength, but also adjusts bone structure and strength to accommodate the various stresses and strains of changing demands of daily living. Increased stress leads to bone formation, strengthening the affected area, whereas weightlessness or limb immobilization leads to mineral loss. Osteocytes entrapped in the bony matrix are thought to function as mechanosensors that signal remodeling through release of prostaglandins, nitric oxide, and growth factors. In animal studies these actions are facilitated and enhanced by PTH.

Actions on Kidney

In the kidney, PTH produces three distinct effects, each of which contributes to the maintenance of calcium homeostasis. In the distal nephron PTH promotes the reabsorption of calcium, and in the proximal tubule it inhibits reabsorption of phosphate and promotes hydroxylation, and hence activation of vitamin D (see below). In producing these effects, PTH binds to G-protein-coupled receptors in both the proximal and distal tubules and stimulates the production of cyclic AMP, diacylglycerol, and IP$_3$. Some cyclic AMP escapes from renal tubular cells and appears in the urine. About one-half of the cyclic AMP found in urine arises in the kidney and is attributable to the actions of PTH.

Calcium Reabsorption

The kidney reacts quickly to changes in PTH concentrations in blood and is responsible for minute-to-minute adjustments in blood calcium. PTH acts directly on the distal portion of the nephron to decrease urinary excretion of calcium well before significant amounts of calcium can be mobilized from bone. About 90% of the filtered calcium is reabsorbed in the proximal tubule and the loop of Henle independently of PTH. Therefore, because its actions are limited to the distal reabsorptive mechanism, PTH can provide only fine-tuning of calcium excretion. Even small changes in the fraction of calcium reabsorbed from the glomerular filtrate, however, can be of great significance. Hypoparathyroid patients whose blood calcium is maintained in the normal range excrete about three times as much urinary calcium as normal subjects.

The cellular mechanisms that account for increased calcium reabsorption in response to PTH are shown in Figure 7. Activation of G-protein–coupled receptors on the basolateral surface of cells in the distal convoluted tubule causes intracellular vesicles that harbor calcium channels to migrate to the luminal surface and fuse with the luminal membrane. Calcium ions in tubular fluid flow passively down their concentration gradient into the cells. In the basolateral membrane PTH activates sodium/calcium exchangers, which exchange three ions of extracellular sodium for one ion of intracellular calcium, a calcium ATPase that pumps calcium across the membrane, and the sodium-potassium ATPase that maintains the electrochemical gradient across the membrane.

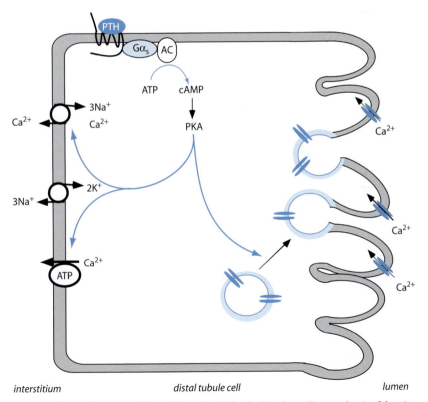

interstitium   *distal tubule cell*   *lumen*

Figure 7 Effects of PTH on calcium reabsorption in the distal nephron. Gα_s, α subunit of the stimulatory G-protein; AC, adenylyl cyclase; cAMP, cyclic adenosine monophosphate; PKA, protein kinase A.

Even when maximally stimulated, this PTH-sensitive mechanism has a low capacity that saturates when the plasma concentration, and hence the amount of calcium that reaches the distal tubule, is high. This circumstance accounts for the paradoxical increase in urinary calcium seen in later phases of PTH action. Regardless of the absolute amount excreted, however, PTH decreases the fraction of filtered calcium that escapes in the urine.

Phosphate Excretion

Parathyroid hormone powerfully inhibits tubular reabsorption of phosphate and thus increases the amount excreted in urine. This effect is seen within minutes after injection of PTH and is exerted in the proximal tubules, where the bulk of phosphate reabsorption occurs. Decreased reabsorption of phosphate results from decreased capacity for sodium–phosphate cotransport across the luminal membrane of tubular cells. In a manner analogous to its effects on calcium reabsorption, PTH decreases the abundance of sodium–phosphate cotransporters in the brush border of proximal tubule cells by stimulating their translocation to intracellular vesicles. This effect also depends on increased production of cyclic AMP (Figure 8).

Effects on Intestinal Absorption

Calcium balance ultimately depends on intestinal absorption of dietary calcium. Calcium absorption is severely reduced in hypoparathyroid patients and dramatically increased in those with hyperparathyroidism. Within a day or two after treatment of hypoparathyroid subjects with PTH, calcium absorption increases. Intestinal uptake of calcium is stimulated by an active metabolite of vitamin D. PTH stimulates the renal enzyme that converts vitamin D to its active form (see below), but has no direct effects on intestinal transport of either calcium or phosphate.

REGULATION OF PTH SECRETION

Chief cells of the parathyroid glands are exquisitely sensitive to changes in extracellular calcium and rapidly adjust their rates of PTH secretion in a manner that is inversely related to the concentration of ionized calcium (Figure 9). The resulting increases or decreases in blood levels of PTH produce either positive or negative changes in the plasma calcium concentration and thereby provide negative feedback signals for regulation of PTH secretion. The activated form of vitamin D, the synthesis of which depends on PTH, is also a negative feedback inhibitor of PTH synthesis (see below). Although blood levels of phosphate are also affected by PTH, high phosphate concentration appears to exert little or no effect on the secretion of PTH, but may exert some direct and indirect effects on

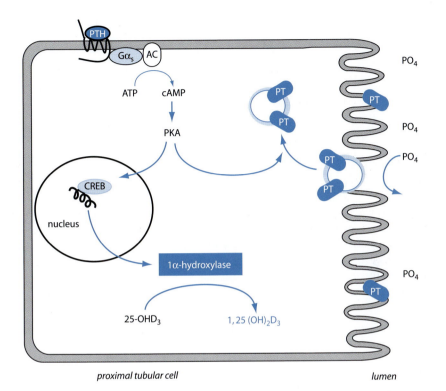

Figure 8 Effects of PTH on proximal tubule cells. PT, Sodium phosphate cotransporter; 25-OHD$_3$ and 1,25(OH)$_2$D$_3$ are metabolic forms of vitamin D (see Figure 13); Gα_s, α subunit of the stimulatory G-protein; AC, adenylyl cyclase; cAMP, cyclic adenosine monophosphate; PKA, protein kinase A; CREB, cyclic AMP response element binding protein.

hormone synthesis. Under experimental conditions the concentration of magnesium in plasma may also influence PTH secretion, but the concentration range in which magnesium inhibits secretion is well beyond that seen physiologically. A decrease in ionized calcium in blood appears to be the only physiologically relevant signal for PTH secretion.

Chief cells are programmed to secrete PTH unless inhibited by extracellular calcium, but secretion is not totally suppressed even when plasma concentrations of calcium are very high. Through mechanisms that are not understood, normal individuals secrete PTH throughout the day in bursts of 1 to 3 pulses per hour. Blood levels of PTH also follow a diurnal pattern, with peak values seen shortly after midnight and minimal values seen in late morning. Diurnal fluctuations appear to arise from endogenous events in the chief cells, and may promote anabolic responses of bone to PTH.

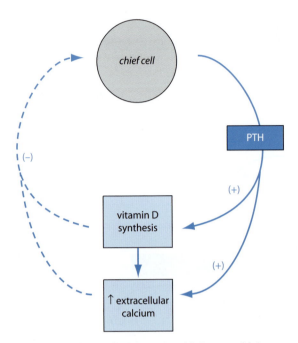

Figure 9 Regulation of PTH secretion. (−), Decrease; (+), increase.

Cellular Mechanisms

The cellular mechanisms by which extracellular calcium regulates PTH secretion are poorly understood. Chief cells are equipped with calcium-sensing receptors in their plasma membranes and can adjust secretion in response to as little as a 2–3% change in extracellular calcium concentration. Calcium-sensing receptors are members of the G-protein-coupled receptor superfamily (Chapter 1) and bind calcium in proportion to its concentration in extracellular fluid. Because they appear to be coupled to adenylyl cyclase through G_i, and to phospholipase C, and perhaps to membrane calcium channels, probably through G_q, several second messengers appear to be involved in governing PTH secretion. Increased extracellular calcium stimulates production of DAG and IP_3 and decreases production of cyclic AMP. Cytosolic calcium rises as a result of IP_3-mediated release from intracellular stores followed by influx through activated membrane channels. Paradoxically, in chief cells, unlike most other secretory cells, increased intracellular calcium is associated with inhibition rather than stimulation of hormone secretion. It is not yet understood how these changes in intracellular messages combine to inhibit PTH secretion. These events are summarized in Figure 10.

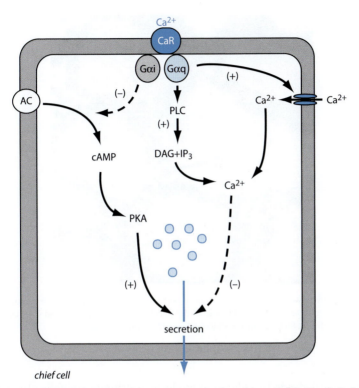

Figure 10 Regulation of parathyroid hormone secretion by calcium (Ca^{2+}). Heptihelical calcium receptors (CaR) on the surface of chief cells communicate with Ca^{2+} channels, adenylyl cyclase (AC), and phospholipase C (PLC) by way of guanosine nucleotide-binding proteins ($G\alpha_i$ and $G\alpha_q$). The resulting increase in Ca^{2+} and inhibition of adenylyl cyclase lowers cyclic AMP (cAMP) and interferes with protein kinase A (PKA)-mediated events that lead to secretion. DAG, Diacylglycerol; IP_3, inositol trisphosphate.

CALCITONIN

CELLS OF ORIGIN

Calcitonin is sometimes also called thyrocalcitonin to describe its origin in the parafollicular cells of the thyroid gland. These cells, which are also called C cells, occur singly or in clusters in or between thyroid follicles. They are larger and stain less densely than follicular cells in routine preparations (Figure 11), and like other peptide hormone-secreting cells, contain membrane-bound storage granules. Parafollicular cells arise embryologically from neuroectodermal cells that migrate to the last branchial pouch, and, in submammalian vertebrates, give rise to the

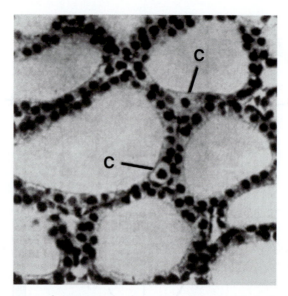

Figure 11   Low-power photomicrograph of a portion of the thyroid gland of a normal dog. Parafollicular (C) cells are located in the walls of the follicles. (From Ham, A.W., and Cormack, D. H., "Histology," 8th Ed., p. 802. Lippincott, Philadelphia, 1979, with permission.)

ultimobranchial glands. In addition to producing calcitonin, parafollicular cells can take up and decarboxylate amine precursors and thus have some similarity to pancreatic alpha cells and cells of the adrenal medulla. Parafollicular cells give rise to a unique neoplasm, medullary carcinoma of the thyroid, which may secrete large amounts of calcitonin.

BIOSYNTHESIS, SECRETION, AND METABOLISM

Calcitonin consists of 32 amino acids and has a molecular weight of about 3,400. Except for a seven-member disulfide ring at the amino terminus, calcitonin has no remarkable structural features. Immunoreactive circulating forms are heterogeneous in size, reflecting the presence of precursors, partially degraded hormone, and disulfide-linked dimers and polymers. The active hormone has a half-life in plasma of about 5–10 minutes and is cleared from the blood primarily by the kidney. The gene that encodes calcitonin also encodes a neuropeptide called calcitonin gene-related peptide (CGRP). This gene contains six exons, but only the first four are represented in the mRNA transcript that codes for the precursor of

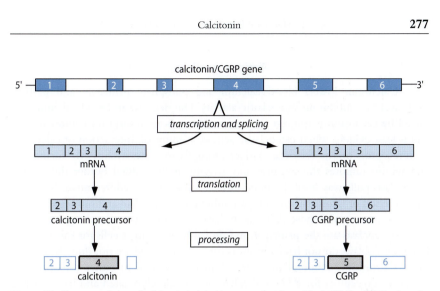

Figure 12 Alternate splicing of calcitonin/calcitonin gene-related peptide (CGRP) mRNA gives rise to either calcitonin or CGRP, with no shared sequence of amino acids.

calcitonin. In the mRNA that codes for CGRP the portion corresponding to exon 4 is deleted and replaced by exons 5 and 6. Because the first exon codes for a non-translated region, and the peptide corresponding to exons 2 and 3 is removed in posttranslational processing, the mature products have no common amino acid sequences (Figure 12).

PHYSIOLOGICAL ACTIONS OF CALCITONIN

Calcitonin escaped discovery for many years because no obvious derange-ment in calcium balance or other homeostatic function results from deficient or excessive production. Thyroidectomy does not produce a tendency toward hyper-calcemia, and thyroid tumors that secrete massive amounts of calcitonin do not cause hypocalcemia. Attention was drawn to the possibility of a calcium-lowering hormone by the experimental finding that direct injection of a concentrated solution of calcium into the thyroid artery caused a more rapid fall in blood calcium than did parathyroidectomy. Indeed, calcitonin promptly and dramatically lowers the blood calcium concentration in many experimental animals. Calcitonin is not a major factor in calcium homeostasis in humans, however, and does not participate in minute-to-minute regulation of blood calcium concentrations. Rather, the importance of calcitonin may be limited to protection against excessive bone resorption.

Actions on Bone

Calcitonin lowers blood calcium and phosphate primarily, and perhaps exclusively, by inhibiting osteoclastic activity. The decrease in blood calcium produced by calcitonin is greatest when osteoclastic bone resorption is most intense and is least evident when osteoclastic activity is minimal. Interaction of calcitonin with receptors on the osteoclast surface promptly increases cyclic AMP formation, and within minutes the expanse and activity of the ruffled border diminishes. Osteoclasts pull away from the bone surface and begin to dedifferentiate. Synthesis and secretion of lysosomal enzymes are inhibited. In less than an hour fewer osteoclasts are present, and those that remain have decreased bone-resorbing activity.

Osteoclasts are the principal, and probably only, target cells for calcitonin in bone. Osteoblasts do not have receptors for calcitonin and are not directly affected by it. Curiously , although they are uniquely expressed in either osteoblasts or osteoclasts, receptors for PTH and calcitonin are closely related and have about a third of their amino acid sequences in common, suggesting they evolved from a common ancestral heptihelical molecule. Because of the coupling phenomenon in the cycle of bone resorption and bone formation discussed earlier, inhibition of osteoclastic activity by calcitonin eventually decreases osteoblastic activity as well. All cell types appear quiescent in histological sections of bone that was chronically exposed to high concentrations of calcitonin. Although osteoclasts express very high numbers of receptors for calcitonin, they quickly become insensitive to the hormone because continued stimulation results in massive down-regulation of receptors.

Actions on Kidney

At high concentrations calcitonin may increase urinary excretion of calcium and phosphorus, probably by acting on the proximal tubules. In humans these effects are small, last only a short while, and are not physiologically important for lowering blood calcium. Renal handling of calcium is not disrupted in patients with thyroid tumors that secrete large amounts of calcitonin. Kidney cells "escape" from prolonged stimulation with calcitonin and become refractory to it, probably as a result of down-regulation of receptors.

REGULATION OF SECRETION

Circulating concentrations of calcitonin are low when blood calcium is in the normal range or below, but they increase proportionately when calcium rises above about 9 mg/dl (Figure 13). Parafollicular cells respond directly to ionized calcium in blood and express the same G-protein-coupled calcium-sensing

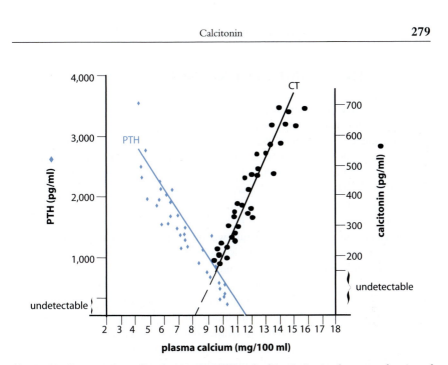

Figure 13 Concentrations of immunoreactive PTH and calcitonin in pig plasma as a function of
plasma calcium. (From Arnaud, C. D., Littledike, T., and Tsao, H. S., *In* "Proceedings of the Symposium
on Calcitonin and C Cells." Heinemann, London, 1969, with permission.)

receptor in their surface membranes as expressed by the parathyroid chief cells.
Both cell types respond to extracellular calcium over the same concentration range,
but their secretory responses are opposite, presumably because of differences in the
secretory apparatus. Although little information is available concerning intracellular
mechanisms in parafollicular cells, the relation between stimulation and secretion
is more typical of endocrine cells than the regulation of PTH secretion.

In addition to the direct stimulation by high concentrations of calcium,
calcitonin secretion may also increase after eating. Gastrin, a hormone produced by
gastric mucosal cells, stimulates parafollicular cells to secrete calcitonin. Other
gastrointestinal hormones that have the same four amino acids at their carboxyl
terminus, including cholecystokinin–pancreozymin, glucagon, and secretin, have
similar effects, but gastrin is the most potent among these agents. Secretion of calcitonin in anticipation of an influx of calcium from the intestine is a feed-forward
mechanism that may guard against excessive concentrations of plasma calcium after
calcium ingestion by decreasing osteoclastic activity. This phenomenon is analogous to the anticipatory secretion of insulin after a carbohydrate-rich meal (see
Chapter 5). Although the importance of this response in humans is not established,
sensitivity of parafollicular cells to gastrin has been exploited clinically as a
provocative test for diagnosing medullary carcinoma of the thyroid.

THE VITAMIN D–ENDOCRINE SYSTEM

A derivative of vitamin D_3, 1,25-dihydroxycholecalciferol [1,25(OH)$_2$D$_3$], is indispensable for maintaining adequate concentrations of calcium in the extracellular fluid and adequate mineralization of bone matrix. Vitamin D deficiency leads to inadequate calcification of bone matrix and severe softening of the skeleton, called *osteomalacia*, and may result in bone deformities and fractures. Osteomalacia in children is called *rickets* and may produce permanent deformities of the weight-bearing bones (bowed legs). Although vitamin D is now often called the vitamin D–endocrine system, when it was discovered as the factor in fish oil that prevents rickets, its hormone-like nature was not suspected. One important distinction between hormones and vitamins is that hormones are synthesized within the body from simple precursors, but vitamins must be provided in the diet. Actually, vitamin D_3 can be synthesized endogenously in humans, but the rate is limited by a nonenzymatic reaction that requires radiant energy in the form of light in the near-ultraviolet range—hence the name "sunshine vitamin." The immediate precursor for vitamin D_3, 7-dehydrocholesterol, is synthesized from acetyl coenzyme A and is stored in skin. Conversion of 7-dehydrocholesterol to vitamin D_3 in skin proceeds spontaneously in the presence of sunlight that penetrates the epidermis, reaching the outer layers of the dermis. Vitamin D deficiency became a significant public health problem as a by-product of industrialization. Urban living, smog, and increased indoor activity limit exposure of the populace to sunshine and hence limit endogenous production of vitamin D_3. This problem is readily addressed by adding vitamin D to foods, particularly milk.

1,25-Dihydroxy vitamin D_3 also fits the description of a hormone in that it travels through the blood in small amounts from its site of production to affect cells at distant sites. Another major difference between a vitamin and a hormone is that vitamins usually are cofactors in metabolic reactions, whereas hormones behave as regulators and interact with specific receptors. 1,25(OH)$_2$D$_3$ produces many of its biological effects in a manner characteristic of steroid hormones (see Chapter 1). It binds to a specific nuclear receptor that is a member of the steroid/thyroid hormone superfamily.

SYNTHESIS AND METABOLISM

The form of vitamin D produced in mammals is called cholecalciferol, or vitamin D_3; it differs from vitamin D_2 (ergosterol), which is produced in plants, only in the length of the side chain. Irradiation of the skin results in photolysis of the bond that links carbons 9 and 10 in 7-dehydrocholesterol, and thus opens the B ring of the steroid nucleus (Figure 14). The resultant cholecalciferol is biologically inert but, unlike its precursor, has a high affinity for a vitamin D-binding

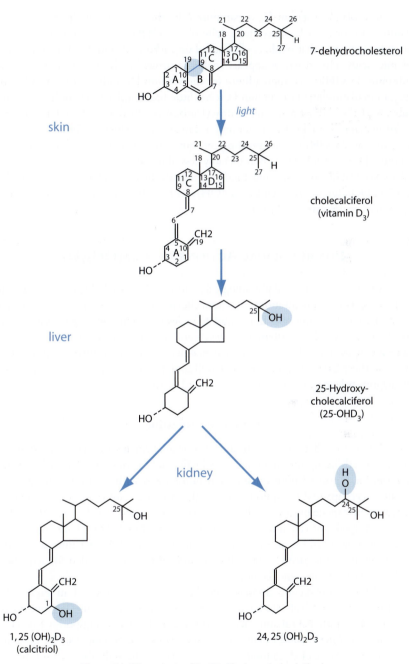

Figure 14 Biosynthesis of 1α,25–dihydroxycholecalciferol.

protein in plasma. Vitamin D_3 is transported by the blood to the liver, where it is oxidized to form 25-hydroxycholecalciferol (25-OHD$_3$) by the same P450 mitochondrial enzyme that oxidizes cholesterol on carbons 26 and 27 in the formation of bile acids. This reaction appears to be controlled only by the availability of substrate. 25-OHD$_3$ has high affinity for the vitamin D-binding protein and is the major circulating form of vitamin D. It has little biological activity. In the proximal tubules of the kidney, a second hydroxyl group is added at carbon 1 by another P450 enzyme to yield the compound 1,25(OH)$_2$D$_3$, which is about 1000 times as active as 25-OHD$_3$, and probably accounts for all the biological activity of vitamin D. 1,25(OH)$_2$D$_3$ is considerably less abundant in blood compared to its 25-hydroxylated precursor and binds less tightly than 25-OHD$_3$ to vitamin D-binding globulin. Consequently, 1,25(OH)$_2$D$_3$ has a half-life in blood of 15 hours, compared to 15 days for 25-OHD$_3$.

Physiological Actions of 1,25(OH)$_2$D$_3$

Overall, the principal physiological actions of 1,25(OH)$_2$D$_3$ increase calcium and phosphate concentrations in extracellular fluid. These effects are exerted primarily on intestine and bone, and to a lesser extent on kidney. Vitamin D receptors are widely distributed, however, and a variety of other actions that are not obviously related to calcium balance have been described or postulated. Because these latter effects are neither well understood nor germane to regulation of calcium balance they will not be discussed further.

Actions on Intestine

Uptake of dietary calcium and phosphate depends on active transport by epithelial cells lining the small intestine. Deficiency of vitamin D severely impairs intestinal transport of both calcium and phosphorus. Although calcium uptake is usually accompanied by phosphate uptake, the two ions are transported by independent mechanisms, both of which are stimulated by 1,25(OH)$_2$D$_3$. Increased uptake of calcium is seen about 2 hours after 1,25(OH)$_2$D$_3$ is given to deficient subjects and is maximal within 4 hours. A much longer time is required when vitamin D is given, presumably because of the time needed for sequential hydroxylations in liver and kidney.

Calcium uptake by duodenal epithelial cells is illustrated in Figure 15. Calcium enters passively down its electrochemical gradient through two novel channels, the epithelial calcium channel (ECaC) and calcium transporter 1 (CaT1). On entry into the cytosol calcium is bound virtually instantaneously by calcium-binding proteins called calbindins and is carried through the cytosol to the basolateral membrane, where it is extruded into the interstitium by calcium ATPase

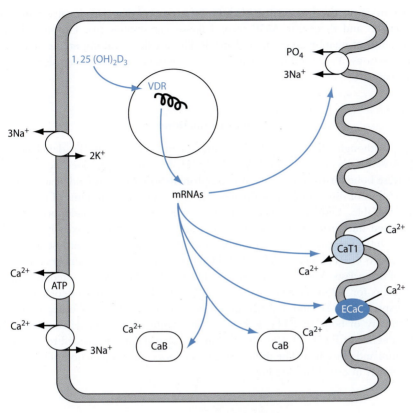

duodenal epithelial cell

Figure 15 Actions of $1,25(OH)_2D_3$ on intestinal transport of calcium. VDR, Vitamin D receptor; CaT1, calcium transporter 1; ECaC, epithelial calcium channel transporter; CaB, calbindin.

(calcium pump) and sodium–calcium antiporters. In addition to shuttling calcium across cells, binding proteins keep the cytosolic calcium concentration low and thus maintain a gradient favorable for calcium influx while affording protection from deleterious effects of high concentrations of free calcium. It appears that the abundance of ECaC and CaT1 in the luminal membrane and at least one of the calbindins in the cytosol depend on $1,25(OH)_2D_3$ through regulation of gene transcription. Similarly, $1,25(OH)_2D_3$ is thought to regulate expression of sodium phosphate transporters in the luminal membrane.

Some evidence obtained in experimental animals and in cultured cells suggests that $1,25,(OH)_2D_3$ may also produce some rapid actions that are not mediated by altered genomic expression. Among these are rapid transport of calcium across the intestinal epithelium by a process that may involve both the IP_3–DAG and the cyclic AMP second messenger systems (see Chapter 1) and activation of membrane calcium channels. The physiological importance of these rapid actions of $1,25,(OH)_2D_3$ and the nature of the receptor that signals them are not known.

Actions on Bone

Although the most obvious consequence of vitamin D deficiency is decreased mineralization of bone, $1,25(OH)_2D_3$ apparently does not directly increase bone formation or calcium phosphate deposition in osteoid. Rather, mineralization of osteoid occurs spontaneously when adequate amounts of these ions are available. Ultimately, increased bone mineralization is made possible by increased intestinal absorption of calcium and phosphate. Paradoxically, perhaps, $1,25(OH)_2D_3$ acts on bone to promote resorption in a manner that resembles the late effects of PTH. Like PTH, $1,25(OH)_2D_3$ increases both the number and activity of osteoclasts. As seen for PTH, osteoblasts rather than mature osteoclasts have receptors for $1,25(OH)_2D_3$. Like PTH, $1,25(OH)_2D_3$ stimulates osteoblastic cells to express M-CSF and RANK ligand as well as a variety of other proteins. Sensitivity of bone to PTH decreases with vitamin D deficiency; conversely, in the absence of PTH, 30–100 times as much $1,25(OH)_2D_3$ is needed to mobilize calcium and phosphate. The molecular sites of cooperative interaction of these two hormones in osteoblasts are not known.

Actions on Kidney

When given to vitamin D-deficient subjects, $1,25(OH)_2D_3$ increases reabsorption of both calcium and phosphate. The effects on phosphate reabsorption are probably indirect. PTH secretion is increased in vitamin D deficiency (see below), and hence tubular reabsorption of phosphate is restricted. Replenishment of $1,25(OH)_2D_3$ decreases the secretion of PTH and thus allows proximal tubular reabsorption of phosphate to increase. Effects of $1,25(OH)_2D_3$ on calcium reabsorption are probably direct. Specific receptors for $1,25(OH)_2D_3$ are found in the distal nephron, probably in the same cells in which PTH stimulates calcium uptake. These cells also express the same vitamin D-dependent proteins that are found in intestinal cells, and are likely to respond to $1,25(OH)_2D_3$ in the same manner as intestinal epithelial cells. It is unlikely that $1,25(OH)_2D_3$ regulates calcium balance on a minute-to-minute basis. Instead, it may support the actions of PTH, which is the primary regulator. The molecular basis for this interaction has not been elucidated.

Actions on the Parathyroid Glands

The chief cells of the parathyroid glands are physiological targets for $1,25(OH)_2D_3$ and respond to it in a manner that is characteristic of negative feedback. In this case, negative feedback is exerted at the level of synthesis rather than secretion. The promoter region of the PTH gene contains a vitamin D response element. Binding of the liganded receptor suppresses transcription of the gene and leads to a rapid decline in the preproPTH mRNA. Because the chief cells store relatively little hormone, decreased synthesis rapidly leads to decreased secretion. In a second negative feedback action, $1,25(OH)_2D_3$ indirectly decreases PTH secretion by virtue of its actions to increase plasma calcium concentration. Consistent with the crucial role of calcium in regulating PTH secretion, the negative feedback effects of $1,25(OH)_2D_3$ on PTH synthesis are modulated by the plasma calcium concentration. Nuclear receptors for $1,25(OH)_2D_3$ in chief cells are down-regulated when the plasma calcium concentration is low and are up-regulated when it is high.

REGULATION OF $1,25(OH)_2D_3$ PRODUCTION

As true of any hormone, the concentration of $1,25(OH)_2D_3$ in blood must be appropriate for prevailing physiological circumstances if it is to exercise its proper role in maintaining homeostasis. Production of $1,25(OH)_2D_3$ is subject to feedback regulation in a fashion quite similar to that of other hormones. PTH increases synthesis of $1,25(OH)_2D_3$, which exerts a powerful inhibitory effect on PTH gene expression in the parathyroid chief cells. The most important regulatory step in $1,25(OH)_2D_3$ synthesis is the hydroxylation of carbon 1 by cells in the proximal tubules of the kidney. The rate of this reaction is determined by the availability of the requisite P450 enzyme, which has a half-life of only about 2–4 hours. In the absence of PTH, the concentration of 1α-hydroxylase in renal cells quickly falls. PTH regulates transcription of the gene that codes for the 1α-hydroxylase enzyme by increasing production of cyclic AMP. Several cyclic AMP response elements (CREs) are present in its promoter region. Activation of protein kinase C through the IP_3-diacylglycerol second messenger system also appears to play some role in up-regulation of this enzyme.

Through a "short" feedback loop, $1,25(OH)_2D_3$ also acts as a negative feedback inhibitor of its own production by rapidly down-regulating 1α-hydroxylase expression. At the same time, $1,25(OH)D_3$ up-regulates the enzyme that hydroxylates vitamin D metabolites on carbon 24 to produce $24,25(OH)_2D_3$ or $1,24,25(OH)_3D_3$. Hydroxylation at carbon 24 is the initial reaction in the degradative pathway that culminates in the production of calcitroic acid, the principal

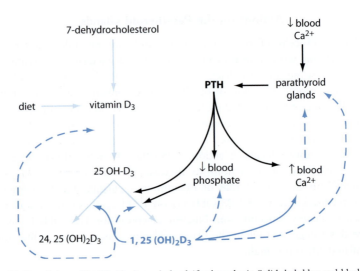

Figure 16 Regulation of 1α,25-dihydroxycholecalciferol synthesis. Solid dark blue and black arrows indicate stimulation; dashed arrows represent inhibition.

biliary excretory product of vitamin D. Up–regulation of the 24 hydroxylase by $1,25(OH)D_3$ is not confined to the kidney, but is also seen in all $1,25(OH)D_3$ target cells. Finally, the results of its actions—increased calcium and phosphate concentrations in blood—directly or indirectly silence the two activators of $1,25(OH)_2D_3$ production, PTH and low phosphate. The regulation of $1,25(OH)_2D_3$ production is summarized in Figure 16.

INTEGRATED ACTIONS OF CALCITROPIC HORMONES

RESPONSE TO A HYPOCALCEMIC CHALLENGE

Because some calcium is always lost in urine, even a short period of total fasting can produce a mild hypocalcemic challenge. More severe challenges are produced by a diet deficient in calcium or anything that might interfere with calcium absorption by the renal tubules or the intestine. The parathyroid glands are exquisitely sensitive to even a small decrease in ionized calcium and promptly increase PTH secretion (Figure 17). Effects of PTH on calcium reabsorption from the glomerular filtrate coupled with some calcium mobilization from bone are evident after about an hour, providing the first line of defense against a hypocalcemic challenge. These actions are adequate only to compensate for a mild or brief

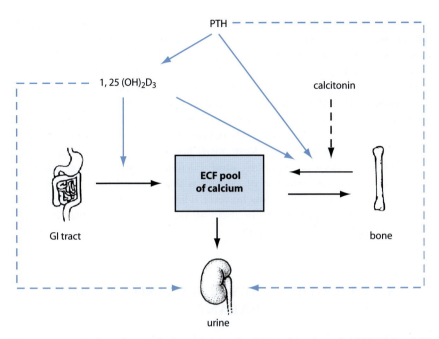

Figure 17   Overall regulation of calcium balance by PTH, calcitonin, and 1,25(OH)$_2$D$_3$. ECF, Extracellular fluid; solid blue arrows indicate stimulation; dashed arrows represent inhibition.

challenge. When the hypocalcemic challenge is large and sustained, additional, delayed responses to PTH are needed. After about 12–24 hours, increased formation of 1,25(OH)$_2$D$_3$ increases the efficiency of calcium absorption from the gut. Osteoclastic bone resorption in response to both PTH and 1,25(OH)$_2$D$_3$ taps the almost inexhaustible reserves of calcium in the skeleton. If calcium intake remains inadequate, skeletal integrity may be sacrificed in favor of maintaining blood calcium concentrations.

RESPONSE TO A HYPERCALCEMIC CHALLENGE

Hypercalcemia is rarely seen under normal physiological circumstances, but it may be a complication of a variety of pathological conditions usually accompanied by increased blood concentrations of PTH or PTHrP. An example of hypercalcemia that might arise under physiological circumstances is the case when a person who has been living for some time on a low-calcium diet ingests calcium-rich food. Under the influence of high concentrations of PTH and 1,25(OH)$_2$D$_3$ that would result from calcium insufficiency, osteoclastic activity transfers bone mineral to the extracellular fluid. In addition, calcium absorptive mechanisms in the intestine and renal tubules are stimulated to their maximal efficiency. Consequently the calcium that enters the gut is absorbed efficiently and blood calcium is increased by a few tenths of a milligram per deciliter. Calcitonin secretion is promptly increased and would provide some benefit through suppression of osteoclastic activity. Although PTH secretion promptly decreases, and its effects on calcium and phosphate transport in renal tubules quickly diminish, several hours pass before hydroxylation of 25-OHD$_3$ and osteoclastic bone resorption diminish. Even after its production is shut down, many hours are required for responses to 1,25(OH)$_2$D$_3$ to decrease. Although some calcium phosphate may crystalize in demineralized osteoid, renal loss of calcium is the principal means of lowering blood calcium. The rate of renal loss, however, is limited to only 10% of the calcium present in the glomerular filtrate, or about 40 mg per hour, even after complete shutdown of PTH-sensitive transport.

OTHER HORMONES AFFECTING CALCIUM BALANCE

In addition to the primary endocrine regulators of calcium balance discussed above, it is apparent that many other endocrine and paracrine factors influence calcium balance. Bone growth and remodeling involve a still incompletely understood interplay of local and circulating cytokines, growth factors, and hormones, including insulin-like growth factor I, growth hormone (see Chapter 9), the cytokines: interleukin-1 (see Chapter 4) interleukin-6, interleukin-11, tumor necrosis factor α, transforming growth factor β, and doubtless many others. The prostaglandins (see Chapter 4) also have calcium-mobilizing activity and stimulate bone lysis. Production of prostaglandins and cytokines is increased in a variety of inflammatory conditions and can lead to systemic or localized destruction of bone.

Many of the systemic hormones directly or indirectly have an impact on calcium balance. Obviously, special demands are imposed on overall calcium balance during growth, pregnancy, and lactation. All of the hormones that govern growth—namely, growth hormone, the insulin-like growth factors, and thyroidal and gonadal hormones (see Chapter 9)—directly or indirectly influence the activity of bone cells and calcium balance. The gonadal hormones, particularly estrogens, play a critical role in maintaining bone mass, which decreases in their absence, leading to *osteoporosis*. This condition is common in postmenopausal women. Osteoblastic cells express receptors for estrogens, which stimulate proliferation of osteoblast progenitors and inhibit production of cytokines such as interleukin-6, which activates osteoclasts. Consequently, in the absence of estrogens osteoclastic activity is increased and osteoblastic activity is decreased, and there is net loss of bone.

Defects in calcium metabolism are also seen in hyperthyroidism and in conditions of excess or deficiency of adrenal cortical hormones. Excessive thyroid hormone accelerates activity of both the osteoclasts and osteoblasts that may result in net bone resorption and a decrease in bone mass. This action may produce a mild hypercalcemia and secondarily suppress PTH secretion and hence $1,25(OH)_2D_3$ production. These hormonal changes result in increased urinary loss of calcium and decreased intestinal absorption. Excessive glucocorticoid concentrations also decrease skeletal mass. Although glucocorticoids stimulate the differentiation of osteoclast progenitors, they decrease proliferation of these progenitor cells, which ultimately leads to a decrease in active osteoblasts. Glucocorticoids also antagonize the actions and formation of $1,25(OH)_2D_3$ by some unknown mechanism, and directly inhibit calcium uptake in the intestine. These changes may increase PTH secretion and stimulate osteoclasts. Conversely, adrenal insufficiency may lead to hypercalcemia, due largely to decreased renal excretion of calcium.

SUGGESTED READING

Brommage, R., and DeLuca, H. F. (1985). Evidence that 1,25-dihydroxyvitamin D_3 is the physiologically active metabolite of vitamin D_3. *Endocr. Rev.* **6**, 491–511.

Brown, E. M., Pollak, M., Seidman, C. E., Seidman, J. G., Chou, Y. H., Riccardi, D., and Hebert, S. C. (1995). Calcium ion-sensing cell-surface receptors, *New Engl. J. Med.* **333**, 234–240.

Diaz, R., Fuleihan, G. E.-H., and Brown, E. M. (2000). Parathyroid hormone and polyhormones: Production and export. *In* "Endocrine Regulation of Water and Electrolyte Balance, Volume 3, Handbook of Physiology, Section 7, The Endocrine System," (J. C. S. Fray, ed.), pp. 607–662. Oxford University Press, New York.

Jones, G., Strugnall, S. A., and DeLuca, H. (1998). Current understanding of the molecular actions of vitamin D. *Physiol. Rev.* **78**, 1193–1231.

Malloy, P. J., Pike, J. W., and Feldman D. (1999). The vitamin D receptor and the syndrome of hereditary 1,25-dihydroxyvitamin D-resistant rickets. *Endocr. Rev.* **20**, 156–188.

Mannstadt, M., Jüppner, H., and Gardella, T. J. (1999). Receptors for PTH and PTHrP: Their biological importance and functional properties. *Am. J. Physiol.* **277**, F665–F675.

Muff, R., and Fischer, J. A. (1992). Parathyroid hormone receptors in control of proximal tubular function. *Annu. Rev. Physiol.* **54**, 67–79.

Nijweide, P. J., Burger, E. H., and Feyen, J. H. M. (1986). Cells of bone: Proliferation, differentiation, and hormonal regulation. *Physiol. Rev.* **66**, 855–886.

Suda, T., Takahashi, N., Udagawa N., Jimi, E., Gillespie, M. T., and Martin, T. J. (1999). Modulation of osteoclast differentiation and function by new members of the tumor necrosis factor receptor and ligand families. *Endocr. Rev.* **20**, 245–397.

CHAPTER 9

Hormonal Regulation of Fuel Metabolism

OVERVIEW

Mammalian survival in a cold, hostile environment demands an uninterrupted supply of metabolic fuels to maintain body temperature, to escape from danger, and to grow and reproduce. A constant supply of glucose and other energy-rich metabolic fuels to the brain and other vital organs must be available at all times despite wide fluctuations in food intake and energy expenditure. Constant availability of metabolic fuel is achieved by storing excess carbohydrate, fat, and protein, principally in liver, adipose tissue, and muscle, and drawing on those reserves when needed. We consider here how fuel homeostasis is maintained minute to minute, day to day, and year to year by regulating fuel storage and mobilization, the mixture of fuels consumed, and food intake. Homeostatic regulation is provided by the endocrine system and the autonomic nervous system. The strategy of hormonal regulation of metabolism during starvation or exercise is to provide sufficient substrate to working muscles while maintaining an adequate concentration of glucose in blood to satisfy the needs of brain and other glucose-dependent cells. When dietary or stored carbohydrate is inadequate, availability of glucose is ensured by (1) gluconeogenesis from lactate, glycerol, and alanine and (2) inhibition of glucose utilization by those tissues that can satisfy their energy needs with other substrates, notably fatty acids and ketone bodies. The principal hormones that govern fuel homeostasis are insulin, glucagon, epinephrine, cortisol, growth hormone (GH), thyroxine (T4), and the newly discovered adipocyte hormone, leptin. The principle target organs for these hormones are adipose tissue, liver, and skeletal muscle.

GENERAL FEATURES OF ENERGY METABOLISM

In discussing how hormones regulate fuel metabolism, we consider first the characteristics of metabolic fuels and the intrinsic biochemical regulatory mechanisms on which hormonal control is superimposed.

Body Fuels

Glucose

Glucose is readily oxidized by all cells; 1 g yields about 4 calories. The average 70 kg man requires approximately 2000 calories per day and therefore would require a reserve supply of approximately 500 g of glucose to ensure sufficient substrate to survive 1 day of food deprivation. If glucose were stored as an isosmolar solution, approximately 10 litres of water (10 kg) would be needed to accommodate a single day's energy needs, and the 70 kg man would have to carry around a storage depot equal to his own weight if he were to survive only 1 week of starvation. Actually, only about 20 g of free glucose is dissolved in extracellular fluids, or enough to provide energy for about 1 hour.

Glycogen

Polymerizing glucose to glycogen eliminates the osmotic requirement for large volumes of water. To meet a single day's energy needs, only about 1.8 kg of "wet" glycogen is required; that is, 500 g of glycogen obligates only about 1.3 liters of water. Glycogen stores in the well-fed 70 kg man are enough to meet only part of a day's energy needs—about 100 g in the liver and about 200 g in muscle.

Protein

Calories can also be stored in somewhat more concentrated form as protein. Storage of protein, however, also obligates storage of some water, and oxidation of protein creates unique by-products: ammonia, which must be detoxified to form urea at metabolic expense, and sulfur-containing acids. The body of a normal 70-kg man in nitrogen balance contains about 10–12 kg of protein, most of which is in skeletal muscle. Little or no protein is stored as an inert fuel depot, so that mobilization of protein for energy necessarily produces some functional deficits. Under conditions of prolonged starvation, as much as one-half of the body protein may be consumed for energy before death ensues, usually from failure of respiratory muscles.

Fat

Triglycerides are by far the most concentrated storage form of high-energy fuel (9 calories/g), and they can be stored essentially without water. The energy needs for 1 day can be met by less than 250 g of triglyceride. Thus a 70-kg man carrying 10 kg of fat maintains an adequate depot of fuel to meet energy needs for more than 40 days. Most fat is stored in adipose tissue, but other tissues, such as muscle, also contain small reserves of triglycerides.

Problems Inherent in the Use of Glucose and Fat as Metabolic Fuels

Glucose and fat are important metabolic fuels, but have associated problems:

1. Fat is the most abundant and efficient energy reserve, but efficiency has its price. When converting dietary carbohydrate to fat, about 25% of the energy is

dissipated as heat. More importantly, synthesis of fatty acids from glucose is an irreversible process. Once the carbons of glucose are converted to fatty acids, virtually no reconversion to glucose is possible. The glycerol portion of triglycerides remains convertible to glucose, but glycerol represents only about 10% of the mass of triglyceride.

2. Limited water solubility of fat complicates transport between tissues. Triglycerides are "packaged" as very low-density lipoproteins (LDLs) or as chylomicrons for transport in blood to storage sites. Uptake by cells follows breakdown to fatty acids by lipoprotein lipase at the external surface or within capillaries of muscle or fat cells. Mobilization of stored triglycerides also requires breakdown to fatty acids, which leave adipocytes in the form of free fatty acids (FFAs). FFAs are not very soluble in water and are transported in blood firmly bound to albumin. Because they are bound to albumin, FFAs have limited access to tissues such as brain; they can be processed to water-soluble forms in the liver, however, which converts them to four-carbon ketoacids (ketone bodies), which can cross the blood–brain barrier.

3. Energy can be derived from glucose without simultaneous consumption of oxygen, but oxygen is required for degradation of fat. Therefore, glucose must be constantly available in the blood to satisfy the needs of red blood cells, which lack mitochondria, and cells in the renal medullae, which function under low oxygen tension. Under basal conditions these cells consume about 50 g of glucose each day and release an equivalent amount of lactate into the blood. Because lactate is readily reconverted to glucose in the liver, however, these tissues do not act as a drain on carbohydrate reserves.

4. In a well-nourished person the brain relies almost exclusively on glucose to meet its energy needs and consumes nearly 150 g per day. The brain does not derive energy from oxidation of FFAs or amino acids. Ketone bodies are the only alternative substrates to glucose, but studies in experimental animals indicate that only certain regions of the brain can substitute ketone bodies for glucose. Total fasting for 4 to 5 days is required before the concentrations of ketone bodies in blood are high enough to provide a significant fraction of the brain's energy needs. Even after several weeks of total starvation, the brain continues to satisfy about one-third of its energy needs with glucose. The brain stores little glycogen and hence must depend on the circulation to meet its minute-to-minute fuel requirements. The rate of glucose delivery depends on its concentration in arterial blood, the rate of blood flow, and the efficiency of extraction. Although an increased flow rate might compensate for decreased glucose concentration, the mechanisms that regulate blood flow in brain are responsive to oxygen and carbon dioxide, rather than to glucose. Under basal conditions the concentration of glucose in arterial blood is about 5 mM (90 mg/dl), of which the brain extracts about 10%. The fraction extracted can double, or perhaps even triple, when the concentration of glucose is low; but when the blood glucose falls below about 30 mg/dl, metabolism

and function are compromised. Thus the brain is exceedingly vulnerable to hypoglycemia, which can quickly produce coma or death.

FUEL CONSUMPTION

The amount of metabolic fuel consumed in a day varies widely and normally is balanced by variations in food intake. The adipose tissue reservoir of triglycerides can shrink or expand to accommodate imbalance in fuel intake and expenditure. Muscle comprises about 50% of body mass and is by far the major consumer of metabolic fuel. Even at rest, muscle metabolism accounts for about 30% of the oxygen consumed. Although normally a 56-kg woman or a 70-kg man consumes about 1600 or 2000 calories in a typical day, daily caloric requirements may range from about 1000 calories with complete bed rest to as much as 6000 with prolonged physical activity. For example, marathon running may consume 3000 calories in only 3 hours. Under basal conditions an individual on a typical mixed diet derives about half of the daily energy needs from the oxidation of glucose, a small fraction from consumption of protein, and the remainder from fat. With starvation or with prolonged exercise, limited carbohydrate reserves are quickly exhausted unless some restriction is placed on carbohydrate consumption by muscle, which has fuel needs far in excess of those of any other tissue and which can be met by increased utilization of fat. In fact, simply providing muscle with fat restricts its ability to consume carbohydrate. Hormonal regulation of energy balance is largely accomplished through adjusting the flux of energy-rich fatty acids and their derivatives to muscle, and the consequent sparing of carbohydrate and protein.

GLUCOSE–FATTY ACID CYCLE

The self-regulating interplay between glucose and fatty acid metabolism is called the glucose–fatty acid cycle. This cycle constitutes an important biochemical mechanism for limiting glucose utilization when alternative substrate is available, and conversely limiting the consumption of stored fat when glucose is available. Fatty acids that are produced in adipose tissue in an ongoing cycle of lipolysis and reesterification may either escape from fat cells to become the free fatty acids, or they may be retained as triglycerides, depending on the availability of α-glycerol phosphate (Figure 1). The only source of α-glycerol phosphate for reesterification of fatty acids is the pool of triose phosphates derived from glucose oxidation, because adipose tissue is deficient in the enzyme required to phosphorylate and hence re-use glycerol released from triglycerides. Consequently, when glucose is abundant, α-glycerol phosphate is readily available, the rate of

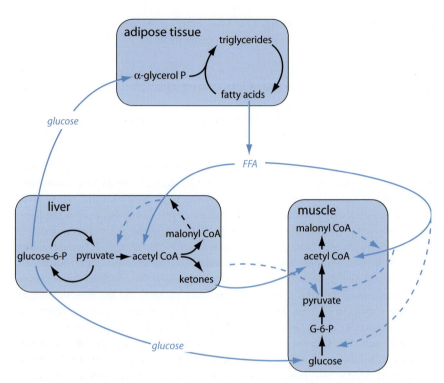

Figure 1 Intraorgan flow of substrate and the competitive regulatory effects of glucose and fatty acids that comprise the glucose–fatty acid cycle. Dashed arrows denote inhibition. See text for details.

reesterification is high relative to lipolysis, and the rate of release of FFAs is low. Conversely, when glucose is scarce, more fatty acids escape and plasma concentrations of FFA increase.

Exposure of muscle to elevated levels of FFAs for several hours decreases transport of glucose across the plasma membrane and phosphorylation to glucose-6-phosphate. The mechanism for this effect is not understood. The resulting decrease of glucose-6-phosphate, which is both a substrate and an allosteric activator of glycogen synthase, results in decreased glycogen formation as well as decreased glucose oxidation by glycolysis. Though somewhat controversial, allosteric effects of products of fatty acid oxidation may further curtail glycolysis by inhibiting of phosphofructokinase. Oxidation of fatty acids or ketone bodies also limits the oxidation of pyruvate to acetyl CoA. It may be recalled (see Chapter 5) that long-chain fatty acids must be linked to carnitine to gain entry into mitochondria, where they are oxidized. Activity of acylcarnitine transferase is increased allosterically by long-chain fatty acid coenzyme A (CoA) and inhibited by malonyl CoA, the formation of which is accelerated when glucose is plentiful.

Oxidation of long-chain fatty acids or ketone bodies to acetyl CoA reduces the cofactor nicotinamide adenine dinucleotide (NAD) to NADH at a rate that exceeds oxidative regeneration in the nonworking muscle. The resulting scarcity of NAD and free CoA limits the breakdown of pyruvate directly, and also activates the mitochondrial enzyme pyruvate dehydrogenase (PDH) kinase that inactivates a key enzyme of pyruvate oxidation. The activity of PDH kinase, in turn, is inhibited by pyruvate. It should be noted that oxidation of pyruvate to acetyl CoA is the reaction that irreversibly removes carbons from the pool of metabolites that are convertible to glucose.

Influx of fatty acids to the liver promotes ketogenesis and gluconeogenesis, largely by the same mechanisms that diminish glucose metabolism in muscle. Metabolism of long-chain fatty acids inhibits the intramitochondrial oxidation of pyruvate to acetyl CoA. Gluconeogenic precursors arriving at the liver in the form of pyruvate, lactate, alanine, or glycerol are thus spared oxidation in the tricarboxylic acid cycle and instead are converted to phosphoenol pyruvate (PEP). Conversely, when glucose is abundant, the concentration of glucose-6-phosphate increases, and gluconeogenesis is inhibited both at the level of fructose-1,6-bisphosphate formation and at the level of pyruvate kinase (Chapter 5, Figure 3). Under these circumstances malonyl CoA formation is increased and fatty acids are restrained from entering the mitochondria and undergoing subsequent degradation.

Through the reciprocal effects of glucose and fatty acids, glucose indirectly regulates its own rate of utilization and production by a negative feedback process that depends on intrinsic allosteric regulatory properties of metabolites and enzymes of the glucose-fatty acid cycle. Hormones may regulate and fine tune metabolism by altering the activities or amounts of enzymes, and by influencing the flow of metabolites. The glucose–fatty acid cycle operates in normal physiology even though the concentration of glucose in blood remains nearly constant. In fact, the contribution of some hormones, notably glucocorticoids and growth hormone (GH), to the maintenance of blood glucose and muscle glycogen stores depends in part on the glucose–fatty acid cycle. Conversely, in addition to stimulating glucose entry into muscle, insulin indirectly increases glucose metabolism by decreasing FFA mobilization from adipose tissue, thereby shutting down the inhibitory influence of the glucose–fatty acid cycle. This effect may be accelerated by a further effect of insulin to increase the formation of malonyl CoA in liver and muscle, thereby diminishing access of fatty acids to the mitochondrial oxidative apparatus.

OVERALL REGULATION OF BLOOD GLUCOSE CONCENTRATION

Despite vagaries in dietary input and large fluctuations in food consumption, the concentration of glucose in blood remains remarkably constant. Its concentration at any time is determined by the rate of input and the rate of removal by the

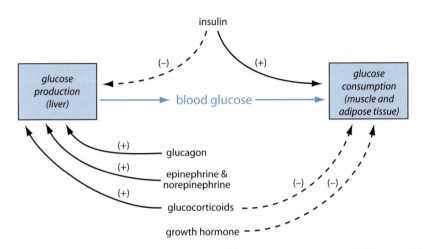

Figure 2 Interaction of hormones to maintain the blood glucose concentration. Solid arrows (+) denote increase; dashed arrows (−) denote decrease.

various body tissues (Figure 2). The rate of glucose removal from the blood varies over a wide range, depending on physical activity and environmental temperature. Even immediately after eating, the rate of input largely reflects activity of the liver, because glucose and other metabolites absorbed from the intestine must pass through the liver before entering the circulation. Liver glycogen is the immediate source of blood glucose under most circumstances. Hepatic gluconeogenesis may contribute to blood glucose directly but is more important for replenishing glycogen stores. The kidneys are also capable of gluconeogenesis, but their role as providers of blood glucose has not been studied to any great extent, except in acidosis, when glucose production from glutamate accompanies renal production and excretion of ammonia. Recent studies in patients undergoing liver transplantation, however, indicate that glucose production by the kidneys immediately after removal of the liver can be substantial, at least for a short time.

SHORT-TERM REGULATION

Minute-to-minute regulation of blood glucose depends on (1) insulin, which, in promoting fuel storage, drives glucose concentrations down, and (2) glucagon, and to a lesser extent catecholamines, which, in mobilizing fuel reserves, drive glucose concentrations up. Effects of these hormones are evident within seconds or minutes and dissipate as quickly. Insulin acts at the level of the liver to inhibit glucose output, and on muscle and fat to increase glucose uptake.

Liver is more responsive to insulin than are muscle and fat, and because of its anatomical location, is exposed to higher hormone concentrations. Smaller increments in insulin concentration are needed to inhibit glucose production than to promote glucose uptake. Glucagon and catecholamines act on hepatocytes to promote glycogenolysis and gluconeogenesis. They have no direct effects on glucose uptake by peripheral tissues, but epinephrine and norepinephrine may decrease the demand for blood glucose by mobilizing alternative fuels—glycogen and fat—within muscle and adipose tissue. Increased blood glucose is perceived directly by pancreatic beta cells, which respond by secreting insulin. Hypoglycemia is perceived not only by the glucagon-secreting alpha cells of pancreatic islets, but also by the central nervous system, which activates sympathetic outflow to the islets and the adrenal medullae. Sympathetic stimulation of pancreatic islets increases secretion of glucagon and inhibits secretion of insulin. In addition, hypoglycemia evokes secretion of the hypothalamic releasing hormones that stimulate ACTH and GH secretion from the pituitary gland (Figure 3). Cortisol, secreted in response to ACTH, and GH act only after a substantial delay and hence are unlikely to contribute to rapid restoration of blood glucose. However, they are important for withstanding a sustained hypoglycemic challenge.

LONG-TERM REGULATION

Long-term regulation, operative on a time scale of hours or perhaps days, depends on direct and indirect actions of many hormones and ultimately ensures (1) that the peripheral drain on glucose reserves is minimized and (2) that liver contains an adequate reservoir of glycogen to satisfy minute-to-minute needs of glucose-dependent cells. To achieve these ends, peripheral tissues, mainly muscle, must be provided with alternate substrate and limit their consumption of glucose. At the same time, gluconeogenesis must be stimulated and supplied with adequate precursors to provide the 150–200 g of glucose needed each day by the brain and other glucose-dependent tissues. Long-term regulation includes all of the responses that govern glucose utilization as well as all those reactions that govern storage of fuel as glycogen, protein, or triglycerides.

INTEGRATED ACTIONS OF METABOLIC HORMONES

Metabolic fuels absorbed from the intestine are largely converted to storage forms in liver, adipocytes, and muscle. It is fair to state that storage is virtually the exclusive province of insulin, which stimulates biochemical reactions that convert simple compounds to more complex storage forms and inhibits fuel mobilization. Hormones that mobilize fuel and defend the glucose concentration of the blood

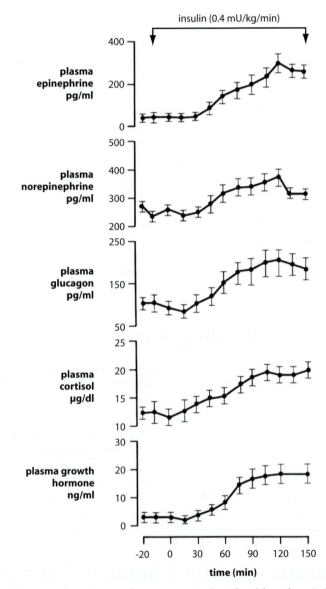

Figure 3 Counterregulatory hormonal responses to insulin-induced hypoglycemia. The infusion of insulin reduced plasma glucose concentration to 50–55 mg/dl. (From Sacca, L., Sherwin, R., Hendler, R., and Felig, P., *J. Clin. Invest.* **63**, 849–857, 1979, with permission.)

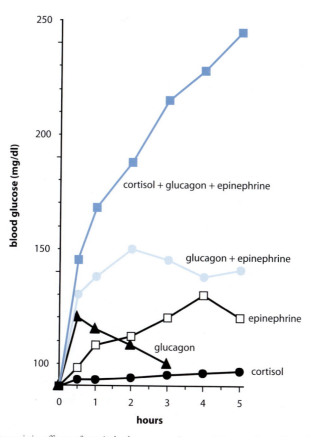

Figure 4 Synergistic effects of cortisol, glucagon, and epinephrine on increasing plasma glucose concentration. Note that the hyperglycemic response to the triple hormone infusion is far greater than the response of each hormone given singly. (Redrawn from data of Eigler, N., Sacca, L., and Sherwin, R. S., *J. Clin. Invest.* **63**, 114–123, 1979.)

are called counterregulatory and include glucagon, epinephrine, norepinephrine, cortisol, and GH. Secretion of most or all of these hormones is increased whenever there is increased demand for energy. These hormones act synergistically and together produce effects that are greater than the sum of their individual actions. In the example shown in Figure 4, glucagon and epinephrine elevated the blood glucose level primarily by increasing hepatic production. When cortisol was given simultaneously, these effects were magnified, even though cortisol had little effect when given alone. Triiodothyronine (T3) must also be considered in this context, because its actions increase the rate of fuel consumption and the sensitivity of target cells to insulin and counterregulatory hormones. Before examining the interactions of these hormones in the whole body, it is useful to summarize their effects on individual tissues.

ADIPOSE TISSUE

The central event in adipose tissue metabolism is the cycle of fatty acid esterification and triglyceride lipolysis (Figure 5). Although reesterification of fatty acids can regulate FFA output from fat cells, regulation of lipolysis and hence the rate at which the cycle spins provides a wider range of control. It has been estimated that under basal conditions 20% of the fatty acids released in lipolysis are reesterified to triglycerides, and that reesterification may decrease to 9–10% during active fuel consumption. Under the same conditions, lipolysis may be varied over a 10-fold range. Catecholamines and insulin, through their antagonistic effects

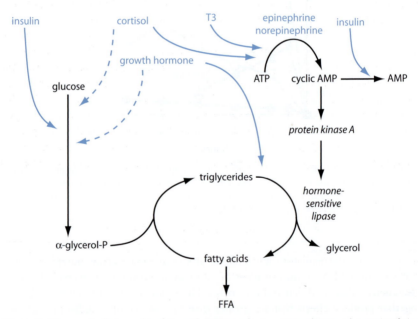

Figure 5 Hormonal effects on free fatty acid (FFA) production. Epinephrine and norepinephrine stimulate hormone-sensitive lipase through a cyclic AMP-mediated process. Insulin antagonizes this effect by stimulating cyclic AMP degradation. T3, cortisol, and growth hormone increase the response of adipocytes to epinephrine and norepinephrine. Growth hormone also directly stimulates lipolysis. Insulin indirectly antagonizes the production of FFA by increasing reesterification. Growth hormone and cortisol increase FFA release by inhibiting reesterification. Solid blue arrows indicate stimulation; dashed arrows indicate inhibition.

on cyclic AMP metabolism, increase or decrease the activity of hormone-sensitive lipase. Responses to these hormones are expressed within minutes. Other hormones, especially cortisol, T3, and GH, modulate the sensitivity of adipocytes to insulin and catecholamines. Modulation is not a reflection of abrupt changes in hormone concentrations but, rather, stems from long-term tuning of metabolic machinery. Finally, GH produces a sustained increase in lipolysis after a delay of about 2 hours. Growth hormone and cortisol also decrease fatty acid esterification by inhibiting glucose metabolism both directly and by decreasing responsiveness to insulin. These hormonal effects on adipose tissue are summarized in Table 1.

MUSCLE

By inhibiting FFA mobilization, insulin promptly decreases plasma FFA concentrations and thus removes a deterrent of glucose utilization in muscle at the same time that it promotes transport of glucose into myocytes. The response to insulin can be divided into two components. Stimulation of glucose transport and glycogen synthesis are direct effects and are seen within minutes. Increased oxidation of glucose that results from release of inhibition requires several hours. Epinephrine and norepinephrine promptly increase cyclic AMP production and glycogenolysis. When the rate of glucose production from glycogen exceeds the need for ATP production, muscle cells release pyruvate and lactate, which can be reconverted to glucose in liver. Growth hormone and cortisol directly inhibit glucose uptake by muscle and indirectly decrease glucose metabolism in myocytes through the agency of the glucose–fatty acid cycle. By indirectly inhibiting glucose metabolism, GH and cortisol decrease glycogen breakdown. The resulting preservation of muscle glycogen has been called the *glycostatic effect* of GH, and is part of

Table 1

Hormonal Effects on Metabolism in Adipocytes

Hormone	Glucose uptake	Lipolysis	Reesterification	Rapid (R) or slow (S)
Insulin	↑	↓	↑	R
Epinephrine and Norepinephrine	↑[a]	↑↑	↑[a]	R
Growth hormone	↓	↑	↓	S
Cortisol	↓	↑[b]	↓	S
T3	↑	↑[b]	↑	S

[a]Increased glucose uptake and reesterification are driven by the increase in lipolysis.
[b]Permissive effects.

Table 2

Hormonal Effects of Glucose Metabolism in Muscle[a]

Hormone	Glucose uptake	Glucose phosphorylation	Glycolysis	Glycogen Storage
Insulin	↑(D)	↑(I)	↑(I)	↑(D)
Epinephrine and norepinephrine	↓(I)[b]	↓(I)[b]	↑	↓(D)
Growth hormone	↓(D, I)	↓(I)	↓(I)	↑(I)
Cortisol	↓(D, I)	↓(I)	↓(I)	↑(I)
T3	↑	↑	↑	↑ or ↓[c]

[a]Directly (D) and indirectly (I) via the glucose fatty acid cycle.
[b]Immediate effect is secondary to glycogenolysis; later effect is secondary to the glucose-fatty acid cycle.
[c]Dependent on the dose; high rates of oxygen consumption may decrease glycogen.

the overall effect of cortisol that gives rise to the term glucocorticoid. Cortisol also inhibits the uptake of amino acids and their incorporation into proteins and simultaneously promotes degradation of muscle protein. As a result, muscle becomes a net exporter of amino acids, which provide substrate for gluconeogenesis in liver. These events are summarized in Table 2. Insulin and GH antagonize the effects of cortisol on muscle protein metabolism.

LIVER

The antagonistic effects of insulin and glucagon on gluconeogenesis, ketogenesis, and glycogen metabolism in hepatocytes are described in Chapter 5. Epinephrine and norepinephrine, by virtue of their effects on cyclic AMP metabolism, share all the actions of glucagon. In addition, these medullary hormones also activate α_1-adrenergic receptors and reinforce these effects through the agency of the diacylglycerol–inositol trisphosphate–calcium system (see Chapter 1). Cortisol is indispensable as a permissive agent for the actions of glucagon and catecholamines on gluconeogenesis and glycogenolysis. In addition, cortisol induces synthesis of a variety of enzymes responsible for gluconeogenesis and glycogen storage. By virtue of its actions on protein degradation in muscle, cortisol is also indispensable for providing substrate for gluconeogenesis. T3 promotes glucose utilization in liver by promoting synthesis of enzymes required for glucose metabolism and lipid formation. Growth hormone is thought to increase hepatic glucose production, probably as a result of increased FFA mobilization, and it also increases ketogenesis largely by increasing mobilization of FFA. These hormonal influences on hepatic metabolism are summarized in Table 3.

Table 3

Hormonal Regulation of Metabolism in Liver[a]

Hormone	Glucose output	Glycogen synthesis (S) or breakdown (B)	Gluconeogenesis	Keto-genesis	Ureogenesis	Lipogenesis
Insulin	↓(D)	S	↓(D)	↓(I, D)	↓(I, D)	↑(D)
Glucagon	↑(D)	B	↑(D)	↑(D)	↑(D)	↓(D)
Epinephrine and norepinephrine	↑(D)	B	↑(D)	↑(I, D)	↑(D)	↓(D)
Growth hormone	↑(I)	(−)	↑(I)	↑(I)	↑(I)	↓(I)[b]
Cortisol	↑(I, D)	S	↑(I, D)	↑(I)	↑(I)	↓(I)
T3	↑(I)	B	↑(I)	↑(I)	↑(I)	↑(D)

[a]Direct (D) and indirect (I) effects.
[b]Growth hormone promotes protein synthesis and hence decreases availability of amino acids for ureogenesis.

PANCREATIC ISLETS

Alpha and beta cells of pancreatic islets are targets for metabolic hormones as well as producers of glucagon and insulin. Glucagon can stimulate insulin secretion, but the physiological significance of such an action is not understood. Insulin inhibits glucagon secretion, and in its absence responsiveness of alpha cells to glucose is severely impaired. Conversely, insulin apparently also exerts autocrine effects on the beta cells and is required to maintain the normal secretory response to increased glucose concentrations. Epinephrine and norepinephrine inhibit insulin secretion and stimulate glucagon secretion. Growth hormone, cortisol, and T3 are required for normal secretory activity of beta cells, which have a reduced capacity for insulin secretion in their absence. The effects of GH and cortisol on insulin secretion are somewhat paradoxical. Although their effects in adipose tissue, muscle, and liver are opposite to those of insulin, GH and cortisol nevertheless increase the sensitivity of beta cells to signals for insulin secretion and exaggerate responses to hyperglycemia (Figure 6). When cortisol or GH is present in excess, higher than normal concentrations of insulin are required to maintain blood glucose in the normal range. Higher concentrations of insulin may contribute to decreased sensitivity by down-regulating insulin receptors in fat and muscle. When either GH or glucocorticoids are present in excess for prolonged periods, diabetes mellitus often results. Approximately 30% of patients suffering from excess GH (acromegaly) and a similar percentage of persons suffering from Cushing's disease

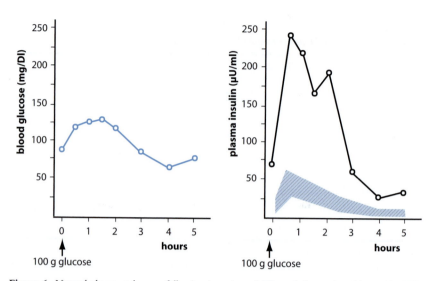

Figure 6 Normal glucose tolerance following ingestion of 100 g of glucose in subject with a GH-secreting pituitary tumor is accompanied by an exaggerated increase in plasma insulin, indicative of decreased insulin sensitivity. The shaded area in the right hand panel represents the plasma insulin response of 43 normal subjects who showed the same changes in glucose concentration after ingestion of 100 g of glucose. (From Daughaday, W. H., and Kipnis, D. M., *Recent Prog. Horm. Res.* **22**, 49–99, 1966, with permission.)

(excess glucocorticoids) experience diabetes mellitus as a complication of their disease. In the early stages diabetes is reversible and disappears when excess pituitary or adrenal secretion is corrected. Later, however, diabetes may become irreversible, and islet cells may be destroyed. This so-called *diabetogenic effect* is an important consideration with chronic glucocorticoid therapy and argues against use of large amounts of GH to build muscle mass in athletes. Hormonal effects on insulin secretion and sensitivity of tissues to insulin are summarized in Table 4.

REGULATION OF METABOLISM DURING FEEDING AND FASTING

POSTPRANDIAL PERIOD

Immediately after eating, metabolic activity is directed toward the processing and sequestration of energy-rich substrates that are absorbed by the intestines. This phase is dominated by insulin, which is secreted in response to three inputs to the beta cells. The *cephalic*, or psychological, aspect of eating stimulates insulin secretion though acetylcholine and vasoactive inhibitory peptide (VIP) released from vagal fibers that innervate islet cells. Food in the small intestine stimulates secretion of

Table 4

Hormonal Effects on Insulin Secretion and Sensitivity of Target Cells to Insulin

Hormone	Insulin secretion by beta cells[a]	Sensitivity of target cells to insulin
Insulin		↓[b]
Glucagon	↑	↓[c]
Epinephrine and norepinephrine	↓	↓[c]
Growth hormone	↑(I)	↓
Cortisol	↑(I)	↓
T3	↑(I)	↑

[a]I, Indirect effect; increases sensitivity to direct stimuli.
[b]Down-regulation of receptors.
[c]Stimulates opposite effect in liver.

intestinal hormones, especially glucagon-like peptide 1 (GLP-1) and glucose-dependent insulinotropic peptide (GIP), which are potent secretagogues for insulin. Finally, the beta cells respond directly to increased glucose and amino acids in arterial blood (see Chapter 5). During the postprandial period the concentration of insulin in peripheral blood may rise from a resting value of about 10 to perhaps as much as 50 μUnits/ml. Glucagon secretion may also increase at this time in response to amino acids in arterial blood. Dietary amino acids may also stimulate GH secretion. Characteristically, the sympathetic nervous system is relatively quiet during the postprandial period, and there is little secretory activity of the adrenal medulla or cortex at this time. Under the dominant influence of insulin, dietary carbohydrates and lipids are transferred to storage depots in liver, adipose tissue, and muscle, and amino acids are converted to proteins in various tissues. Extrahepatic tissues use dietary glucose and fat to meet their needs instead of glucose derived from hepatic glycogen or fatty acids mobilized from adipose tissue. Hepatic glycogen increases by an amount equivalent to about half of the ingested carbohydrate. Fatty acid mobilization is inhibited by the high concentrations of insulin and glucose in blood. Of course, the composition of the diet profoundly affects postprandial responses. Obviously, a diet rich in carbohydrate elicits quantitatively different responses from one that is mainly composed of fat.

POSTABSORPTIVE PERIOD

Several hours after eating, when metabolic fuels have largely been absorbed from the intestine, the body begins to draw on fuels that were stored during the postprandial period. During this period insulin secretion returns to relatively low basal rates and is governed principally by the concentration of glucose in blood, which has returned to about 5 mM (90 mg/dl). About 75% of the glucose secreted by the liver derives from glycogen, and the remainder comes from gluconeogenesis, driven principally by glucagon. Although the rate of glucagon secretion is relatively low at this time, the decline in insulin concentration enables the actions of glucagon to prevail. Growth hormone and cortisol are also secreted at relatively low basal rates in the postabsorptive period. About 75% of the glucose consumed by extrahepatic tissues during this period is taken up by brain, blood cells, and other tissues that consume fuels independently of insulin. Muscle and adipose tissue, which are highly dependent on insulin, account for the remaining 25%. FFAs gradually increase as adipose tissue is progressively relieved of the restraint imposed by high levels of insulin during the postprandial period. Blood glucose remains constant during this period, but glucose metabolism in muscle decreases as the restrictive effects of the glucose–fatty acid cycle become operative. Liver gradually depletes its glycogen stores and begins to rely more heavily on gluconeogenesis from amino acids and glycerol to replace glucose consumed by extrahepatic tissues.

FASTING

More than 24 hours after the last meal, an individual can be considered to be fasting. At this time, circulating insulin concentrations decrease further, and glucagon and GH increase. Cortisol secretion follows its basal diurnal rhythmic pattern (see Chapter 4) unaffected by fasting at this early stage, but basal concentrations of cortisol play their essential permissive role in allowing gluconeogenesis and lipolysis to proceed. Glucocorticoids and GH also exert a restraining influence on glucose metabolism in muscle and adipose tissue. With the further decrease in insulin concentration, there is a further decrease in restraint of lipolysis. The lipolytic cycle speeds up, fatty acid esterification decreases, and FFA mobilization is accelerated. This effect is supported and accelerated by GH and cortisol. Decreased insulin permits a net breakdown of muscle protein, and the amino acids that consequently leave muscle, mainly as alanine, provide the substrate for gluconeogenesis. Fuel consumption after 24 hours of fasting is shown in Figure 7.

With prolonged fasting of 3 days or more, increased GH and decreased insulin result in even greater mobilization of FFAs. Ketogenesis becomes significant, driven by the almost unopposed action of glucagon. By about the third day of starvation, ketone bodies in blood reach concentrations of 2 to 3 mM and begin to provide for an appreciable fraction of the brain's metabolic needs. Urinary nitrogen excretion decreases to the postabsorptive level or below as the pool of rapidly turning over proteins diminishes. During subsequent weeks of total starvation, nitrogen excretion continues at a low but steady rate, with carbon skeletons from amino acids providing substrate for gluconeogenesis and the intermediates needed to maintain the tricarboxylic acid cycle. Glycerol liberated from triglycerides

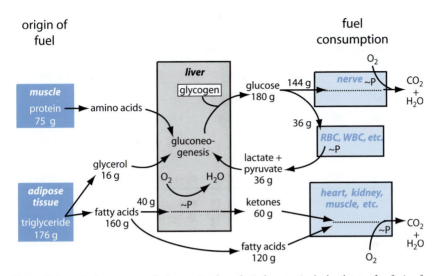

Figure 7 Quantitative turnover of substrates in a hypothetical person in the basal state after fasting for 24 hours (−1800 calories). (From Cahill, G. F., Jr., *N. Engl. J. Med.* **282**, 668–675, 1970, by permission of The New England Journal of Medicine.)

provides the other major substrate for gluconeogenesis. Renal gluconeogenesis from glutamate accompanies production of ammonia stimulated by ketoacidosis. Virtually all other energy needs are met by oxidation of fatty acids and ketones until triglyceride reserves are depleted. In the terminal stages of starvation, proteins may become the only remaining substrate and are rapidly broken down to amino acids. Gluconeogenesis briefly increases once again until cumulative protein loss precludes continued survival. Curiously, continued slow loss of protein during starvation of the extremely obese individual may result in death from protein depletion even before fat depots are depleted.

Table 5 gives some representative values for hormone concentrations in blood in the transition from the fed to the fasting state. Values for cortisol remain unchanged or might even decrease somewhat until late in starvation. Concentrations of cortisol shown in Table 5 represent morning values and change with the time of day in a diurnal rhythmic pattern that is not altered by fasting (see Chapter 4). Even though its concentration does not increase during fasting, cortisol nevertheless is an essential component of the survival mechanism. In its absence, mechanisms for producing and sparing carbohydrates are virtually inoperative, and death from hypoglycemia is inevitable. The role of glucocorticoids in fasting is a good example of permissive action, in which a hormone maintains the instruments of metabolic adjustments so that other agents can manipulate those instruments effectively. Hypoglycemia or perhaps nonspecific stress may account for increased cortisol in the terminal stages of starvation.

Table 5

Representative Values for Some Blood Constituents during Fasting

Sampling time	Glucose (mg/dl)	Insulin (mU/ml)	Glucagon (pg/ml)	Cortisol (ng/ml)	GH (ng/ml[a])	T3 (ng/ml)
Postprandial	150	50	120		1.9	1.2
Postabsorptive	90	15	100	120	—	1.15
Day 1	80	10–12	120	120	5.4	1.15
Day 3	70	8	150	110	—	0.70
Day 5	70	7	150	110	6.1	0.60

[a]Values for GH are from Ho et al., J. Clin. Invest. **81**, 968–975 (1988), and represent the integrated hormone concentration calculated from blood samples taken every 20 minutes over a 24-hour period.

The decrease in plasma concentrations of T3 is not indicative of decreased secretion of TSH or thyroid hormone, but rather reflects decreased conversion of plasma T4 to T3. At least during the first few days of fasting, T4 concentrations in plasma remain constant. The slight decline in T4 seen with more prolonged fasting probably reflects a decrease in plasma binding proteins. Recall that T3, which is formed mostly in extrathyroidal tissue, is the biologically active form of the hormone (see Chapter 3). Deiodination of T4 can lead to the formation of T3 or the inactive metabolite rT3. With starvation, the concentration of rT3 in plasma increases, suggesting that metabolism of T4 shifted from the formation of the active to the inactive metabolite. Some of this increase may also be accounted for by a somewhat slower rate of degradation of rT3. Decreased production of T3 results in an overall decrease in metabolic rate and can be viewed as an adaptive mechanism for conservation of metabolic fuels.

Secretion of GH follows a pulsatile pattern that is exaggerated during starvation (see Chapter 10). Fasting increases both the frequency of secretory pulses and their amplitude. The values for GH shown in Table 5 represent concentrations present in a mixed sample of blood that was drawn at 20 minute intervals over a 24-hour period. The metabolic changes produced by an increase in GH secretion are similar to those that result from a decrease in insulin secretion. Growth hormone increases lipolysis, decreases glucose utilization in muscle and fat, and increases glucose production by the liver. These effects of GH are relatively small compared to the effects of diminished insulin secretion. However, persons suffering from a deficiency of GH may become hypoglycemic during fasting, and treatment with GH helps to maintain their blood glucose (Figure 8). In the non-fasting individual, GH stimulates the liver and other tissues to secrete insulin-like growth factor I (IGF-I), which stimulates protein synthesis. Curiously, the liver

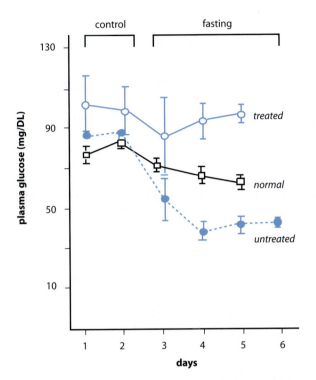

Figure 8 Plasma concentrations of glucose in the plasma of normal subjects and patients suffering from isolated deficiency of GH (shown in blue) during control days of normal food intake and while fasting. Some GH-deficient patients were untreated and others were given 5 mg of hGH per day (treated). The fast began after the collection of blood on day 2. (From Merimee, T. J., Felig, P., Marliss, E., Fineberg, S. E., and Cahill, G. F., Jr., *J. Clin. Invest.* **50**, 574–582, 1971, with permission.)

becomes insensitive to this effect of GH during fasting, and plasma concentrations of IGF-I fall dramatically. This, too, may be an adaptive mechanism that maximizes availability of amino acids for gluconeogenesis and replenishment of critical proteins.

HORMONAL INTERACTIONS DURING EXERCISE

During exercise, overall oxygen consumption may increase 10- to 15-fold in a well-trained young athlete. The requirements for fuel are met by mobilization of reserves within muscle cells and from extramuscular fuel depots. Rapid uptake of glucose from blood can potentially deplete, or at least dangerously lower, glucose concentrations and hence can jeopardize the brain unless some physiological controls are operative. We can consider two forms of exercise: short-term maximal effort, characterized by sprinting for a few seconds, and sustained aerobic work, characterized by marathon running.

SHORT-TERM MAXIMAL EFFORT

For the few seconds of the 100-yard dash, endogenous ATP reserves in muscle—creatine phosphate and glycogen—are the chief sources of energy. For short-term maximal effort, energy must be released from fuel before circulatory adjustments can provide the required oxygen. Breakdown of glycogen to lactate provides the needed ATP and is activated in part through intrinsic biochemical mechanisms that activate glycogen phosphorylase and phosphofructokinase. For example, calcium released from the sarcoplasmic reticulum in response to neural stimulation not only triggers muscle contraction but also activates glycogen phosphorylase. These intrinsic mechanisms are reinforced by epinephrine and nor-epinephrine released from the adrenal medullae and sympathetic nerve endings in response to central activation of the sympathetic nervous system.

The endocrine system is important primarily for maintaining or replenishing fuel reserves in muscle. Through the actions of hormones and the glucose–fatty acid cycle already discussed, glycogen reserves in resting muscle are sustained at or near capacity, so that muscle is always prepared to respond to demands for maximal effort. During the recovery phase lactate released from working muscles is converted to glucose in liver and can be exported back to muscle in the classic *Cori cycle*. Insulin secreted in response to increased dietary intake of glucose or amino acids promotes reformation of glycogen.

SUSTAINED AEROBIC EXERCISE

Glucose taken up from the blood or derived from muscle glycogen is also the most important fuel in the early stages of moderately intense exercise, but with continued effort, dependence on fatty acids increases. Although fat is a more efficient fuel than glucose from a storage point of view, glucose is more efficient than fatty acids from the perspective of oxygen consumption, and yields about 20% more energy per liter of oxygen. Table 6 shows the changes in fuel consumption with time in subjects exercising at 30% of their maximal oxygen consumption. For reasons that are not fully understood, working muscles, even in the trained athlete, cannot derive more than about 70% of their energy from oxidation of fat. Hypoglycemia and exhaustion occur when muscle glycogen is depleted. With sustained exercise, the decline in insulin and the increase in all of the counterregulatory hormones contribute to supplying fat to the working muscles and maximizing gluconeogenesis (Figure 9).

Table 6

Fuels Consumed by Leg Muscles of Man during Mild Prolonged Exercise[a]

Period of exercise (minutes)	Contribution to oxygen uptake (%)		
	Plasma glucose	Plasma free fatty acids	Muscle glycogen
40	27	37	36
90	41	37	22
180	36	50	14
240	30	62	8

[a]Modified from Ahlborg et al., J. Clin. Invest. **53**, 1080–1090 (1974).

Anticipation of exercise may be sufficient to activate the sympathetic nervous system, which is of critical importance not only for supplying the fuel for the working muscles but for making the cardiovascular adjustments that maintain blood flow to carry fuel and oxygen to muscle, gluconeogenic precursors to liver, and heat to sites of dissipation. Insulin secretion is shut down by sympathetic activity. This removes the major inhibitory influence on production of glucose by the liver, glycogen breakdown in muscle, and FFA release from adipocytes. At first glance decreasing insulin secretion might seem deleterious for glucose consumption in muscle. However, the decrease in insulin concentration decreases glucose uptake only by nonworking muscles. Mobilization of GLUT 4 and transport of glucose across the sarcolemma are stimulated by muscular contractions independently of insulin. Glucose metabolism in working muscles is therefore not limited by membrane transport but, rather, by phosphofructokinase, which in turn is responsive to a variety of intracellular metabolites that coordinate its activity with energy demand.

Increased hepatic glucose production results primarily from the combined effects of the decrease in insulin secretion and the increase in glucagon secretion augmented with some contribution from catecholamines. The contributions of the increased secretion of GH and cortisol to this effect are unlikely to be important initially, but with sustained exercise the contributions of both are likely to increase. Actions of both hormones increase the output of FFA and glycerol and decrease glucose utilization by adipocytes and nonworking muscles. Additionally, the increased cortisol would be expected to increase the expression of gluconeogenic enzymes in the liver.

Glycogen reserves of nonworking muscles may provide an important source of carbohydrate for working muscles during sustained exercise and for restoring muscle glycogen after exercise. Epinephrine and norepinephrine stimulate

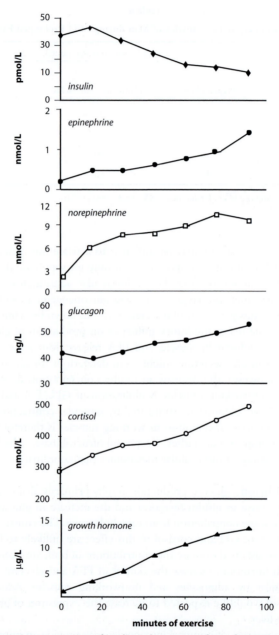

Figure 9 Changes in concentration of insulin and counterregulatory hormones during prolonged moderate exercise. Values shown are the means obtained for eight young men exercising on a bicycle ergometer at ~50% of maximum oxygen consumption. (Drawn from data of Davis, S. N., Galassetti, P., Wasserman, D. H., and Tate, D., *J. Clin. Endocrinol. Metab.* **85**, 224–230, 2000.)

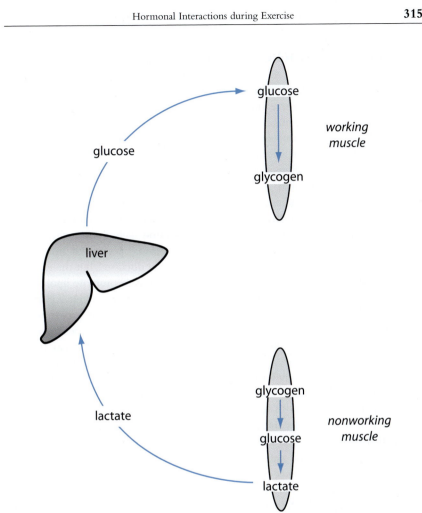

Figure 10   Postulated interaction between previously exercising muscle and previously resting muscle via the Cori cycle during recovery from prolonged arm exercise. (From Ahlborg, G., Wahren, J., and Felig, P., *J. Clin. Invest.* **77**, 690–699, 1986, with permission.)

glycogenolysis in nonworking as well as in working muscles. Glucose-6-phosphate produced from glycogen can be completely broken down to carbon dioxide and water in working muscles, but nonworking muscles convert it to pyruvate and lactate, which escape into the blood. Liver then reconverts these three-carbon acids to glucose, which is returned to the circulation and selectively taken up by the working muscles (Figure 10).

LONG-TERM REGULATION OF FUEL STORAGE

Adipose tissue diffusely scattered throughout the body has an almost limitless capacity for fuel storage. Increasing or decreasing the total amount of fat stored is achieved primarily by changes in the volume of the fat droplets within each adipocyte by the mechanisms of fat deposition and mobilization already discussed. In addition, the number of adipocytes is not fixed and may increase throughout life by multiplication and differentiation of precursor cells. Conversely, fat cells may become depleted of their triglyceride stores and undergo dedifferentiation and apoptosis. The process of adipogenesis depends on locally produced growth factors and cytokines as well as hormones. Insulin and cortisol promote differentiation of human preadipocytes, and tend to favor accumulation of body fat, whereas GH inhibits both differentiation and storage. These chronic actions of insulin and GH are consistent with the short-term actions already discussed, but the effects of cortisol are unexpected in light of its short-term actions to promote lipolysis and decrease fatty acid reesterification. However, removal of the adrenal glands in experimental animals prevents or reverses all forms of genetic or experimentally induced obesity. Additionally, chronic excess production of glucocorticoids in humans is associated with increased body fat, especially in the torso (truncal obesity), the face (moon face), and between the scapulae (buffalo hump).

For most people, body fat reserves are maintained at a nearly constant level throughout adult life despite enormous variations in daily food consumption and energy expenditure. Figure 11 summarizes the findings of five independent studies of changes in body weight and fat mass with aging in about 12,000 individuals. Although total body fat increased with increasing age, the increase averaged less than a gram per day when averaged over a period of 50 years and corresponds to a daily positive energy balance of about 6 calories. Assuming that daily energy consumption averages about 2000 calories, the intake of fuel in a mixed diet matched the rate of energy utilization with an error of 0.3%. However, affluence, ready access to high-calorie foods, and technology that fosters sedentary activities in contemporary society have so distorted the balance between caloric intake and energy expenditure that 30% of the American population is now classified as obese.

Understanding of the mechanisms that govern long-term fuel storage requires understanding regulation of energy expenditure as well as food intake. Physical activity accounts for only about 30% of daily energy expenditure: 60% is expended at rest for maintenance of ion gradients, renewal of cellular constituents, neuronal activity, and to support cardiopulmonary work. The remaining 10% is dissipated as the thermogenic effect of feeding and the consequent processes of assimilation. Neither resting nor thermogenic energy expenditure is fixed, but each is adjustable in a manner that tends to keep body fat reserves constant. In the experiment illustrated in Figure 12, normal human subjects were overfed or underfed in order to increase or decrease body weight by 10%. They were then given

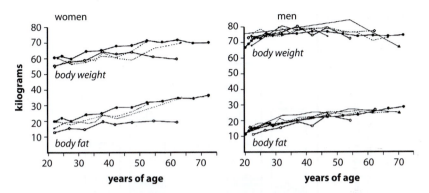

Figure 11 Cross-sectional data obtained in five independent studies showing changes in body weight and fat content with aging. (From Forbes, G. B., and Reina, J. C., *Metabolism* **19**, 653–663, 1970, with permission.)

just enough food each day to maintain their new weight at a constant level. Energy utilization increased disproportionally in the overfed subjects and decreased disproportionally in the underfed subjects. These compensatory changes in energy expenditure opposed maintenance of the change from initial body fat content. How such changes in energy expenditure are brought about is not understood. One possibility is that metabolic efficiency may be regulated by adjusting expression of genes that encode proteins that uncouple ATP generation from oxygen consumption (see Chapter 3).

HYPOTHALAMIC CONTROL OF APPETITE AND FOOD INTAKE

Studies such as those illustrated in Figures 11 and 12 and many older observations gave rise to the idea that the mass of the fat storage depot is monitored, and is maintained at a nearly constant set point through feedback mechanisms that regulate food consumption and energy expenditure (Figure 13). Clinical observations and studies in experimental animals established that the hypothalamus coordinates the drive for food intake with energy-consuming processes such as temperature regulation, growth, and reproduction. Various injuries to the hypothalamus can produce either insatiable eating behavior accompanied by severe obesity, or food avoidance and lethal starvation. A complex neural network interconnects "satiety centers" in the medial hypothalamus and "hunger centers" in the lateral hypothalamus with each other, with autonomic integrating centers in the hypothalamus and brain stem, and with neurons in the arcuate and paraventricular nuclei that regulate secretion of hypophysiotropic hormones (see Chapter 2). In recent years important advances have been made in understanding the complex

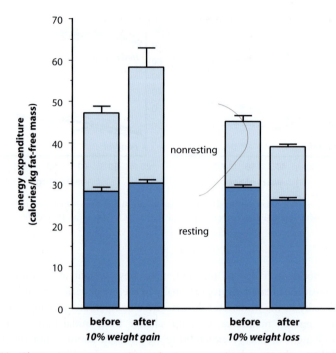

Figure 12   Changes in energy expenditure after increase or decrease of body weight. Thirteen normal subjects were overfed a defined diet until their body weight increased by 10%. Eleven normal subjects were underfed until their body weight decreased by 10%. Both groups were then fed just enough to maintain their new weights for 2 weeks, at which time energy expenditure and lean body mass were measured. (Drawn from data of Leibel, R. L., Rosenbaum, M., and Hirsch, J. *N. Engl. J. Med.* **332**, 673–674, 1995.)

process of regulating food intake, including discovery of the adipocyte hormone *leptin* and some of the neuropeptide transmitters associated with appetite control and their receptors.

LEPTIN

Leptin, which means "thin," is expressed primarily but not exclusively in adipocytes. Inactivating mutations of the gene that encodes leptin or its receptor result in hyperphagia (excess food consumption), obesity, diabetes, impaired temperature regulation, and infertility in rodents. In the very rare cases that have been reported in humans, mutation of the genes that code either for leptin or for its

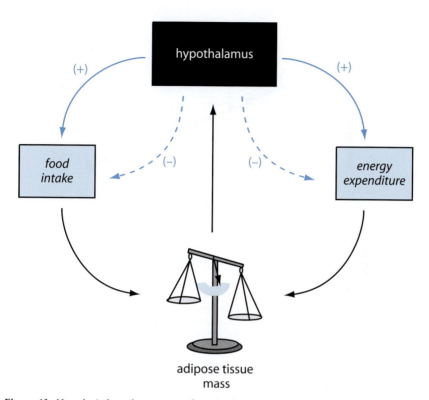

Figure 13 Hypothetical regulatory system for maintaining constancy of adipose mass in which the mass of total stored fat is monitored. Adjustments in energy intake and expenditure are made to maintain constancy. Solid lines (+) denote increase; dashed arrows (−) denote decrease.

receptor results in hyperphagia, morbid obesity, and impaired sexual development. When administered to obese, leptin-deficient mice, leptin decreases body weight by reducing food intake and increasing energy utilization (Figure 14).

Leptin concentrations in blood correlate positively with body fat content (Figure 15), suggesting that leptin might provide a means for monitoring fat stores. Blood levels of leptin also reflect changes in nutritional state. Within hours after initiation of fasting, leptin concentrations decrease sharply (Figure 16), and, conversely, sustained overfeeding increases plasma levels. Because leptin concentrations change to a far greater extent in response to nutritional status than to changes in adipose mass, it has been suggested that a decrease in blood leptin concentration may act as a starvation signal to increase food intake and initiate energy conservation.

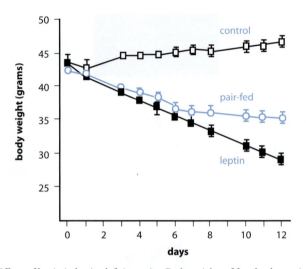

Figure 14   Effects of leptin in leptin-deficient mice. Body weights of female obese mice treated with saline (control) or 270 μg of leptin/day were compared to body weights of obese mice treated with saline, but fed an amount of food equal to that consumed by the leptin-treated mice. Note that the loss of body weight produced by leptin was not accounted for simply by decreased food intake. (From Levin, N., Nelson, C., Gurney, A., Vandlen, R., and de Sauvage, F., *Proc. Natl. Acad. Sci. U.S.A.* **93**, 1726–1730, 1996, by permission of author.)

Biosynthesis, Secretion, and Effects

Leptin is encoded as a 167–amino-acid prohormone in the *ob* gene located on chromosome 7. Its tetra helical structure resembles that of the class of cytokines and hormones that includes GH and prolactin. Because mRNA levels correlate with circulating leptin concentrations and with low levels of the hormone present in adipocytes, secretion of leptin is thought to be regulated at the level of gene transcription. Little is known of the cellular events that are associated with leptin secretion or of the mechanisms that regulate its synthesis. Studies of isolated adipocytes indicate that enlargement of the lipid storage droplet increases leptin mRNA production, but the cellular mechanisms that are activated by fat cell enlargement are not understood. Insulin and cortisol act synergistically to increase leptin synthesis and secretion, whereas GH, norepinephrine, or increased activity of the sympathetic nervous system decreases leptin production. Plasma concentrations of leptin appear to follow a circadian pattern, with highest levels found at night. Frequent spikes in leptin concentration in blood are indicative of synchronized pulsatile secretion, but how secretion by diffusely distributed adipocytes is

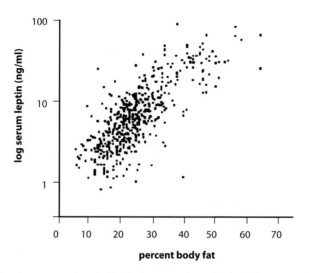

Figure 15 Leptin concentrations in blood plasma correlate with body fat content in 500 human study participants. (From Caro, J. F., Sinha, M. K., Kolaczynski, J. W., Zhang, P. L., and Considine, R. V., *Diabetes* **45**, 1455–1462, 1996, reprinted with permission from *The American Diabetes Association.*)

coordinated is not understood. More than 40% of the leptin in blood is bound to a soluble fragment of the leptin receptor. Leptin is cleared from the blood primarily by the kidney.

The leptin receptor, like receptors for GH and prolactin, belongs to the class of transmembrane cytokine receptors that signals by activation of the cytosolic tyrosine kinase JAK-2 (see Chapters 1 and 10). Multiple splice variants of leptin receptor mRNA give rise to different isoforms, including one that circulates as a soluble protein. Other isoforms have truncated cytoplasmic domains, but only the form with the full-length cytoplasmic tail appears to be capable of signaling. Truncated forms, which are expressed in vascular endothelium and the choroid plexus, may serve a transport function to facilitate passage of leptin across blood vessels in the brain to breech the brain barrier and thus deliver leptin to target cells in the central nervous system.

Hypothalamic Neuronal Targets and Their Peptide Products

The principal targets for leptin are neurons in the arcuate nuclei of the hypothalamus, but leptin receptors are also found in neurons of the paraventricular,

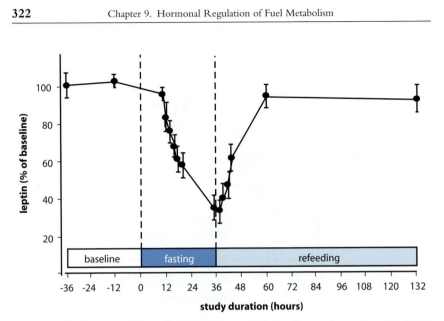

Figure 16 Fasting and refeeding effects on leptin concentrations in the plasma of normal human subjects. (From Kolaczynski, J. W., Considine, R. V., Ohannesian, J., Marco, C., Opentanova, I., Nyce, M. R., Myint, M., and Caro, J. F., *Diabetes* **45**, 1511–1515, 1996, reprinted with permission from *The American Diabetes Association*.)

ventromedial, and dorsomedial nuclei and neurons in the lateral hypothalamus. Neuropeptide transmitters of some of these neurons have been identified along with their receptors and the sites of their expression have been located. These neurons project to hypothalamic and brain stem autonomic integrating centers. Neurons from the arcuate and paraventricular nuclei project to the median eminence of the hypothalamus, where they release hypophysiotropic hormones into the hypophysial–portal circulation (see Chapter 2). Through these neural connections leptin integrates nutritional status with adrenal, gonadal, and thyroidal function and with GH and prolactin secretion. Some of the relevant neuropeptides are briefly described in the following list:

1. *Neuropeptide Y* (NPY). This 36-amino-acid peptide is abundantly expressed in neurons of the arcuate nuclei of the hypothalamus that have axons projecting to the paraventricular nuclei and the lateral hypothalamic area. Its expression is increased during fasting. When administered into the hypothalamus of rodents, NPY stimulates food intake, lowers energy expenditure, and produces obesity. Expression of NPY is increased in leptin deficiency and suppressed by leptin administration.

2. *Proopiomelanocortin* (POMC). This precursor of ACTH in anterior pituitary cells is also expressed in the arcuate nuclei of the hypothalamus, where

posttranslational processing (see Figure 3 in Chapter 2) gives rise to α-melanocyte-stimulating hormone.

   3. *α-melanocyte-stimulating hormone* (MSH). As a neuropeptide, α-MSH is a potent negative regulator of food intake and activates melanocortin receptors in neurons in the dorsomedial and paraventricular nuclei of the hypothalamus and the lateral hypothalamic area. Expression of α-MSH in rodents is increased by leptin and suppressed by overfeeding. Pharmacological blockade or genetic depletion of brain melanocortin receptors results in obesity.

   4. *Agouti-related protein* (AGRP). In the skin, α-MSH increases expression of a black pigment in hair follicles. A protein, called agouti, competes with α-MSH for the melanocortin receptor, and under its influence, a yellow pigment is expressed. The observation that a mutation that results in ubiquitous inappropriate expression of the agouti gene in mice also produces obesity led to the discovery that neurons in the arcuate nucleus express a similar protein, the agouti-related protein (AGRP), which also competes with α-MSH for binding to the melanocortin receptor. AGRP and neuropeptide Y are coexpressed in arcuate neurons, and their combined actions provide a strong drive for food intake. Expression of AGRP is increased in leptin deficiency and decreased by leptin treatment.

   5. *Melanin-concentrating hormone* (MCH). Originally described as the factor that opposes the melanophore-dispersing activity of MSH in fish, it is now known that MCH is also expressed in neurons in the lateral hypothalamus. When administered to rodents, MCH stimulates feeding behavior and antagonizes the inhibitory effects of α-MSH. Unlike AGRP, MCH does not bind to the same receptor as α-MSH. Fasting increases MCH expression, and absence of the MCH gene in mice results in reduction in body fat content.

   6. *Cocaine- and amphetamine-regulated transcript* (CART). This is another appetite-suppressing peptide that was discovered during studies of drugs of abuse. It is widely expressed in the brain and other endocrine tissues, including some neurons in the arcuate nuclei that also express α-MSH. Administration of CART inhibits the feeding response to NPY, and mice that are deficient in CART become obese.

Other neuropeptides and amine transmitters that originate in neurons in various brain loci and the gastrointestinal tract also participate in the complex regulation of feeding behavior. A complete "wiring diagram" cannot yet be drawn, but some of the relationships are shown in Figure 17.

Other Effects of Leptin

   Through its actions on neurons associated with autonomic and anterior pituitary functions, leptin affects temperature regulation, reproduction, and adrenal cortical function. Leptin receptors are also found in many cells outside of the

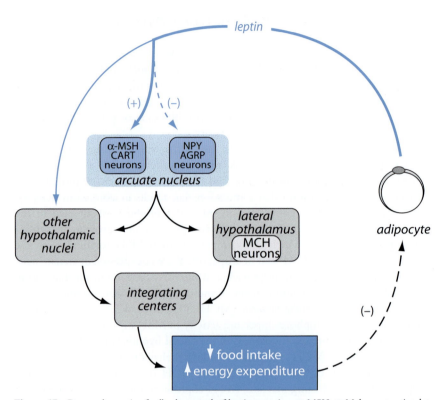

Figure 17   Proposed negative feedback control of leptin secretion. α-MSH, α-Melanocyte-stimulating hormone; CART, cocaine- and amphetamine-regulated transcript; NPY, neuropeptide Y; AGRP, agouti-related peptide; MCH, melanin-concentrating hormone.

central nervous system. Adipocytes express leptin receptors and respond to leptin in an autocrine manner with an increase in lipolysis. Leptin acts directly on pancreatic beta cells to inhibit insulin secretion and thus forms one arm of a negative feedback arrangement between beta cells and adipocytes in which stimulation of leptin secretion by insulin leads to inhibition of insulin secretion by leptin. The presence of leptin receptors in the gonads suggests that peripheral actions of leptin may complement the fertility-promoting effects exerted at the hypothalamic level. Other peripheral effects of leptin include actions in bone marrow to promote hematopoiesis and actions in capillary endothelium to increase angiogenesis (blood vessel formation). Leptin is also produced by the placenta and may play a role in fetal development.

SUGGESTED READING

Ahima, R. S., and Flier, J. S. (2000). Leptin. *Annu. Rev. Physiol.* **62**, 413–437.

Felig, P., Sherwin, R. S., Soman, V., Wahren, J., Hendler, R., Sacca, L., Eigler, N., Goldberg, D., and Walesky, M. (1979). Hormonal interactions in the regulation of blood glucose. *Recent Prog. Horm. Res.* **35**, 501–528.

Jefferson, L. S. and Cherrington A. D. (2001). "The Endocrine Pancreas and Regulation of Metabolism, Handbook of Physiology, Section 7, Volume II." Oxford University Press, New York. This work contains detailed treatment of many of the topics discussed in this chapter.

Randle, P. J., Kerbey, A. L., and Espinal, J. (1988). Mechanisms decreasing glucose oxidation in diabetes and starvation: Role of lipid fuels and hormones. *Diabetes/Metab. Rev.* **4**, 623–638.

Ruderman, N. B., Saha, A. K., Vavvas, D., and Witters, L. A. (1999). Malonyl-CoA, fuel sensing, and insulin resistance. *Am. J. Physiol.* **276**, E1–E18.

Wasserman D. H. and Cherrington A. D. (1996). Regulation of extramuscular fuel sources during exercise. *In* "Exercise: Regulation and Integration of Multiple Systems, Handbook of Physiology, Section 12" (L. B. Rowell, and J. T. Shepherd, eds.), pp. 1036–1074. Oxford University Press, New York.

Hormonal Control of Growth

OVERVIEW AND GENERAL CONSIDERATIONS

The simple word "growth" describes a variety of processes, both living and nonliving, that share the common feature of increase in mass. For purposes of this chapter we limit the definition of growth to mean the organized addition of new tissue that occurs normally in development from infancy to adulthood. This process is complex and depends on the interplay of genetic, nutritional, and environmental influences as well as actions of the endocrine system. Growth of an individual or an organ involves increases both in cell number and in cell size, differentiation of cells that perform highly specialized functions, and tissue remodeling, which may require apoptosis as well as new cell formation. Most of these processes depend on locally produced growth factors that operate through paracrine or autocrine mechanisms. Many continue to operate throughout life, providing not only for cell renewal but also for adaptations to meet changing physiological demands. Dozens of families of growth factors have been described and an unknown number of others await discovery. Regulation of growth by the endocrine system can be viewed as coordination of local growth processes with overall development of the individual and with external environmental influences. This chapter describes the hormones that play important roles in growth and their interactions at critical times in development.

Growth is most rapid during prenatal life. In only 9 months body length increases from just a few micrometers to almost 30% of final adult height. The growth rate decelerates after birth but during the first year of life is rapid enough that the infant increases half again in height, to about 45% of final adult stature. Thereafter growth decelerates and continues at a slower rate, about 2 inches per year until puberty. Steady growth during this juvenile period contributes the largest fraction, about 40%, to final adult height. With the onset of sexual development, growth accelerates to about twice the juvenile rate and contributes about 15–18% of final adult height before stopping altogether (Figures 1 and 2). Our understanding of hormonal influences on growth is limited largely to the juvenile and adolescent periods, but emerging information is providing insight into regulation of prenatal growth, which is largely independent of the classical hormones. During the juvenile period, the influence of growth hormone (GH) is preeminent, but appropriate secretion of thyroid and adrenal hormones and insulin is essential for optimal growth. The adolescent growth spurt reflects the added input of androgens and estrogens, which speed up growth and the maturation of bone that brings growth to a halt.

GROWTH HORMONE

Consequences of Deficiency or Excess

Growth hormone, which is also called somatotropin, is the single most important hormone required for normal growth. Attainment of adult size is absolutely dependent on GH; in its absence growth is severely limited. *Pituitary dwarfism* is the failure of growth that results from lack of GH during childhood.

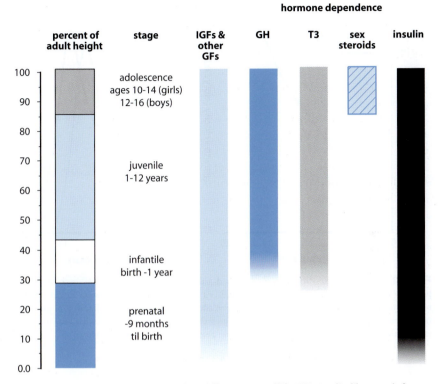

Figure 1 Hormonal regulation of growth at different stages of life. IGFs, Insulin-like growth factors; GH, growth hormone; T3, triiodothyronine.

Pituitary dwarfs typically are of normal weight and length at birth and grow rapidly and nearly normally during early infancy. Before the end of the first year, however, growth is noticeably below normal and continues slowly for many years; left untreated, children with this hormone defect may reach heights of around 4 feet. Typically the pituitary dwarf retains a juvenile appearance because of the retention of "baby fat" and the disproportionally small size of maxillary and mandibular bones.

Pituitary dwarfism is not a single entity and may encompass a range of defects. The deficiency in GH may be accompanied by deficiencies of several or all other anterior pituitary hormones (*panhypopituitarism*), as might result from defects in pituitary development (see Chapter 2). Alternatively, traumatic injury to the pituitary gland or a tumor that either destroys pituitary cells or their connections to the hypothalamus might also interfere with normal pituitary function. Individuals with such trauma do not mature sexually and suffer from inadequacies

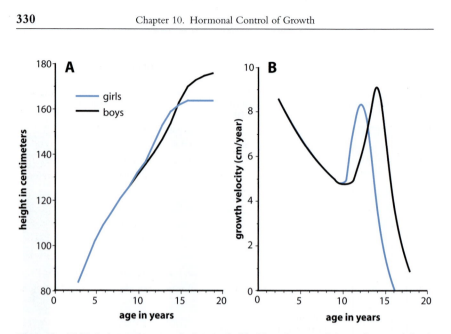

Figure 2   (A) Typical growth curves for boys and girls. Note that growth is not linear and that it proceeds at the same rate in juvenile boys and girls. At puberty, which begins earlier in girls than in boys, there is a spurt in growth that immediately precedes growth arrest. (B) Nonlinearity of growth is more clearly evident when plotted as changes in growth velocity over time. Note that growth is very rapid in the newborn, but slows during the juvenile period and accelerates at puberty.

of thyroid and adrenal glands as well. The lack of GH might also be an isolated inherited defect, with no abnormalities in other pituitary hormones. Aside from their diminutive height, such individuals are normal in all respects and can reproduce normally. Causes of isolated GH deficiency are multiple, and include derangements in synthesis, secretion, and end–organ responsiveness.

    Overproduction of GH may occur either as a result of some derangement in mechanisms that control secretion by normal pituitary cells or from tumor cells that secrete GH autonomously. Overproduction of GH during childhood results in *gigantism*, in which an adult height in excess of 8 feet has occasionally been reported. Overproduction of GH during adulthood, after the growth plates of long bones have fused (see below), produces growth only by stimulation of responsive osteoblastic progenitor cells in the periosteum. There is thickening of the cranium, the mandible, and enlargement of some facial bones and bones in the hands and feet. Growth and deformities in these acral parts give rise to the term *acromegaly* to describe this condition. Persistence of responsive cartilage progenitor cells in the costochrondral junctions leads to elongation of the ribs, giving a typical barrel-chested appearance. In acromegalic patients there is also thickening of the skin and disproportionate growth of some soft tissues, including spleen and liver.

When thinking about giants and dwarfs, it is important to keep in mind the limitations of GH action. The pediatric literature makes frequent use of the term *genetic potential* in discussions of diagnosis and treatment of disorders of growth. Predictions of how tall a child will be as an adult are usually based on the average of parental height plus 2.5 inches for boys or minus 2.5 inches for girls. We can think of GH as the facilitator of expression of genetic potential for growth rather than as the primary determinant. The entire range over which GH can influence adult stature is only about ~30% of genetic potential. A person destined by genetic makeup to attain a final height of 6 feet will attain a height of about 4 feet even in the absence of GH and is unlikely to exceed 8 feet in height even with massive overproduction of GH from birth. We do not understand what determines genetic potential for growth, but it is clear that although both arise from a single cell, a hypopituitary elephant is enormously larger than a giant mouse. Within the same species, something other than aberrations in GH secretion accounts for the large differences in size of miniature and standard poodles or chihuahuas and great danes.

SYNTHESIS, SECRETION, AND METABOLISM

Although the anterior pituitary gland produces at least six hormones, more than one-third of its cells synthesize and secrete GH. In humans the 5 to 10 mg of stored GH make it the most abundant hormone in the pituitary, accounting for almost 10% of the dry weight of the gland. More than 10 times as much GH is produced and stored as any other pituitary hormone. Of the GH produced by somatotropes, 90% is composed of 191 amino acids and has a molecular weight of about 22,000. The remaining 10%, called 20K GH, has a molecular weight of 20,000 and lacks the 15-amino-acid sequence corresponding to residues 32 to 46. Both forms are products of the same gene and result from alternate splicing of the RNA transcript. Both forms of hormone are secreted and have similar growth-promoting activity, although metabolic effects of the 20K form are reduced.

About half of the GH in blood circulates bound to a protein that has the same amino acid sequence as the extracellular domain of the GH receptor (see below). In fact, the plasma GH-binding protein is a product of the same gene that encodes the GH receptor and originates by proteolytic cleavage of the receptor at the outer surface of target cells. It is thought that the binding protein provides a reservoir of GH that prolongs its half-life and buffers changes in free hormone concentration. Free GH can readily cross capillary membranes, but bound hormone is restricted to the vascular compartment. The half-life of GH in blood is about 20 minutes. GH that crosses the glomerular membrane is reabsorbed and destroyed in the kidney, which is the major site of GH degradation. Less than 0.01% of the hormone secreted each day reaches the urine in recognizable form. GH is also degraded in its various target cells following uptake by receptor-mediated endocytosis.

MODE OF ACTION

Like other peptide and protein hormones GH binds to its receptor on the surface of target cells. The GH receptor is a glycoprotein that has a single membrane-spanning region and a relatively long intracellular tail that has no catalytic activity and does not interact with G proteins. The GH receptor binds to a cytosolic enzyme called Janus kinase-2 (JAK-2), which catalyzes the phosphorylation of the receptor and other proteins on tyrosine residues (see Chapter 1). Growth hormone activates a signaling cascade by binding sequentially to two GH receptor molecules to form a receptor dimer that sandwiches the hormone between the two receptor molecules. Such dimerization of receptors is also seen for other hormone and cytokine receptors of the superfamily to which the GH receptor belongs. Dimerization of receptors brings the bound JAK-2 enzymes into favorable alignment to promote tyrosine phosphorylation and activation of their catalytic sites. In addition, dimerization may also recruit JAK-2 molecules to unoccupied binding sites on the receptors. Tyrosine phosphorylation provides docking sites for other proteins and facilitates their phosphorylation. One group of target proteins, called Stat proteins, involved in signal transduction and activation of transcription, can migrate to the nucleus and activate gene transcription (see Chapter 1). Another target group, the mitogen-activated protein (MAP) kinases, is also thought to have a role in promoting gene transcription. Activation of the GH receptor also results in an influx of extracellular calcium through voltage-regulated channels, which may further promote transcription of target genes. All in all, GH produces its effects in various cells by stimulating the transcription of specific genes.

PHYSIOLOGICAL ACTIONS OF GH

Effects on Skeletal Growth

The ultimate height attained by an individual is determined by the length of the skeleton—in particular, the vertebral column and long bones of the legs. Growth of these bones occurs by a process called *endochondral ossification*, in which proliferating cartilage is replaced by bone. The ends of long bones are called *epiphyses* and arise from ossification centers that are separate from those responsible for ossification of the *diaphysis*, or shaft. In the growing individual the epiphyses are separated from the diaphysis by cartilaginous regions called *epiphyseal plates*, in which continuous production of chondrocytes provides the impetus for diaphyseal elongation. Chondrocytes in epiphyseal growth plates are arranged in orderly columns in parallel with the long axis of the bone (Figure 3). Frequent division of small, flattened cells in the germinal zone at the distal end of the growth plate provides for continual elongation of columns of chondrocytes. As they grow and

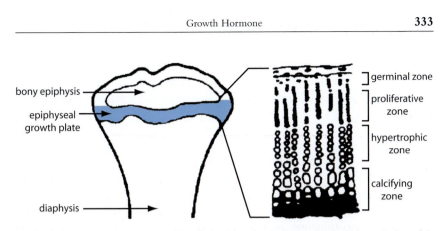

Figure 3 Schematic representation of the tibial epiphyseal growth plate. (From Ohlsson, C., Isgaard, J., Törnell, J., Nilsson, A., Isaksson, O. G. P., and Lindahl, A., *Acta Paediatr. Suppl.* **391**, 33–40, 1993, with permission.)

mature, chondrocytes produce the mucopolysaccharides and collagen, which constitute the cartilage matrix. Cartilage cells hypertrophy, become heavily vacuolated, and degenerate as the surrounding matrix becomes calcified. Ingrowth of blood vessels and migration of osteoblast progenitors from the marrow result in replacement of calcified cartilage with true bone. Proliferation of chondrocytes at the epiphyseal border of the growth plate is balanced by cellular degeneration at the diaphyseal end, so that in the normally growing individual the thickness of the growth plate remains constant as the epiphyses are pushed further and further outward by the elongating shaft of bone.

Eventually, progenitors of chondrocytes are either exhausted or lose their capacity to divide. As remaining chondrocytes go through their cycle of growth and degeneration, the epiphyseal plate becomes progressively narrower and is ultimately obliterated when diaphyseal bone fuses with the bony epiphyses. At this time, the epiphyseal plates are said to be closed, and the capacity for further growth is lost. In the absence of GH there is severe atrophy of the epiphyseal plates, which become narrow as proliferation of cartilage progenitor cells slows markedly. Conversely, after GH is given to a hypopituitary subject, resumption of cellular proliferation causes columns of chondrocytes to elongate and epiphyseal plates to widen. This characteristic response has been used as the basis of a biological assay for GH in experimental animals.

Growth of bone requires that diameter as well as length increase. Thickening of long bones is accomplished by proliferation of osteoblastic progenitors from the connective tissue sheath (*periosteum*) that surrounds the diaphysis. As it grows, bone is also subject to continual reabsorption and reorganization, with the incorporation of new cells that originate in both the periosteal and endosteal regions. Remodeling, which is an intrinsic property of skeletal growth, is accompanied by destruction and replacement of calcified matrix, as described in

Chapter 8. Treatment with GH often produces a transient increase in urinary excretion of calcium, phosphorus, and hydroxyproline, reflecting bone remodeling. Hydroxyproline derives from breakdown and replacement of collagen in bone matrix.

SOMATOMEDIN HYPOTHESIS

The epiphyseal growth plates are obviously stimulated after GH is given to hypophysectomized animals, but little or no stimulation of cell division, protein synthesis, or incorporation of radioactive sulfur into mucopolysaccharides of cartilage matrix was observed when epiphyseal cartilage taken from hypophysectomized rats was incubated with GH. In contrast, when cartilage taken from the same rats was incubated with blood plasma from hypophysectomized rats that had been treated with GH, there was a sharp increase in matrix formation, protein synthesis, and DNA synthesis. Blood plasma obtained from normal rats produced similar effects, but plasma from hypophysectomized rats had little effect unless the rats were first treated with GH. These experiments gave rise to the hypothesis that GH may not act directly to promote growth but, instead, stimulates the liver to produce an intermediate, blood-borne substance that activates chondrogenesis and perhaps other GH-dependent growth processes in other tissues. This substance was later named *somatomedin* (because it is a somatotropin mediator), and on subsequent purification was found to consist of two closely related substances that also produce the insulin-like activity that persists in plasma after all the authentic insulin is removed by immunoprecipitation. These substances are now called *insulin-like growth factors*, or IGF-I and IGF-II. Of the two, IGF-I appears to be the more important mediator of the actions of GH, and has been studied more thoroughly. Although some aspects of the original somatomedin hypothesis have been discarded (see below), the crucial role of IGF-I as an intermediary in the growth-promoting action of GH is now firmly established.

In general, plasma concentrations of IGF-I reflect the availability of GH or the rate of growth. They are higher than normal in blood of persons suffering from acromegaly and are very low in GH-deficient individuals. Children whose growth is more rapid than average have higher than average concentrations of IGF-I, whereas children at the lower extreme of normal have lower values. When GH is injected into GH-deficient patients or experimental animals, IGF-I concentrations increase after a delay of about 6 to 8 hours and remain elevated for more than a day. Children or adults who are resistant to GH because of a receptor defect have low concentrations of IGF-I in their blood despite high concentrations of GH. Growth of these children is restored to nearly normal rates following daily administration of IGF-I (Figure 4). Disruption of the IGF-I gene in mice causes severe growth retardation despite high concentrations of GH in their blood. Daily

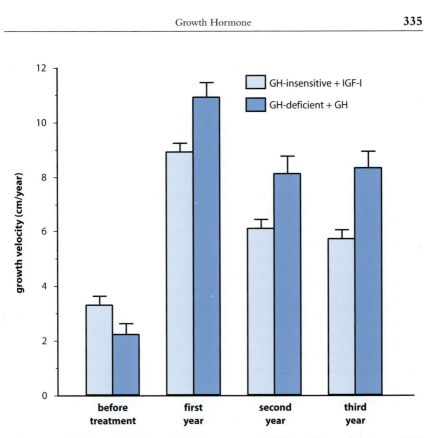

Figure 4 Insulin-like growth factor-I (IGF-I) treatment of children with growth hormone (GH) insensitivity due to a receptor deficiency, compared to GH treatment of children with GH deficiency. (Plotted from data of Guevara-Aguirre, J., Rosenbloom, A. L., Vasconez, O., Martinez, V, Gargosky, S., Allen, L., and Rosenfeld, R., *J. Clin. Endocrinol. Metab.* **82**, 629–633, 1997.)

treatment with large doses of GH does not accelerate their growth. Similarly, a child with a homozygous deletion of the IGF-I gene suffered severe pre- and postnatal growth retardation that was partially corrected by daily treatment with IGF-I.

Although overwhelming evidence indicates that IGF-I stimulates cell division in cartilage and many other tissues and accounts for much and perhaps all of the growth-promoting actions of GH, the somatomedin hypothesis as originally formulated is inconsistent with recent experimental findings. Production of IGF-I is not limited to the liver, and may be increased by GH in many tissues, including cells in the epiphyseal growth plate. Direct infusion of small amounts of GH into epiphyseal cartilage of the proximal tibia in one leg of hypophysectomized rats was found to stimulate tibial growth of that limb, but not of the contralateral limb.

Only a direct action of GH on osteogenesis can explain such localized stimulation of growth, because IGF-I that arises in the liver is equally available in the blood supply to both hind limbs. It is now apparent that GH stimulates prechondrocytes and other cells in the epiphyseal plates to synthesize and secrete IGFs that act locally in an autocrine or paracrine manner to stimulate cell division, chondrocyte maturation, and bone growth. Evidence to support this conclusion includes findings of receptors for both GH and IGF-I in cells in the epiphyseal plates along with the GH-dependent increase in mRNA for IGFs. Thus growth of the long bones might be stimulated by IGF-I that reaches the bones either through the circulation or by diffusion from local sites of production, or some combination of the two.

A genetic engineering approach was adopted to evaluate the relative importance of locally produced and blood-borne IGF-I. A line of mice was developed in which the IGF-I gene was selectively disrupted only in hepatocytes. Concentrations of IGF-I in the blood of these animals were severely reduced, but their growth and body proportions were no different from those of control animals that produced normal amounts of IGF-I in their livers and had normal blood levels of IGF-I. These findings indicate that locally produced IGF-I is sufficient to account for normal growth and that IGF-I in the circulation plays only a minor role, if any, in stimulating growth. However, the average concentration of GH in the blood of these genetically altered mice was considerably increased, consistent with the negative feedback effect of IGF-I on GH secretion. The current view of the relationship between GH and IGF-I is summarized in Figure 5. GH acts directly on both the liver and its peripheral target tissues to promote IGF-I production. Liver is the principal source of IGF in blood, but target tissues also make some contributions. Stimulation of growth is provided primarily by locally produced IGF-I acting in an autocrine/paracrine manner, but some IGF-I produced in liver or elsewhere probably makes a small contribution. The major function of blood-borne IGF is to regulate GH secretion.

Properties of the Insulin-like Growth Factors

IGF-I and IGF-II are small, unbranched peptides that have molecular masses of about 7500 Da. They are encoded in separate genes located on chromosomes 12 and 11, respectively, and are expressed in a wide variety of cells. Although the regulatory elements and exon/intron architecture of their genes differ significantly, protein structures of the IGFs are very similar to each other and to proinsulin (see Chapter 5), both in terms of amino acid sequence and in the arrangement of disulfide bonds. The IGFs share about 50% amino acid identity with insulin. In contrast to insulin, however, the region corresponding to the connecting peptide is retained in the mature form of the IGFs, which also have a C-terminal extension. Both IGF-I and IGF-II are present in blood at relatively high

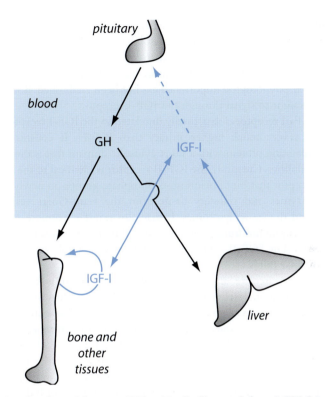

Figure 5 The roles of growth hormone (GH) and insulin-like growth factor-I (IGF-I) in promoting growth. GH stimulates IGF-I production in liver and epiphyseal growth plates. Epiphyseal growth is stimulated primarily by autocrine/paracrine actions of IGF-I. Hepatic production of IGF-I acts primarily as a negative feedback regulator of GH secretion. Dashed arrow signifies inhibition. Liver is the principal source of IGF-I in blood, but other GH target organs may also contribute to the circulating pool.

concentrations throughout life, although the absolute amounts differ at different stages of life. The concentration of IGF-II is usually about three times higher than that of IGF-I.

Two receptors for the IGFs have been identified. The IGF-I receptor, which binds IGF-I with greater affinity than IGF-II, is remarkably similar to the insulin receptor. Like the insulin receptor (see Chapter 5), it is a tetramer that consists of two membrane-spanning beta subunits connected by disulfide bonds to two extracellular alpha subunits, which contain the IGF-binding domain. As in the insulin receptor, the beta subunits have intrinsic protein tyrosine kinase activity that is activated by ligand binding and catalyzes the phosphorylation of some of its own tyrosine residues and tyrosines on many of the same proteins that mediate the intracellular responses to insulin: the insulin receptor substrates, phosphatidylinositol 3-kinase, etc. Cells that coexpress both insulin and IGF-I receptors may also produce "hybrid receptors" that have one alpha and one beta subunit of the insulin receptor coupled to one alpha one beta subunit of the IGF-I receptor. These receptors behave more like IGF-I receptors than like insulin receptors. Their physiological importance has not been established. Both IGF-I and IGF-II are thought to signal through the IGF-I receptor.

The IGF-II receptor is structurally unrelated to the IGF-I receptor and binds IGF-II with a very much higher affinity than IGF-I . It consists of a single membrane-spanning protein with a short cytosolic domain that is thought to lack signaling capabilities. Curiously, the IGF-II receptor is identical to the mannose-6-phosphate receptor that binds mannose-6-phosphate groups on newly synthesized lysosomal enzymes and transfers them from the trans-Golgi vesicles to the endosomes and thence to lysosomes. It may also transfer mannose-6-phosphate-containing glycoproteins from the extracellular fluid to the lysosomes by an endocytotic process. The IGF-II receptor plays an important role in clearing IGF-II from extracellular fluids.

The IGFs circulate in blood tightly bound to IGF binding proteins (IGFBPs). Six different closely related IGFBPs, each the product of a separate gene, are found in mammalian plasma and extracellular fluids. Their affinities for both IGF-I and IGF-II are considerably higher than are the affinities of the IGF-I receptors for either IGF-I or IGF-II. The combined binding capacity of all the IGFBPs in plasma is about twice that needed to bind all of the IGFs in blood. IGFBP-3, whose synthesis is stimulated by GH, IGF-I, and insulin, is the most abundant form and is complexed with most of the IGF-I and IGF-II in plasma. IGFBP-3 and its cargo of IGFs form a large 150-kDa ternary complex with a third protein, the acid-labile subunit (ALS), whose synthesis is also stimulated by GH. Consequently, the concentrations of both proteins are quite low in the blood of GH-deficient subjects and increase on treatment with GH. The remaining IGFs in plasma are distributed among the other IGFBPs that do not bind to ALS, and hence form complexes that are small enough to escape across the capillary endothelium. Of these, IGFBP-2 is the most important quantitatively. Its concentration in blood is increased in plasma of GH-deficient patients and is decreased by GH, but rises dramatically after administration of IGF-I.

Normally the binding capacity of IGFBP-3 is saturated, whereas the other IGFBPs have free binding sites. Consequently the IGFs do not readily escape from the vascular compartment and have half-lives in blood of about 15 hours. Proteolytic "clipping" of IGFBP-3 by proteases present in plasma lowers its binding affinity and releases IGF-I, which may then form lower molecular weight complexes with other IGFBPs and escape to the extracellular fluid. The major functions of the IGF binding proteins in blood are to provide a plasma reservoir of IGF-I and IGF-II, to slow their degradation, and to regulate their bioavailability.

The IGFBPs are synthesized locally in conjunction with IGF in a wide variety of cells and are widely distributed in extracellular fluid. Their biology is complex and not completely understood. It may be recalled that the IGFs mediate localized growth in response to a variety of signals in addition to GH. Many different cells both produce and respond to IGF-I, which is a small and readily diffusible molecule. The IGFBPs may provide a means of restricting the extent of cell growth to the precise location dictated by physiological demand. Because their affinity for both IGFs is so much greater than the affinity of the IGF-I receptor, the IGFBPs can successfully compete for binding free IGF and thus restrict its bioavailability. Conversely, IGFBPs may also enhance the actions of IGF-I. Some of the IGFBPs bind to extracellular matrices, where they may provide a localized reservoir of IGFs that might be released by proteolytic modification of the IGFBPs. Binding to the cell surface lowers the affinity of some of the IGFBPs and thus provides a means of targeted delivery of free IGF-I to receptive cells. Some evidence also suggests that IGFBPs may produce biological effects that are independent of the IGFs.

Effects of GH/IGF-I on Body Composition

Growth hormone-deficient animals and human subjects have a relatively high proportion of fat, compared to water and protein, in their bodies. Treatment with GH changes the proportion of these bodily constituents to resemble the normal juvenile distribution. Body protein stores increase, particularly in muscle, and there is a relative decrease in fat. Despite their relatively higher fat content, subjects who are congenitally deficient in GH or unresponsive to it actually have fewer total adipocytes than do normal individuals. Their adiposity is due to an increase in the amount of fat stored in each cell. Treatment with GH restores normal cellularity by increasing proliferation of fat cell precursors through autocrine stimulation of by IGF-I. Curiously, however, GH also restrains the differentiation of fat cell precursors into mature adipocytes. The overall decrease in body fat produced by GH results from decreased deposition of fat and accelerated mobilization and increased reliance of fat for energy production (see Chapter 9).

Most internal organs grow in proportion to body size, except liver and spleen, which may be disproportionally enlarged by prolonged treatment with GH. The heart may also be enlarged in acromegalic subjects, in part from stimulation of cardiac myocyte growth by GH or IGF, and in part from hypertension, which is frequently seen in these individuals. Conversely, GH deficiency beginning in childhood is associated with decreased myocardial mass due to decreased thickness of the ventricular walls. Treatment of these individuals with GH leads to increased myocardial mass and improved cardiac performance. Skin and the underlying

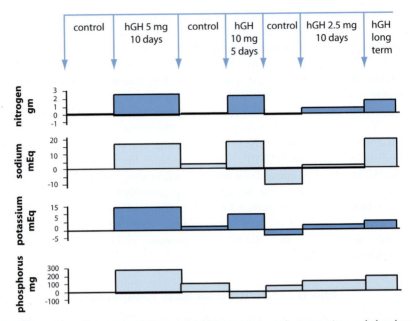

Figure 6   Effects of human growth hormone (hGH) on nitrogen, sodium, potassium, and phosphorus balances in an 11.5-year-old girl with pituitary dwarfism. Changes above the control base line represent retention of the substance; changes below the base line represent loss. (From Hutchings, J. J., Escamilla, R. F., Deamer, W. C., *et al., J. Clin. Endocrinol. Metab.* **19**, 759–764, 1959, by permission of The Endocrine Society.)

connective tissue also increase in mass, but GH does not appear to influence growth of the thyroid, gonads, or reproductive organs.

Changes in body composition and organ growth have been monitored by studying changes in the biochemical balance of body constituents (Figure 6). When human subjects or experimental animals are given GH repeatedly for several days, there is net retention of nitrogen, reflecting increased protein synthesis. Urinary nitrogen is decreased, as is the concentration of urea in blood. Net synthesis of protein is increased without an accompanying change in the net rate of protein degradation. Increased retention of potassium reflects the increase in intracellular water that results from increased cell size and number. An increase in sodium retention and the consequent expansion of extracellular volume is characteristic of GH replacement and may result from activation of sodium channels in the distal portions of the nephron. Increased phosphate retention reflects expansion of the cellular and skeletal mass and is brought about in part by activation of sodium phosphate cotransporters in the proximal tubules and activation of the 1α-hydroxylase that catalyzes production of calcitriol (Chapter 8).

Comparison of the Actions of GH and IGF-I

Although IGF-I figures prominently as an intermediary in the actions of GH, not all of the effects of GH are mediated by the IGFs. For example, hepatocytes and adipocytes lack IGF-I receptors, but GH nevertheless modifies expression of a variety of their genes. Clinical experience indicates that IGF-I is less effective than GH in promoting skeletal growth (see Figure 4), but more effective than GH in stimulating growth of some soft tissues. The physiological importance of IGF-I-independent actions of GH is unknown, but treatment with IGF-I reproduces most of the effects of GH on body size and composition in patients who are insensitive to GH because of a receptor defect.

REGULATION OF GH SECRETION

Growth is a slow, continuous process that takes place over more than a decade. It might be expected, therefore, that concentrations of GH in blood would be fairly static. In contrast to such expectations, however, frequent measurements of GH concentrations in blood plasma throughout the day reveal wide fluctuations, indicative of multiple episodes of secretion. Because metabolism of GH is thought to be invariant, changes in plasma concentration imply changes in secretion. In male rats, GH is secreted in regular pulses every 3.0 to 3.5 hours in what has been called an *ultradian* rhythm. In humans, GH secretion is also pulsatile, but the pattern of changes in blood concentrations is usually less obvious than in rats. Frequent bursts of secretion occur throughout the day, with the largest being associated with the early hours of sleep (Figure 7). In addition, stressful changes in the internal and external environments can produce brief episodes of hormone secretion. Little information or diagnostic insight can therefore be obtained from a single random measurement of the GH concentration in blood. Because secretory episodes last only a short while, multiple, frequent measurements are needed to evaluate functional status or to relate GH secretion to physiological events. Alternatively, it is possible to withdraw small amounts of blood continuously over the course of a day, and, by measuring GH in the pooled sample, to obtain a 24-hour integrated concentration of GH in blood.

The possible physiological significance of intermittent as compared to constant secretion of GH has received much attention experimentally. Pulsatile administration of GH to hypophysectomized rats is more effective in stimulating growth than is continuous infusion of the same daily dose. However, similar findings have not been made in human subjects, whose rate of growth, like that of experimental animals, can be restored to normal or near normal with single daily or every-other-day injections of GH. Although expression of some hepatic genes appears to be sensitive to the pattern of changes in plasma GH concentrations in

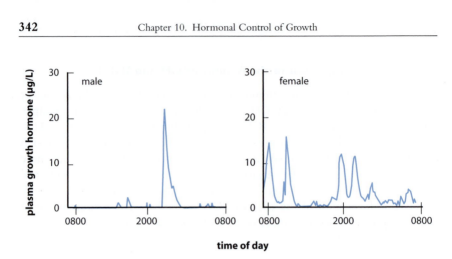

Figure 7   Growth hormone concentrations in blood sampled at 10-minute intervals over a 24-hour period in a normal man and a normal woman. The large pulse in the man coincides with the early hours of sleep. Note that the pulses of secretion are more frequent and of greater amplitude in the woman. (From Asplin, C. M., Faria, H. C. S., Carlsen, E. C., *et al.*, *J. Clin. Endocrinol. Metab.* **69**, 239–245, 1989, by permission of The Endocrine Society.)

rodents, there is neither evidence for comparable effects in humans nor an obvious relationship of the affected genes to growth of rodents. In normal human adults the same total amount of GH given either as a constant infusion or in eight equally spaced brief infusions over 24 hours increased expression IGF-I and IGFBP-3 to the same extent.

EFFECTS OF AGE

Using the continuous sampling method, it was found that GH secretion, though most active during the adolescent growth spurt, persists throughout life, long after the epiphyses have fused and growth has stopped (Figure 8). In midadolescence the pituitary secretes between 1 and 2 mg of GH per day. Between ages 20 and 40 years, the daily rate of secretion gradually decreases in both men and women, but remarkably, even in middle age the pituitary continues to secrete about 0.1 mg of GH every day. Low rates of GH secretion in the elderly may be related to loss of lean body mass in later life. Changes in GH secretion with age primarily reflect changes in magnitude of secretory pulses (Figure 9).

REGULATORS OF GH SECRETION

In addition to spontaneous pulses, secretory episodes are induced by metabolic signals such as a rapid fall in blood glucose concentration or an increase in certain

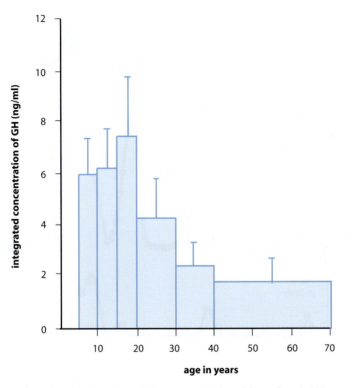

Figure 8   Relation between the integrated plasma concentration of GH and age in 173 normal males and females. (From Zadik, Z., Chalew, S. A., McCarter, R. J., Jr., Meistas, M., and Kowarski, A. A., *J. Clin. Endocrinol. Metab.* **60**, 513–516, 1985, by permission of The Endocrine Society.)

amino acids, particularly arginine and leucine. The physiological significance of these changes in GH secretion is not understood, but provocative tests using these signals are helpful for judging competence of the GH secretory apparatus (Figure 10). Traumatic and psychogenic stresses are also powerful inducers of GH secretion in humans and monkeys, but whether increased secretion of GH is beneficial for coping with stress is not established and is not universally seen in mammals. In rats, for example, GH secretion is inhibited by the same signals that increase it in humans. However, regardless of their significance, these observations indicate that GH secretion is under minute-to-minute control by the nervous system. That control is expressed through the hypothalamo–hypophyseal portal circulation, which delivers two hypothalamic neuropeptides to the somatotropes: GH-releasing hormone (GHRH) and somatostatin. It is possible that a third hormone, ghrelin (see Chapter 2), also plays a role in this regard, but data

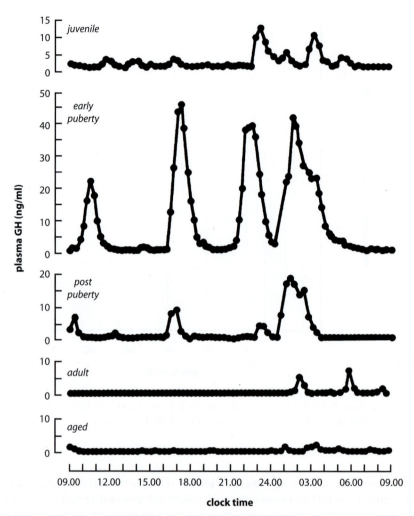

Figure 9 Changing patterns of GH secretion with age. (Modified from Robinson, I. C. A. F., and Hindmarsh, P. C., "Handbook of Physiology, Section 7, The Endocrine System, Volume V: Hormonal Control of Growth," pp. 329–396, Oxford University Press, New York, 1999, with permission.)

to support this premise are not yet in hand. GHRH provides the primary drive for GH synthesis and secretion. In its absence, or when a lesion interrupts hypophyseal portal blood flow, secretion of GH ceases. Somatostatin reduces or blocks secretion of GH in response to GHRH, but has little or no influence on GH synthesis. Somatostatin and GHRH also exert reciprocal inhibitory influences on GHRH and somatostatin neurons (Figure 11).

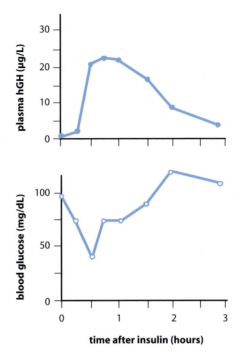

Figure 10　Acute changes in plasma growth hormone concentration (upper panel) in response to insulin-induced hypoglycemia (lower panel). (Reprinted with permission from Roth, J., Glick, S. M., Yalow, R. S., and Berson, S., *Science* **140**, 987–989, copyright 1963 American Association for the Advancement of Science.)

Defective hypothalamic production or secretion of GHRH may be a more common cause of GH deficiency than are defects in the pituitary gland. GH concentrations in plasma are restored to normal in many GH-deficient individuals after treatment with GHRH, suggesting that their somatotropes are competent but not adequately stimulated. Defects in somatostatin synthesis or secretion are not known to be responsible for disease states, but long-acting analogs of somatostatin are used to decrease GH secretion in patients with acromegaly.

In addition to neuroendocrine mechanisms that adjust secretion in response to changes in the internal or external environment, secretion of GH is under negative feedback control. As for other negative feedback systems, products of GH action, principally IGF-I, act as inhibitory signals (Figure 12). IGF-I acts primarily on the pituitary to decrease GH secretion in response to GHRH. Some evidence suggests that IGF-I may also increase somatostatin secretion. Increased concentrations of FFA or glucose, which are also related to GH action, exert inhibitory effects, probably by increasing somatostatin secretion, but fasting, which is also

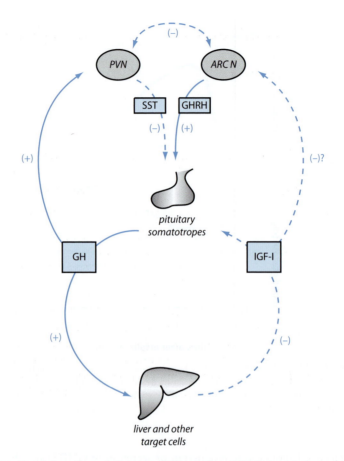

Figure 11 Regulation of growth hormone (GH) secretion. PVN, Periventricular nuclei; ARCN, arcuate nuclei; SST, somatostatin; GHRH, growth hormone-releasing hormone; IGF-I, insulin-like growth factor-I; (+), stimulation; (–), inhibition.

associated with increased FFA, inhibits somatostatin secretion. Growth hormone exerts a short-loop negative feedback effect on its own secretion by inhibiting GHRH secretion and stimulating somatostatin secretion.

Negative feedback control sets the overall level of GH secretion by regulating the amounts of GH secreted in each pulse. The phenomenon of pulsatility and the circadian variation that increases the magnitude of the secretory pulses at night are entrained by neural mechanisms. Pulsatility appears to be the result of reciprocal intermittent secretion of both GHRH and somatostatin. It appears that bursts of GHRH secretion are timed to coincide with interruptions

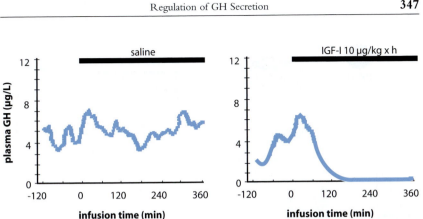

Figure 12 Effects of insulin-like growth factor-I (IGF-I) on growth hormone (GH) secretion in normal fasted men. Values shown are averages for the same 10 men given infusions of either physiological saline (control) or IGF-I for the periods indicated. Note: IGF-I completely blocked GH secretion after a lag period of 1 hour. (Redrawn from Hartman, M. L., Clayton, P. E., Johnson, M. L., Celniker, A., Perlman, A. J., Alberti, K. G., and Thorner, M. O., *J. Clin. Invest.* **91**, 2453–2462, 1993, with permission.)

in somatostatin secretion. Experimental evidence obtained in rodents indicates that GHRH-secreting neurons in the arcuate nuclei communicate with somatostatin-secreting neurons in the periventricular nuclei either directly or through interneurons, and conversely that somatostatinergic neurons communicate with GHRH neurons. However, understanding of how reciprocal changes in secretion of these two neurohormones are brought about is still incomplete. GHRH neurons express receptors for ghrelin, but the role of ghrelin in the regulation of GH secretion remains to be uncovered.

ACTIONS OF GHRH, SOMATOSTATIN, AND IGF-I ON THE SOMATOTROPE

Receptors for GHRH, somatostatin, and IGF-I are present on the surface of somatotropes. The complex interplay of these hormones is illustrated in Figure 13. Receptors GHRH and somatostatin are coupled to several G proteins and express their antagonistic effects on GH secretion in part through their opposing influences on cyclic AMP production, cyclic AMP action, and cytosolic calcium concentrations. GHRH activates adenylyl cyclase through a typical G_s-linked mechanism (see Chapter 1). Cyclic AMP activates protein kinase A, some of which migrates to the nucleus and phosphorylates the cyclic AMP response element binding protein (CREB). Activation of CREB promotes the expression of the transcription factor, pit-1, which in turn increases transcription of genes for both

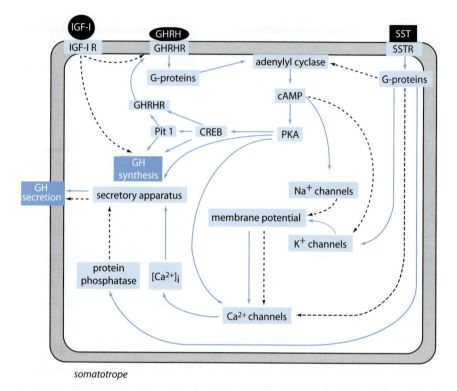

somatotrope

Figure 13 Effects of growth hormone-releasing hormone (GHRH), insulin-like growth factor-I (IGF-I), and somatostatin (SST) on the somatotrope. IGF-IR, GHRHR, and SSTR are receptors for IGF-I, GHRH, and SST. CAMP, Cyclic adenosine monophosphate; CREB, cAMP response element binding protein; PKA, protein kinase A; Pit-1, pituitary-specific transcription factor-1; $[Ca^{2+}]_i$, intracellular free calcium concentration. Solid blue arrows indicate increase or stimulation; dashed black arrows indicate decrease or inhibition. See text for discussion.

GH and the GHRH receptor. In addition, cyclic AMP-dependent phosphorylation of voltage-sensitive calcium channels is thought to lower their threshold and increase their probability of opening. Voltage-sensitive calcium channels are activated by a G-protein-dependent mechanism that depolarizes the somatotrope membrane by activating sodium channels and inhibiting potassium channels. The resulting increase in cytosolic calcium concentration ($[Ca^{2+}]_i$) triggers exocytosis of GH. Increased $[Ca^{2+}]_i$ also limits the secretory event by inhibiting voltage-sensitive calcium channels and restoring membrane polarity by activating potassium channels. Somatostatin acts through the inhibitory guanine nucleotide binding protein (G_i) to antagonize activation of adenylyl cyclase. Somatostatin receptors also inhibit calcium channels and activate potassium channels through a

G-protein-mediated mechanism. Activation of potassium channels hyperpolarizes the plasma membrane, which prevents activation of voltage-sensitive calcium channels. Somatostatin also activates protein phosphatase through a G-protein-dependent mechanism and thereby antagonizes activation of the secretory apparatus by protein kinase A.

The negative feedback effects of IGF-I are slower in onset than the G-protein-mediated effects of GHRH and somatostatin and require tyrosine phosphorylation-initiated changes in gene expression that down-regulate GHRH receptors and GH synthesis. Somatotropes also express G-protein-coupled receptors for ghrelin. In cultured cells, the activated ghrelin receptor signals through the inositol trisphosphate/diacylglycerol (IP_3/DAG) second messenger system (Chapter 1). Both the release of calcium from intracellular stores in response to IP_3 and the DAG-dependent activation of protein kinase C complement the actions of GHRH and enhance GH secretion. Because the physiology of ghrelin remains to be elaborated, it is not included in Figure 13.

THYROID HORMONES

As already mentioned in Chapter 4, growth is stunted in children suffering from unremediated deficiency of thyroid hormones. Treatment of hypothyroid children with thyroid hormone results in rapid "catch up" growth and accelerated maturation of bone. Conversely, hyperthyroidism in childhood increases the rate of growth, but because of early epiphyseal closure, the maximum height attained is not increased. Thyroidectomy of juvenile experimental animals produces nearly as drastic an inhibition of growth as hypophysectomy, and restoration of normal amounts of triiodothyronine (T3) or thyroxine (T4) promptly reinitiates growth. Young mice grow somewhat faster than normal after treatment with thyroxine, but although final size is attained earlier, adult size is no greater than normal.

The effects of thyroid hormones on growth are intimately entwined with GH. T3 and T4 have little, if any, growth-promoting effect in the absence of GH. Plasma concentrations of both GH and IGF-I are reduced in hypothyroid children and adults and are restored by treatment with thyroid hormone (Figure 14). This decrease is due to decreased amplitude of secretory pulses and possibly also to a decrease in frequency consistent with impairments at both the hypothalamic and pituitary levels. Insulin-induced hypoglycemia and other stimuli for GH secretion produce abnormally small increases in the concentration of GH in plasma of hypothyroid subjects. Such blunted responses to provocative signals probably reflect decreased sensitivity to GHRH as well as depletion of GH stores. Curiously, GH secretion in response to a test dose of GHRH is decreased when there is either a deficiency or an excess of thyroid hormone (Figure 15).

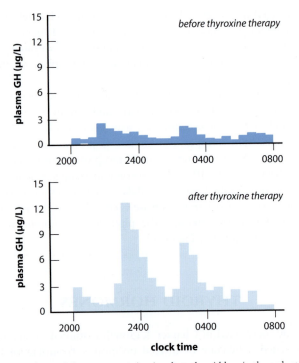

Figure 14 Nocturnal growth hormone secretion in a hypothyroid boy in the early stages of puberty before and 2 months after treatment with thyroxine. Each bar represents the average plasma GH concentration during a 30-minute period of continuous slow withdrawal of blood. Plasma IGF-I was increased more than four fold during treatment. (From Chernausek, S. D., and Turner, R., *J. Pediatr.* **114**, 968–972, 1989, with permission.)

DEPENDENCE OF GH SYNTHESIS AND SECRETION ON T3

The promoter region of the rodent GH gene contains a thyroid hormone response element and its transcriptional activity is enhanced by T3. Furthermore, T3 increases the stability of the GH messenger RNA transcripts. GH synthesis comes to an almost complete halt and the somatotropes become severely depleted of GH only a few days after thyroidectomy. The human GH gene lacks the thyroid hormone response element, and its transcription is not directly activated by T3. However, thyroid hormones affect synthesis of human GH indirectly. Blunted responses to GHRH in hypothyroid children and adults probably result from decreased expression of GHRH receptors by somatotropes. Thyroidectomy also decreases the abundance of GHRH receptors in rodent somatotropes, and hormone replacement restores both the number of receptors and receptor mRNA.

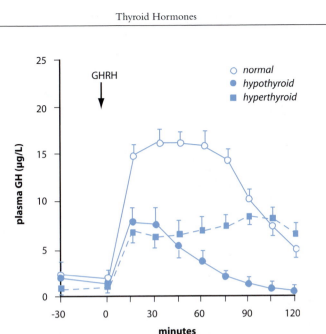

Figure 15   Decreased growth hormone secretion in response to a test dose of growth hormone-releasing hormone (GHRH) in hypothyroid and hyperthyroid individuals. Data shown represent average responses from 30 normal, 25 hypothyroid, and 38 hyperthyroid adult patients. (From Valcavi, R., Zini, M., and Portioli, M., *J. Endocrinol. Invest.* **15**, 313–330, 1992, copyright Editrice Kurtis.)

IMPORTANCE OF T3 FOR EXPRESSION OF GH ACTIONS

Failure of growth in thyroid-deficient individuals is largely due to a deficiency of GH, which may be compounded by a decrease in sensitivity to GH. Treatment of thyroidectomized animals with GH alone can reinitiate growth, but even large amounts cannot sustain a normal rate of growth unless some thyroid hormone is also given. In rats that were both hypophysectomized and thyroidectomized, T4 decreased the amount of GH needed to stimulate growth (increased sensitivity) and exaggerated the magnitude of the response (increased efficacy). Thyroid hormones increase expression of GH receptors in rodent tissues. T3 and T4 also potentiate effects of GH on the growth of long bones and increase its effects on protein synthesis in muscle and liver. IGF-I concentrations are reduced in the blood of hypothyroid individuals partly because of decreased circulating GH and partly because of decreased hepatic responsiveness to GH. In addition, tissues isolated from thyroidectomized animals are less responsive to IGF-I.

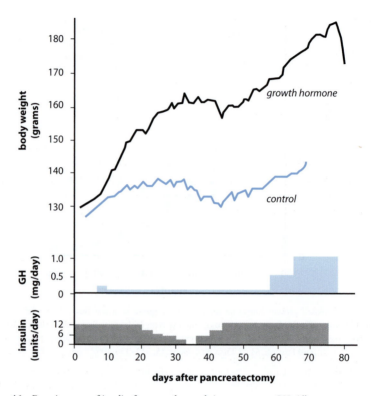

Figure 16   Requirement of insulin for normal growth in response to GH. All rats were pancreatectomized 3 to 7 weeks before the experiment was begun. Each rat was fed 7 g of food per day. The
treated group was injected with the indicated amounts of GH daily. Note the failure to respond to GH
in the period between 20 and 44 days, coincident with the decrease in daily insulin dose, and the
resumption of growth when the daily dose of insulin was restored. (From Scow, R. O., Wagner, E. M.,
and Ronov, E., *Endocrinology* **62**, 593–604, 1958, by permission of The Endocrine Society.)

INSULIN

Although neither GH nor thyroxine appears to be an important determinant
of fetal growth, many investigators have suggested that insulin may serve as a
growth-promoting hormone during the fetal period. Infants born of diabetic
mothers are often larger than normal, especially when the diabetes is poorly
controlled. Because glucose readily crosses the placenta, high concentrations of
glucose in maternal blood increase fetal blood glucose and stimulate the fetal
pancreas to secrete insulin. In the rare cases of congenital deficiency of insulin that
have been reported, fetal size is below normal. Structurally, insulin is closely related
to IGF-I and IGF-II, and when present in adequate concentrations can activate

IGF-I receptors, which are closely related structurally to the insulin receptor. We do not know if the effects of insulin on fetal growth are mediated by the insulin receptor or IGF-I receptors.

Optimal concentrations of insulin in blood are required to maintain normal growth during postnatal life, but it has been difficult to obtain a precise definition of the role of insulin. Because life cannot be maintained for long without insulin, dramatic effects of sustained deficiency on final adult size are not seen. However, growth is often retarded in insulin-dependent diabetic children, particularly in the months leading up to the diagnosis of full-blown disease. Studies in pancreatectomized rats indicate a direct relation between the effectiveness of GH and the dose of insulin administered. Treatment with GH sustained a rapid rate of growth so long as the daily dose of insulin was adequate, but growth progressively decreased as the dose of insulin was reduced (Figure 16). Conversely, insulin cannot sustain a normal rate of growth in the absence of GH.

The effects of insulin on postnatal growth cannot be attributed to changes in GH secretion, which, if anything, is increased in human diabetics. Although insulin was sometimes used diagnostically to provoke GH secretion, it is the resulting hypoglycemia, rather than insulin per se, that stimulates GH release. Expression of IGF-I mRNA in liver and other tissues is decreased in diabetes and in low-insulin states such as fasting or caloric restriction, consistent with the possibility that insulin is permissive for growth. Insulin stimulates protein synthesis and inhibits protein degradation, and in its absence, protein breakdown is severe. Consequently, without insulin, normal responses to GH are not seen; the anabolic effects of GH on body protein either cannot be expressed or are masked by simultaneous, unchecked catabolic processes.

GONADAL HORMONES

PUBERTAL GROWTH SPURT

Awakening of the gonads at the onset of sexual maturation is accompanied by a dramatic acceleration of growth. The adolescent growth spurt, like other changes at puberty, is attributable to steroid hormones of the gonads and perhaps the adrenals. Because the development of pubic and axillary hair at the onset of puberty is a response to increased secretion of adrenal androgens, this initial stage of sexual maturation is called *adrenarche*. The physiological mechanisms that trigger increased secretion of adrenal androgens and the awakening of the gonadotropic secretory apparatus are poorly understood; they are considered further in Chapter 11. At the same time that gonadal steroids promote linear growth, they accelerate closure of the epiphyses and therefore limit the final height that can be attained. Children who undergo early puberty and hence experience their growth spurt while their contemporaries continue to grow at the slower prepubertal rate are likely to be the tallest and most physically developed in grade school or junior high, but are among the shortest in their high school graduating class. Deficiency of gonadal hormones, if left untreated, delays epiphyseal closure, and despite the absence of a pubertal growth spurt, such hypogonadal individuals tend to be tall and have unusually long arms and legs.

Effects of Estrogens and Androgens

In considering the relationship of the gonadal hormones to the pubertal growth spurt it is important to understand several caveats:

- Androgens and estrogens are produced and secreted by both the ovaries and the testes.
- Androgens are precursors of estrogens and are converted to estrogens in a reaction catalyzed by the enzyme P450 aromatase in the gonads and extragonadal tissues.
- Estrogens produce their biological effects at hormone concentrations that are more than 1000 times lower than the concentrations at which androgens produce their effects.

Until recently it was generally accepted that androgens produce the adolescent growth spurt in both sexes. This idea was rooted in the observations that, even at the relatively low doses used therapeutically in women, administration of estrogens inhibits growth, and administration of androgens stimulates growth. Some "experiments of nature" that have come to light in recent years have challenged this idea and lead to the opposite conclusion. Girls with ovarian agenesis (congenital absence of ovaries) have short stature and do not experience an adolescent growth surge. Their growth is increased by treatment with doses of estrogen that are below the threshold needed to cause breast development. In normal girls the adolescent growth spurt usually occurs before estrogen secretion is sufficient to initiate growth of breasts.

Patients of either sex who have a homozygous disruption of the P450 aromatase gene do not experience a pubertal growth spurt despite supranormal levels of androgens, and continue to grow at the juvenile rate well beyond the time of normal epiphyseal fusion unless estrogens are given. Although sexual development cannot occur in girls with this extremely rare defect, affected males develop normally. A man with a homozygous disruption of the α-estrogen receptor similarly had normal sexual development, but failed to experience an adolescent growth spurt. At age 28, when he was diagnosed, he was 6 feet 8 inches tall and his epiphyses had still not closed. In contrast, patients with nonfunctional androgen receptors experience a normal pubertal growth spurt and their epiphyses close at the normal time. These observations established that estrogen rather than androgen is responsible for both acceleration of growth at puberty and maturation of the epiphyseal plates.

It is noteworthy that estrogen concentrations increase in the plasma of both boys and girls early in puberty, and reach similar concentrations at the onset of the growth spurt. The well-established growth-promoting effect of androgens administered to children whose epiphyses are not yet fused is likely attributable to androgen conversion to estrogen. Synthetic androgens that are chemically modified in ways that prevent aromatization are ineffective in promoting growth, even though other aspects of androgen activity (Chapter 11) are fully evident. Nevertheless, the maximal rate of growth achieved in adolescence is greater in males than in females, and a supportive growth-promoting role for androgens is not ruled out. In addition, androgens stimulate growth of muscle, particularly in the upper body. Androgen secretion during puberty in boys produces a doubling of muscle mass by increasing the size and number of muscle cells. Such growth of muscle can occur in the absence of GH or thyroid hormones and is mediated by the same androgen receptors that are expressed in other androgen-sensitive tissues (see Chapter 11). Stimulation of muscle growth by androgens is most pronounced in androgen-deficient or hypopituitary subjects, and only small effects, if any, are seen in men with normal testicular function, except perhaps when very large doses of so-called anabolic steroids are used.

Effects on GH Secretion and Action

Most, if not all, of the increase in height stimulated by estrogens or androgens at puberty is due to increased secretion of GH (Figure 17). During the pubertal growth spurt or when androgens are given to prepubertal children, there is an increase in both frequency and amplitude of secretory pulses of GH (Figure 18), and GHRH concentrations are increased in peripheral blood of boys and girls during puberty. The concentration of IGF-I in blood also increases during the pubertal growth spurt or after androgens are given to prepubertal children. This increase is probably a consequence of increased secretion of GH.

We still do not understand the basis for either the stimulatory or inhibitory effects of estrogen on linear growth. At the same concentrations that inhibit growth, estrogens increase GH secretion, and, as we have seen (Figure 7), plasma concentrations of GH are higher in women. In addition, the GH secretory apparatus tends to be more sensitive to environmental influences in women than in men, and the circulating concentrations of GH tend to rise more readily in women in response to provocative stimuli. Although we do not understand its molecular basis, inhibitory effects of estrogens on growth appear to result from interference with the action of GH at the level of its target cells. Estrogens act directly on the epiphyseal plates, which lose their capacity to replenish cartilage progenitor cells. Estrogens, which are not catabolic, also antagonize effects of GH

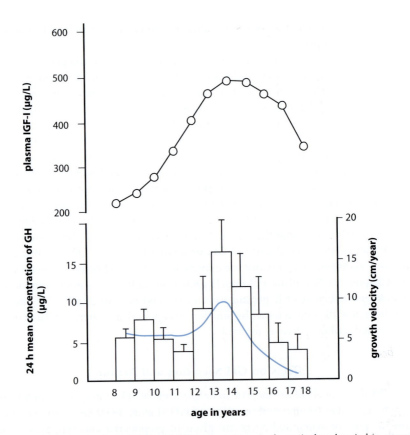

Figure 17   Changes in plasma IGF-I and GH concentrations in the peripubertal period in normal boys. Bars in the lower panel indicate 24–hour integrated concentrations of GH. The blue curve is the idealized growth velocity curve for North American boys. (Top, from Juul, A., Dalgaard, P., Blum, W. F., Bang, P., Hall, K., Michaelsen, K. F., Muller, J., and Skakkebaek, N. E., *J. Clin. Endocrinol. Metab.* **80**, 2534–2542, 1995, by permission of The Endocrine Society; bottom, from Martha, P. M., Jr., Rogol, A. D., Veldhuis, J. D., Kerrigan, J. R., Goodman, D. W., and Blizzard, R. M., *J. Clin. Endocrinol. Metab.* **69**, 563–570, 1989, by permission of The Endocrine Society.)

on nitrogen retention and minimize the increase in IGF-I in blood of hypophysectomized or hypopituitary individuals treated with GH. Neither estrogens nor androgens affect responsiveness to IGF-I.

GLUCOCORTICOIDS

Normal growth requires secretion of glucocorticoids, which have widespread effects in promoting optimal function of a variety of organ systems

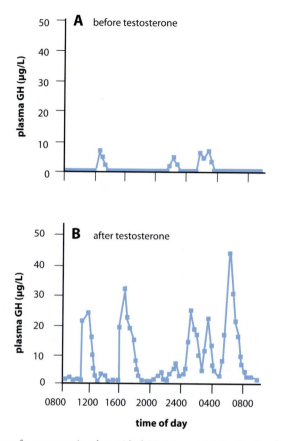

Figure 18 Effects of testosterone in a boy with short stature and delayed puberty. (A) Initial staus, before testosterone. (B) During therapy with long-acting testosterone. Note the increase in frequency and amplitude of growth hormone secretory episodes in the treated subjects. (From Link, K., Blizzard, R. M., Evans, W. S., Kaiser, D. L., Parker, M. W., and Rogol, A. D., *J. Clin. Endocrinol. Metab.* **62**, 159–164, 1986, by permission of The Endocrine Society.)

(see Chapter 4), a sense of health and well-being, and normal appetite. Glucocorticoids are required for synthesis of GH and have complex effects on GH secretion. When given acutely, they may enhance GH gene transcription and increase responsiveness of somatotropes to GHRH. However, GH secretion is reduced by excessive glucocorticoids, probably as a result of increased somatostatin production. Children suffering from overproduction of glucocorticoids (Cushing's disease) experience some stunting of their growth. In a recently reported case involving identical twins, the unaffected twin was 21 cm taller at age 15 than her sister, whose disease was untreated until around the time of puberty. Similar

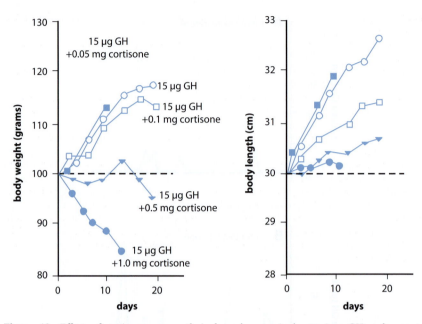

Figure 19 Effects of cortisone on growth in hypophysectomized rats given GH replacement. The growth-promoting response to GH, measured as a change in either body weight or length, decreased progressively as the dose of cortisone was increased from 0.1 to 1.0 mg per day. The decrease in body weight seen when 1.0 mg per day of cortisone was given probably results from net breakdown of muscle mass (see Chapter 4). (From Soyka, L. F., and Crawford, J. D., *J. Clin. Endocrinol. Metab.* **25**, 469–475, 1965, by permission of The Endocrine Society.)

impairment of growth is seen in children treated chronically with high doses of glucocorticoids to control asthma or inflammatory disorders. Consistent with their catabolic effects in muscle and lymphoid tissues, glucocorticoid also antagonize the actions of GH. Hypophysectomized rats grew less in response to GH when cortisone was given simultaneously (Figure 19). Glucocorticoids similarly blunt the response to GH administered to hypopituitary children. The cellular mechanisms for this antagonism are not yet understood. IGF-I production may be reduced by treatment with glucocorticoids, but we do not know if this is a cause or an effect of the decreased action of GH.

SUGGESTED READING

Baumann, G. (1991). Growth hormone heterogeneity: Genes, isohormones, variants and binding proteins. *Endocr. Rev.* **12**, 424–449.

Clemmons, D. R. (2001). Use of mutagenesis to probe IGF-binding protein structure/function relationships. *Endocr. Rev.* **22**, 800–817.

Frank, S. J. (2002). Receptor dimerization in GH and erythropoietin action—It takes two to tango, but how? *Endocrinology* **143**, 2–10.

Giustina, A., and Veldhuis, J. D. (1998). Pathophysiology of the neuroregulation of growth hormone secretion in experimental animals and the human. *Endocr. Rev.* **19**, 717–797.

Jones, J. I., and Clemmons, D. R. (1995). Insulin-like growth factors and their binding proteins: Biological actions. *Endocr. Rev.* **16**, 3–34.

Kostyo, J. L. (ed.) (1999). "Handbook of Physiology, Section 7: The Endocrine System. Volume V: Hormonal Control of Growth," Oxford University Press, New York. (This volume contains many excellent chapters that are relevant to topics considered here.)

LeRoith, D. E., Bondy, C., Yakar, S., Liu J. L., and Butler, A. (2001). The somatomedin hypothesis: 2001. *Endocr. Rev.* **22**, 53–74.

Ohlsson, C., Bengtsson, B.-Å., Isaksson, O. G. P., Andreassen, T. T., and Slootweg, M. C. (1998). Growth hormone and bone. *Endocr. Rev.* **19**, 55–79.

Rosenfeld, R. G., Rosenbloom, A. L., and Guevara-Aguirre, J. (1994). Growth hormone (GH) insensitivity due to primary GH receptor deficiency. *Endocr. Rev.* **15**, 369–390.

Smit, L., Meyer, D. J., Argetsinger, L. S., Schwartz, J., and Carter-Su, C. (1999). Molecular events in growth hormone-receptor interaction and signaling. *In*: "Handbook of Physiology, Section 7: The Endocrine System, Volume V: Hormonal Control of Growth," (J. L. Kostyo, ed.), pp. 445–480. Oxford University Press, New York.

CHAPTER 11

Hormonal Control of Reproduction in the Male

OVERVIEW

The testes serve the dual function of producing sperm and hormones. The principal testicular hormone is the steroid testosterone, which has an intratesticular role in sperm production and an extratesticular role in promoting delivery of sperm to the female genital tract. In this respect, testosterone promotes development and maintenance of accessory sexual structures responsible for nurturing gametes and ejecting them from the body, development of secondary sexual characteristics that make men attractive to women, and those behavioral characteristics that promote successful procreation. Testicular function is driven by the pituitary through the secretion of two gonadotropic hormones: follicle-stimulating hormone (FSH) and luteinizing hormone (LH). Secretion of these pituitary hormones is controlled by (1) the central nervous system through intermittent secretion of the hypothalamic hormone, gonadotropin-releasing hormone (GnRH), and (2) the testes through the secretion of testosterone and inhibin. Testosterone, its potent metabolite 5α-dihydrotestosterone, and an additional testicular secretion called antimüllerian hormone also function as determinants of sexual differentiation during fetal life.

MORPHOLOGY OF THE TESTES

The testes are paired ovoid organs located in the scrotal sac outside the body cavity. The extraabdominal location is coupled with vascular countercurrent heat exchangers and muscular reflexes that retract the testes to the abdomen, permitting testicular temperature to be maintained constant at about 2°C below body temperature. For reasons that are not understood, this small reduction in temperature is crucial for normal *spermatogenesis* (sperm production). Failure of the testes to descend into the scrotum results in failure of spermatogenesis, although production of testosterone may be maintained. The two principal functions of the testis—sperm production and steroid hormone synthesis—are carried out in morphologically distinct compartments. Sperm are formed and develop within *seminiferous tubules*, which comprise the bulk of testicular mass. Testosterone is produced by the *interstitial cells of Leydig*, which lie between the seminiferous tubules (Figure 1). The entire testis is encased in an inelastic fibrous capsule consisting of three layers of dense connective tissue and some smooth muscle.

Blood reaches the testes primarily through paired spermatic arteries and is first cooled by heat exchange with returning venous blood in the *pampiniform plexus*. This complex tangle of blood vessels is formed by the highly tortuous and convoluted artery intermingling with equally tortuous venous branches that converge to form the spermatic vein. This arrangement provides a large surface area for warm arterial blood to transfer heat to cooler venous blood across thin vascular walls. Rewarmed venous blood returns to the systemic circulation primarily through the internal spermatic veins.

LEYDIG CELLS AND SEMINIFEROUS TUBULES

Leydig cells are embedded in loose connective tissue that fills the spaces between seminiferous tubules. They are large polyhedral cells with an extensive

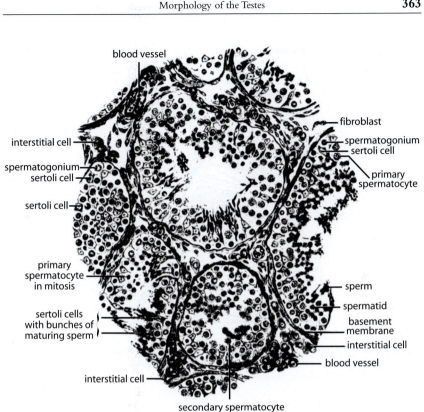

blood vessel

fibroblast

interstitial cell

spermatogonium
sertoli cell

spermatogonium
sertoli cell

primary
spermatocyte

sertoli cell

primary
spermatocyte
in mitosis

sperm

spermatid

sertoli cells
with bunches of
maturing sperm

basement
membrane

interstitial cell

blood vessel

interstitial cell

secondary spermatocyte

Figure 1 Histological section of human testis. The transected tubules show various stages of
spermatogenesis. (From Fawcett, D. W., "A Textbook of Histology," 11th Ed., p. 804. W. B. Saunders,
Philadelphia, 1986, with permission.)

smooth endoplasmic reticulum characteristic of steroid-secreting cells. Although
extensive at birth, Leydig cells virtually disappear after the first 6 months of
postnatal life, only to reappear more than a decade later with the onset of puberty.

Seminiferous tubules are highly convoluted loops that range from about 120
to 300 μm in diameter and from 30 to 70 cm in length. They are arranged in
lobules bounded by fibrous connective tissue. Each testis has hundreds of such
tubules that are connected at both ends to the *rete testis* (Figure 2). It has been
estimated that, if laid end to end, the seminiferous tubules of the human testes
would extend more than 250 meters. The seminiferous epithelium that lines the
tubules consists of three types of cell: *spermatogonia*, which are stem cells; *spermatocytes* in the process of becoming sperm; and *Sertoli cells*, which nurture developing
sperm and secrete a variety of products into the blood and the lumens of seminiferous tubules. Seminiferous tubules are surrounded by a thin coating of peritubular

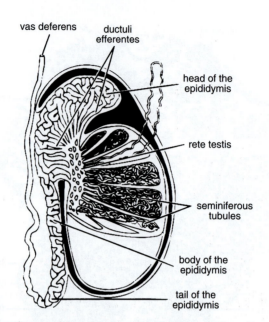

vas deferens

ductuli
efferentes

head of the
epididymis

rete testis

seminiferous
tubules

body of the
epididymis

tail of the
epididymis

Figure 2 Cutaway diagram of the architecture of the human testis. (From Fawcett, D. W., "A Textbook of Histology," 11th Ed., p. 797. W. B. Saunders, Philadelphia, 1986, with permission; modified from "Textbook of Human Anatomy." Macmillan, London, 1957.)

epithelial cells, which in some species are contractile and help propel the nonmotile sperm through the tubules toward the rete testis.

MALE REPRODUCTIVE TRACT

Spermatogenesis goes on continuously from puberty to senescence along the entire length of the seminiferous tubules. Though a continuous process, spermatogenesis can be divided into three discrete phases: (1) mitotic divisions, which replenish the spermatogonia and provide the cells destined to become mature sperm; (2) meiotic divisions, which reduce the chromosome number and produce a cluster of haploid spermatids; and (3) transformation of spermatids into mature sperm (spermiogenesis), a process involving the loss of most of the cytoplasm and the development of flagella. These events occur along the length of the seminiferous tubules in a definite temporal and spatial pattern. A spermatogenic cycle includes all of the transformations from spermatogonium to spermatozoan and requires about 64 days. As the cycle progresses, germ cells move from the basal portion of the germinal epithelium toward the lumen.

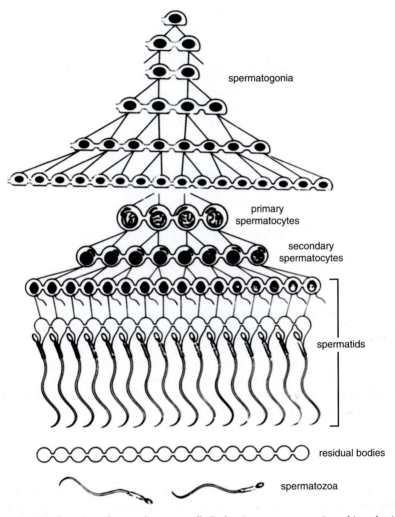

Figure 3   The formation of mammalian germ cells. Each primary spermatogonium ultimately gives rise to 64 sperm cells. Cytokinesis is incomplete in all but the earliest spermatogonial divisions, resulting in expanding clones of germ cells that remain joined by intercellular bridges. (From Fawcett, D. W., "A Textbook of Histology," 11th Ed., p. 815. W. B. Saunders, Philadelphia, 1986, with permission.)

    Successive cycles begin before the previous one has been completed, so that at any given point along a tubule different stages of the cycle are seen at different depths of the epithelium (Figure 3). Spermatogenic cycles are synchronized in adjacent groups of cells, but the cycles are slightly advanced in similar groups of cells located immediately upstream, so that cells at any given stage of the spermato-genic cycle are spaced at regular intervals along the length of the tubules. This spatial organization is called the spermatogenic wave. This complex series of events ensures that mature spermatozoa are produced continuously. About 2 million spermatogonia, each giving rise to 64 sperm cells, begin this process in each testis every day. Hundreds of millions of spermatozoa are thus produced daily throughout six or more decades of reproductive life.

Sertoli cells are remarkable polyfunctional cells; their activities are intimately related to many aspects of the formation and maturation of spermatozoa. They extend through the entire thickness of the germinal epithelium, from basement membrane to lumen, and in the adult take on exceedingly irregular shapes that are determined by the changing conformation of the developing sperm cells embedded in their cytoplasm (Figure 4). Differentiating sperm cells are isolated from the bloodstream and must rely on Sertoli cells for their sustenance. Adjacent Sertoli cells arch above the clusters of spermatogonia that nestle between them at the level of the basement membrane. Adjacent Sertoli cells form a series of tight junctions that limit passage of physiologically relevant molecules into or out of seminiferous tubules. This so-called blood–testis barrier actually has selective permeability that allows rapid entry of testosterone, for example, but virtually completely excludes cholesterol. The physiological significance of the blood–testis barrier has not been established, but it is probably of some importance that spermatogonia are located on the blood side of the barrier, whereas developing spermatids are restricted to the luminal side. In addition to harboring developing sperm, Sertoli cells secrete a watery fluid that transports spermatozoa through the seminiferous tubules and into the *epididymis*, where 99% of the fluid is reabsorbed.

The remaining portion of the male reproductive tract consists of modified excretory ducts that ultimately deliver sperm to the exterior along with secretions of accessory glands that promote sperm survival and fertility. Sperm leave the testis through multiple *ductuli efferentes*, which have ciliated epithelia that facilitate passage from the rete testis into the highly convoluted and tortuous duct of the epididymis. The epididymis is the primary area for maturation and storage of sperm, which remain viable within its confines for months.

Sperm are advanced through the epididymis, particularly during sexual arousal, by rhythmic contractions of circular smooth muscle surrounding the duct. At ejaculation, sperm are expelled into the *vas deferens* and ultimately through the *urethra*. An accessory storage area for sperm lies in the ampulla of the vas deferens, posterior to the *seminal vesicles*. These elongated, hollow evaginations of the defer-ential ducts secrete a fluid rich in citric acid and fructose that provides nourishment for the sperm after ejaculation. Metabolism of fructose provides the energy for sperm motility. Additional citrate and a variety of enzymes are added to the ejaculate by the *prostate*, which is the largest of the accessory secretory glands.

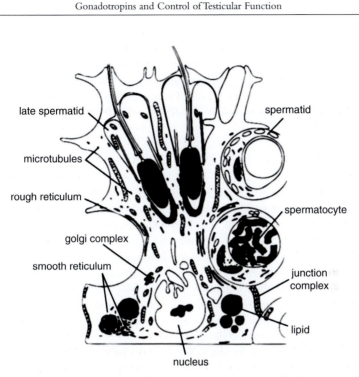

Figure 4   Ultrastructure of the Sertoli cell and its relation to the germ cells. The spermatocytes and early spermatids occupy niches in the sides of the columnar supporting cell, whereas late spermatids reside in deep recesses in its apex. (From Fawcett, D. W., "A Textbook of Histology," 11th Ed., p. 834. W. B. Saunders, Philadelphia, 1986, with permission.)

Sperm and the combined secretions of the accessory glands make up the *semen*, of which less than 10% is sperm.

GONADOTROPINS AND CONTROL OF TESTICULAR FUNCTION

Physiological activity of the testis is governed by two pituitary gonadotropic hormones, follicle-stimulating hormone (FSH) and luteinizing hormone (LH) (see Chapter 2). These same gonadotropic hormones are also produced in pituitary glands of women, and because their physiology has been studied more extensively in women, the names that have been adopted for them describe their activity in the ovary (see Chapters 12 and 13). FSH and LH are closely related glycoprotein hormones that consist of a common alpha subunit and unique beta subunits that confer FSH or LH specificity. The three subunits are the products of three genes that are regulated independently. Both gonadotropins are synthesized and secreted by a single class of pituitary cells, the gonadotrope. Their sites of stimulation of testicular function, however, are discrete: LH acts on the Leydig cells and FSH acts on the Sertoli cells in the germinal epithelium.

LEYDIG CELLS

The principal role of Leydig cells is synthesis and secretion of testosterone in response to stimulation by LH. Testosterone is an important paracrine regulator of intratesticular functions as well as a hormonal regulator of a variety of extra-testicular cells. In addition to stimulating steroidogenesis, LH controls the availability of its own receptors (down-regulation) and governs growth and differentiation of Leydig cells. After hypophysectomy of experimental animals, Leydig cells atrophy and lose their extensive smooth endoplasmic reticulum where the bulk of testosterone synthesis takes place. LH restores them to normal and can produce frank hypertrophy if given in excess. Leydig cells, which are abundant in newborn baby boys, regress and die shortly after birth. Secretion of LH at the onset of puberty causes dormant Leydig cell precursors to proliferate and differentiate into mature steroidogenic cells. In the fetus, growth and development of Leydig cells depend initially on the placental hormone, chorionic gonadotropin, which is present in high concentrations and which stimulates LH receptors, and later on LH secreted by the fetal pituitary gland.

The LH receptor is a member of the superfamily of receptors that are coupled to heterotrimeric G-proteins. LH stimulates the formation of cyclic AMP (see Chapter 1), which in turn activates protein kinase A and the subsequent phosphorylation of proteins that promote steroidogenesis, gene transcription, and other cellular functions. As with the adrenal cortex (see Chapter 4), the initial step in the synthesis of testosterone is the conversion of cholesterol to pregnenolone. This reaction requires mobilization of cholesterol from storage droplets and its translocation from cytosol to the intramitochondrial compartment, where cleavage of the side chain occurs (desmolase reaction). Access of cholesterol to the P450scc enzyme in the mitochondria is governed by a still to be defined action of the steroid acute regulatory protein (StAR), the expression and phosphorylation of which are accelerated by cyclic AMP. Stored cholesterol may derive either from *de novo* synthesis within the Leydig cell or from circulating cholesterol, which enters the cell by receptor-mediated endocytosis of low-density lipoproteins.

The biochemical pathway for testosterone biosynthesis is shown in Figure 5. Neither LH nor cyclic AMP appears to accelerate the activity of any of the four enzymes responsible for conversion of the 21-carbon pregnenolone to testosterone. It may be recalled that the rate of steroid hormone secretion as well as synthesis is determined by the rate of conversion of cholesterol to pregnenolone.

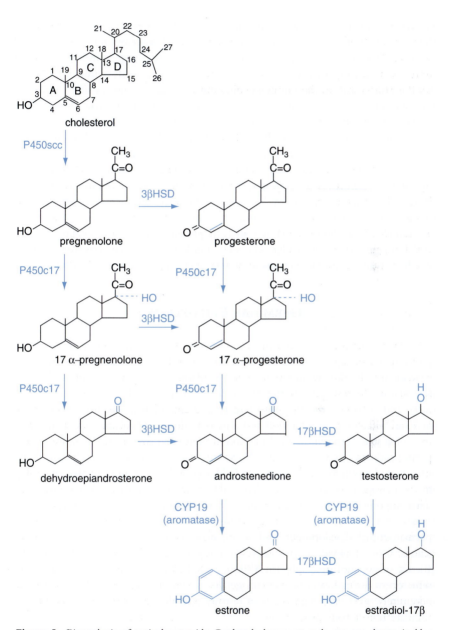

Figure 5 Biosynthesis of testicular steroids. Catalyzed changes at each step are shown in blue. Testosterone comprises more than 99% of testicular steroid hormone production. 3βHSD, 3β-Hydroxysteroid dehydrogenase.

However, in maintaining the functional integrity of the Leydig cells, LH maintains the levels of all of the steroid transforming enzymes. Transcription of the gene that encodes P450c17, the enzyme responsible for the two-step conversion of 21-carbon steroids to 19-carbon steroids, appears to be especially sensitive to cyclic AMP. Testosterone released from Leydig cells may diffuse into nearby capillaries for transport in the general circulation, or may diffuse to nearby seminiferous tubules where it performs its essential role in spermatogenesis.

The testes also secrete small amounts of estradiol and some androstenedione, which serves as a precursor for extratesticular synthesis of estrogens. The Leydig cell is the chief source of testicular estrogens, but immature Sertoli cells have the capacity to convert testosterone to estradiol. In addition, developing sperm cells express the enzyme P450arom (aromatase) and may also convert androgens to estrogens. Estradiol is present in seminal fluid and may be essential for reabsorption of seminal tubular fluid in the rete testis. The presence of estrogen receptors in the epididymis and several testicular cells, including Leydig cells, suggests that estradiol may have other important actions in normal sperm formation and maturation.

GERMINAL EPITHELIUM

The function of the germinal epithelium is to produce large numbers of sperm that are capable of fertilization. The Sertoli cells, which are interposed between the developing sperm and the vasculature, harbor and nurture sperm as they mature. Sertoli cells are the only cells that express FSH receptors in human males and therefore are the only targets of FSH. In the immature testis FSH increases Sertoli cell proliferation and differentiation and probably maintains their functional state throughout life. In its absence testicular size is severely reduced and sperm production, which is limited by Sertoli cell availability, is severely restricted. It has been known for many years that FSH, LH, and testosterone all play vital roles in spermatogenesis. It is likely that FSH indirectly regulates development of spermatogonia by stimulating Sertoli cells to produce both growth and survival factors that prevent germ cell apoptosis. Withdrawal of FSH and LH arrests spermatogonial development, and is the major rate-limiting step in spermato-genesis. Once formed, spermatocytes progress through meiosis, normally in the absence of gonadotropic support. Although FSH and testosterone are required for initiation of normal rates of spermatogenesis, sperm formation can be maintained indefinitely with very high doses of testosterone alone or with sufficient LH to stimulate testosterone production.

Sertoli cells lack receptors for LH but are richly endowed with androgen receptors, indicating that the actions of LH on Sertoli cell function are indirect and are mediated by testosterone, which reaches Sertoli cells in high concentration by diffusion from adjacent Leydig cells, and perhaps also by peptide factors produced by Leydig cells. Testosterone readily passes through the blood–testis barrier and is found in high concentrations in seminiferous fluid. However, the absence of androgen receptors in developing human sperm cells indicates that support of sperm cell development by testosterone is also exerted indirectly by way of the Sertoli cells. Although testosterone is critically important for spermatogenesis, it is ineffective in this regard when administered in amounts sufficient to restore normal blood concentrations. For reasons that are not understood, the intratesticular concentration needed to support spermatogenesis is many times higher than needed to saturate androgen receptors. The concentration of testosterone in testicular venous blood is 40 to 50 times that found in peripheral blood and its concentration in aspirates of human testicular fluid is more than 100 times greater than the concentration found in blood plasma.

The FSH receptor is closely related to the LH receptor, and, when stimulated, activates adenylyl cyclase through the agency of the stimulatory G-protein alpha subunit ($G\alpha_s$; see Chapter 1). The resulting activation of protein kinase A catalyzes phosphorylation of proteins that regulate the cytoskeletal elements that maintain the tortuous shape of these cells, production of the membrane glycoproteins that govern adherence to developing sperm, and expression of specific genes that code for proteins that directly and indirectly regulate germ cell development (Figure 6). Some of the proteins that Sertoli cells secrete into the seminiferous tubules are thought to facilitate germ cell maturation in the epididymis and perhaps more distal portions of the reproductive tract. On stimulation by FSH, Sertoli cells may also secrete paracrine factors that enhance Leydig cell responses to LH.

FSH and testosterone have overlapping actions on Sertoli cells and act synergistically, but the precise actions of each remain unknown. Recently described "experiments of nature" have shed some light on the relative importance of FSH and testosterone in spermatogenesis. Inactivating mutations of the beta subunit of LH in humans or rodents result in total failure of spermatogenesis, despite normal prenatal sexual development (see below), suggesting that testosterone is indispensable. In contrast, inactivating mutations of the gene for the FSH receptor did not prevent affected men or mice from fathering offspring. Thus FSH apparently is not absolutely required for spermatogenesis, but the patients and rodents with inactive FSH receptors had small testes, low sperm counts, and a preponderance of defective sperm. Thus stimulation by FSH at some period of life is required for production of normal amounts and quality of sperm.

TESTOSTERONE

SECRETION AND METABOLISM

Testosterone is the principal androgen secreted by the mature testis. The normal young man produces about 7 mg each day, of which less than 5% is derived from adrenal secretions. This amount decreases somewhat with age, so that by the

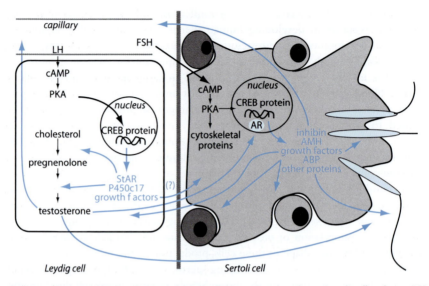

Figure 6   Actions of FSH and LH on the testis. FSH acts directly only on Sertoli cells, whereas LH acts directly solely on Leydig cells. Paracrine cross-talk mediated by growth factors likely takes place between Sertoli and Leydig cells and between Sertoli cells and germ cells. cAMP, Cyclic adenosine monophosphate; PKA, protein kinase A; CREB, cyclic AMP response element binding protein; StAR, steroid acute regulatory protein; P450c17, 17α-hydroxylase/lyase; AR, androgen receptor; AMH, antimüllerian hormone; ABP, androgen-binding protein.

seventh decade and beyond testosterone production may have decreased to 4 mg per day; in the absence of illness or injury, there is no sharp drop in testosterone production akin to the abrupt cessation of estrogen production in the postmenopausal woman. As with the other steroid hormones, testosterone in blood is largely bound to plasma protein, with only about 2–3% present as free hormone. About 50% is bound to albumin, and about 45% to sex hormone-binding globulin (SHBG), which is also called testosterone–estradiol-binding globulin (TeBG). This glycoprotein binds both estrogen and testosterone, but its single binding site has a higher affinity for testosterone. Its concentration in plasma is decreased by androgens. Consequently, SHBG is more than twice as abundant in the circulation of women, compared to men. In addition to its functions as a carrier protein, SHBG may also act as an enhancer of hormone action. Persuasive evidence has been amassed to indicate that SHBG binds to specific receptors on cell membranes, and increases the formation of cyclic AMP when its steroid-binding site is occupied. The nature of the receptor has not been characterized, nor has the physiological importance of this action been established. Although testosterone decreases expression of SHBG in hepatocytes, both testosterone and FSH increase transcription of the same gene in Sertoli cells; the protein product of this was given the name androgen-binding protein (ABP) before its identity with SHBG was known. ABP is secreted into the lumens of seminiferous tubules.

Testosterone that is not bound to plasma proteins diffuses out of capillaries and enters nontarget as well as target cells. In some respects, testosterone can be considered to be a prohormone, because it is converted in extratesticular tissues to other biologically active steroids. Testosterone may be reduced to the more potent androgen, 5α-dihydrotestosterone, in the liver in a reaction catalyzed by the enzyme 5α-reductase type I, and then is returned to the blood. This enzyme is a component of the steroid hormone degradative pathway and also reduces 21-carbon adrenal steroids. Testosterone is also reduced to dihydrotestosterone in the cytoplasm of its target cells mainly through the catalytic activity of 5α-reductase type II, whose abundance in these cells is increased by the actions of testosterone. Dihydrotestosterone is only about 5% as abundant in blood as testosterone and is derived primarily from extratesticular metabolism.

Some testosterone is also metabolized to estradiol (Figure 7) in both androgen target and nontarget tissues. A variety of cells, including some in brain, breast, and adipose tissue, can convert testosterone and androstenedione to estradiol and estrone, which produce cellular effects that are different from, and sometimes opposite to, those of testosterone. The concentration of estrogens in blood of normal men is similar to that of women in the early follicular phase of the menstrual cycle (see Chapter 12). About two-thirds of these estrogens are formed from androgen outside of the testis. Although less than 1% of the peripheral pool of testosterone is converted to estrogens, it is important to recognize that estradiol produces its biological effects at concentrations that are far below those needed for androgens to produce their effects. In other tissues, including liver, reduction catalyzed by 5β-reductase destroys androgenic potency. The liver is the principal site of degradation of testosterone and releases water-soluble sulfate or glucuronide conjugates into blood for excretion in the urine.

MECHANISM OF ACTION

Like other steroid hormones, testosterone penetrates target cells and stimulates their growth and function. Androgen target cells generally convert testosterone to 5α-dihydrotestosterone before it binds to the androgen receptor. The androgen receptor is a ligand-dependent transcription factor that belongs to the nuclear receptor superfamily (see Chapter 1). It binds both testosterone and dihydrotestosterone, but the dihydro form dissociates from the receptor much more slowly than testosterone and therefore is the predominant androgen

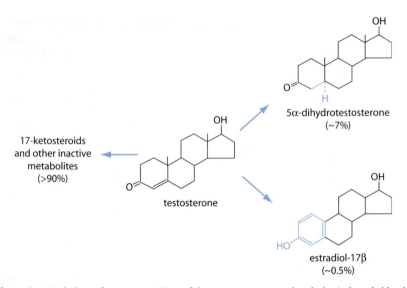

Figure 7   Metabolism of testosterone. Most of the testosterone secreted each day is degraded by the liver and other tissues by reduction of the A ring, oxidation of the 17-hydroxyl group, and conjugation with polar substituents. Conversion to 5α-dihydrotestosterone takes place in target cells catalyzed mainly by the type II dehydrogenase and in nontarget cells mainly but not exclusively by the type I dehydrogenase. Aromatization of testosterone to estradiol may occur directly or after conversion to androstenedione. Note that 5α-dihydrotestosterone cannot be aromatized or reconverted to testosterone.

associated with DNA. It is likely that the higher affinity of dihydrotestosterone for the androgen receptor accounts for its greater biological potency compared to testosterone. On binding testosterone or dihydrotestosterone the liganded receptor complex binds to androgen response elements in specific target genes, and along with a cell-specific array of transcription factors and coactivators, regulates expression of a cadre of genes that are characteristic of each particular target cell. These events are summarized in Figure 8. Some "nongenomic" actions of testosterone have also been described and include rapid increases in intracellular calcium. We do not understand the cellular mechanisms of these effects or their physiological importance.

Dihydrotestosterone in the prostate stimulates expression of a protein, the so-called prostate-specific antigen (PSA), which is found in blood in high concentrations in patients afflicted with prostate cancer. It's abundance in plasma is now widely used diagnostically as a marker of prostate cancer. PSA is a serine protease that is synthesized in the columnar cells of the glandular epithelium and is secreted into the semen. Cleavage of seminogelin by PSA causes liquifaction of the ejaculate and is thought to increase sperm mobility.

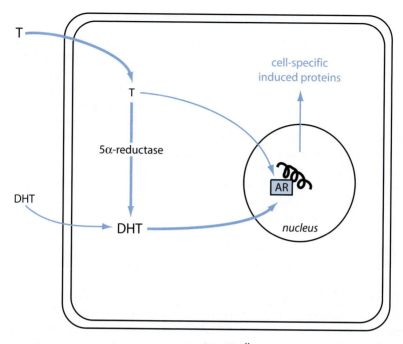

target cells

Figure 8 Action of testosterone. Testosterone (T) enters its target cell and binds to its nuclear androgen receptor (AR) either directly or after it is converted to dihydrotestosterone (DHT). The thickness of the arrows reflects the quantitative importance of each reaction. The hormone–receptor complex binds to DNA along with a variety of cell-specific nuclear regulatory proteins to induce formation of the RNA that encodes the proteins that express effects of the hormone. Not shown: Testosterone may also bind to membrane receptors and initiate rapid ionic changes that may reinforce its genomic effects. Testosterone may also produce rapid changes in cyclic AMP through the binding of the sex hormone-binding globulin to surface receptors.

PHYSIOLOGICAL EFFECTS

Male Genital Tract

Testosterone promotes growth, differentiation, and function of accessory organs of reproduction. Its effects on growth of the genital tract begin early in embryonic life and are not completed until adolescence, after an interruption of more than a decade. Maintenance of normal reproductive function in the adult also depends on continued testosterone secretion. The secretory epithelia of the seminal vesicles and prostate atrophy after castration but can be restored with injections of androgen.

Secondary Sexual Characteristics

In addition to its effect on organs directly related to transport and delivery of sperm, testosterone affects a variety of other tissues and thus contributes to the morphological and psychological components of masculinity. These characteristics are clearly an integral part of reproduction, for they are related to the attractiveness of the male to the female. During early adolescence androgens that arise from the adrenals and later from the testes stimulate growth of pubic hair. Growth of chest, axillary, and facial hair is also stimulated, but scalp hair is affected in the opposite manner. Recession of hair at the temples is a typical response to androgen, and adequate amounts allow expression of genes for baldness. Growth and secretion of sebaceous glands in the skin are also stimulated, a phenomenon undoubtedly related to the acne of adolescence. Dihydrotestosterone is the important androgen for recession of scalp hair and stimulation of the sebaceous glands.

Androgen secretion at puberty stimulates growth of the larynx and thickening of the vocal chords and thus lowers the pitch of the voice. At this time also the characteristic adolescent growth spurt results from an interplay of testosterone and growth hormone (see Chapter 10) that promotes growth of the vertebrae and long bones. Development of the shoulder girdle is pronounced. This growth is self-limiting, as androgens, after extragonadal conversion to estrogens, accelerate epiphyseal closure (see Chapter 10). Androgens promote growth of muscle, especially in the upper torso. Indeed, men have almost half again as much muscle mass as women. In some animals the temporal and masseter muscles are particularly sensitive to androgenic stimulation. Growth and nitrogen retention, of course, are also related to stimulation of appetite and increased food intake. Accordingly, androgens bring about increased physical vigor and a feeling of well-being. Testosterone also stimulates red blood cell production by direct effects on bone marrow and by stimulating secretion of the hormone erythropoietin from the kidney. This action of androgens accounts for the higher hematocrit in men than women. In both men and women androgens increase sexual drive (libido).

SEXUAL DIFFERENTIATION

Primordial components of both male and female reproductive tracts are present in early embryos of both sexes, and their development is either stimulated or suppressed by humoral factors arising in the testes. The indifferent gonads present in the early embryo differentiate into testes under the influence of the product of the sex-determining gene (called *SRY*, for sex-determining region of the Y chromosome). By about the seventh week of embryonic life the medulla of the primitive gonad becomes distinguishable as a testis, with the appearance of cords of cells that give rise to seminiferous tubules. Leydig cells appear about

10 days later and undergo rapid proliferation for the next 6 to 8 weeks in response to chorionic gonadotropin, which is produced by the placenta in large amounts at this time, and perhaps also in response to LH secreted by the fetal pituitary gland. Fetal Leydig cells secrete sufficient testosterone to raise blood concentrations to the same levels as those seen in adult men. Testosterone accumulation is enhanced by an additional effect of the *SRY* gene product, which blocks expression of aromatase and thus prevents conversion of testosterone to estrogens.

DEVELOPMENT OF INTERNAL REPRODUCTIVE DUCTS AND THEIR DERIVATIVES

Regardless of its genetic sex, the embryo has the potential to develop phenotypically either as male or female. The pattern for female development is expressed unless overridden by secretions of the fetal testis. The early embryo develops two sets of ducts that are the precursors of either male or female internal genitalia (Figure 9). Seminal vesicles, epididymes, and vasa deferentia arise from primitive

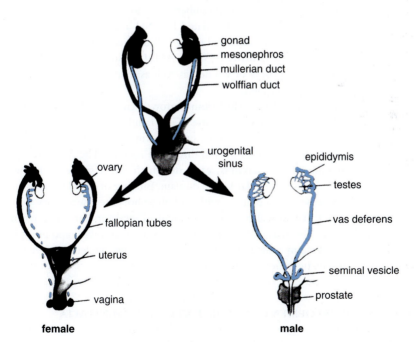

Figure 9 Development of the male and female internal genitalia. (From Jaffe, R. B., "Reproductive Endocrinology," 4th Ed., p. 364. W. B. Saunders, Philadelphia, 1999, with permission.)

mesonephric, or wolffian, ducts. Internal genitalia of the female, including the uterus, fallopian tubes, and upper vagina, develop from the paramesonephric, or müllerian, ducts. When stimulated by testosterone the wolffian ducts differentiate into male reproductive structures, but in the absence of androgen, they regress and disappear. In contrast, müllerian ducts develop into female reproductive structures unless actively suppressed. Testosterone does not stimulate müllerian regression. Under the influence of the *SRY* gene product and specific transcription factors, Sertoli cells in newly differentiated seminiferous tubules secrete a glycoprotein called antimüllerian hormone (AMH), which initiates apoptosis of tubular epithelial cells and reabsorption of the müllerian ducts. In experiments in which only one testis was removed from embryonic rabbits, the müllerian duct regressed only on the side with the remaining gonad, indicating that antimüllerian hormone must act locally as a paracrine factor. The wolffian duct regressed on the opposite side, suggesting that testosterone too must act locally to sustain the adjacent wolffian duct, because the amounts that reached the contralateral duct through the circulation were inadequate to prevent its regression (Figure 10).

Sertoli cell production of AMH is not limited to the embryonic period, but continues into adulthood. AMH is present in adult blood serum and in seminal plasma, where it binds to sperm and may increase their motility. Plasma concentrations of AMH are highest in the prepubertal period and fall as testosterone concentrations rise. Its secretion is stimulated by FSH and strongly inhibited by testosterone. In the testis AMH inhibits Leydig cell differentiation and expression of steroidogenic enzymes, particularly P450c17. AMH is also expressed in the adult ovary and is found in the plasma of women as well as men. No extragonadal role for AMH has yet been established.

AMH is a member of the transforming growth factor β (TGFβ) family of growth factors. As for other members of the TGFβ family, AMH signals by way of membrane receptors that have intrinsic protein kinase activity that catalyzes phosphorylation of proteins on serine and threonine residues. The AMH receptor consists of two nonidentical subunits, each of which has a single membrane-spanning region and an intracellular kinase domain. Binding of AMH to its specific primary receptor causes it to complex with and phosphorylate a secondary signal-transducing subunit that may also be a component of receptors for other agonists of the TGFβ family. The activated receptor complex associates with and phosphorylates cytosolic proteins called Smads which enter the nucleus and activate transcription of specific genes (Figure 11).

DEVELOPMENT OF THE EXTERNAL GENITALIA

The urogenital sinus and the genital tubercle are the primitive structures that give rise to the external genitalia in both sexes. Masculinization of these structures

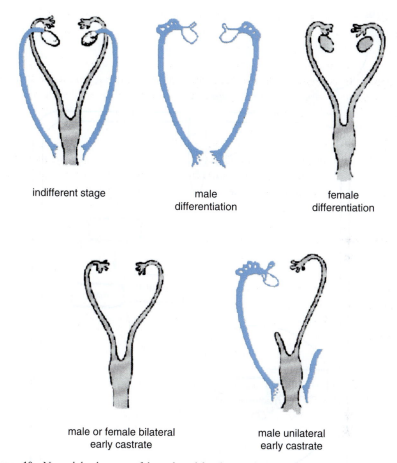

indifferent stage

male
differentiation

female
differentiation

male or female bilateral
early castrate

male unilateral
early castrate

Figure 10 Normal development of the male and female reproductive tracts. Tissues destined to form the male tract are shown in blue; tissues that develop into the female tract are shown in gray. Bilateral castration of either male or female embryos results in development of the female pattern. Early unilateral castration of male embryos results in development of the normal male duct system on the side with the remaining gonad, but female development on the contralateral side. This pattern develops because both testosterone and antimüllerian hormone act as paracrine factors. (Modified from Jost, A., "Hermaphroditism, Genital Anomalies and Related Endocrine Disorders," 2nd Ed., p. 16. Williams & Wilkins, Baltimore, 1971.)

to form the penis, scrotum, and prostate gland depends on secretion of testosterone by the fetal testis. Unless stimulated by androgen, these structures develop into female external genitalia. When there is insufficient androgen in male embryos, or too much androgen in female embryos, differentiation is incomplete and the external genitalia are ambiguous. Differentiation of the masculine external genitalia

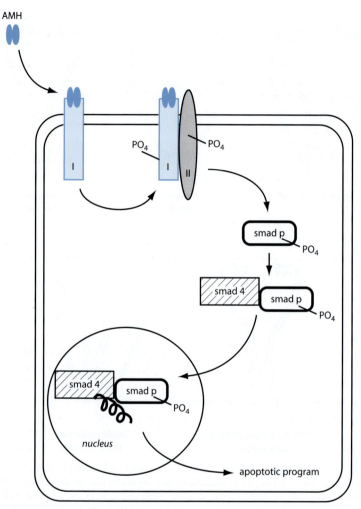

müllerian tubular cell

Figure 11 Antimüllerian hormone (AMH) signaling pathway. AMH binds to its specific primary receptor (I), which then forms a heterodimer with and phosphorylates the secondary signal-transducing subunit (II). The activated receptor complex then catalyzes phosphorylation of Smad proteins on serine and threonine residues, causing them to bind Smad 4, which carries them into the nucleus where transcription of specific genes results in expression of an apoptotic program and resorption of the müllerian duct cells.

depends on dihydrotestosterone rather than testosterone. The 5α–reductase type II responsible for conversion of testosterone to dihydrotestosterone is present in tissues destined to become external genitalia even before the testis starts to secrete testosterone. In contrast, this enzyme does not appear in tissues derived from the wolffian ducts until after they differentiate, indicating that testosterone rather than dihydrotestosterone was the signal for differentiation of the wolffian derivatives.

The importance of androgen action in sexual development is highlighted by a fascinating human syndrome called testicular feminization, which can be traced to an inherited defect in the single gene on the X chromosome that encodes the androgen receptor. Afflicted individuals have the normal female phenotype, but have sparse pubic and axillary hair and no menstrual cycles. Genetically, they are male and have intraabdominal testes and circulating concentrations of testosterone and estradiol that are within the range found in normal men, but their tissues are totally unresponsive to androgens. Their external genitalia are female because, as already mentioned, the primordial tissues develop in the female pattern unless stimulated by androgen. Because AMH production and responsiveness are normal and their wolffian ducts are unable to respond to androgen, both of these duct systems regress and neither male nor female internal genitalia develop. Secondary sexual characteristics including breast development appear at puberty in response to estrogens formed extragonadally from testosterone.

POSTNATAL DEVELOPMENT

Aside from a brief surge in androgen production during the immediate neonatal period, testicular function enters a period of quiescence, and further development of the male genital tract is arrested until the onset of puberty. Increased production of testosterone at puberty promotes growth of the penis and scrotum and increases pigmentation of the genitalia as well as the depth of rugal folds in scrotal skin. Further growth of the prostate, seminal vesicles, and epididymes also occurs at this time. Although differentiation of the epididymes and seminal vesicles was independent of dihydrotestosterone during the early fetal period, later acquisition of 5α-reductase type II makes this more active androgen the dominant form stimulating growth and secretory activity during the pubertal period. Increased secretion of FSH at puberty stimulates multiplication of Sertoli cells and growth of the seminiferous tubules, which constitute the bulk of the testicular mass.

The importance of some of the foregoing information is highlighted by another interesting genetic disorder that has been described as "penis at twelve." Affected individuals have a deletion or inactivating mutation in the gene that codes for 5α-reductase type II, and hence they cannot convert testosterone to dihydrotestosterone in derivatives of the genital tubercule. Although testes and wolffian derivatives develop normally, the prostate gland is absent, and external genitalia at birth are ambiguous or overtly feminine. Affected children have been raised as females. With the onset of puberty there is an increase in testosterone production and an increase in the expression of 5α-reductase type I in the skin. Significant growth of the penis occurs at this time in response to 5α-dihydrotestosterone produced in the liver and skin by the catalytic activity of 5α-reductase type I.

REGULATION OF TESTICULAR FUNCTION

Testicular function, as we have seen, depends on stimulation by two pituitary hormones, FSH and LH. Without them, the testes lose spermatogenic and steroidogenic capacities, and either atrophy or fail to develop. Secretion of these hormones by the pituitary gland is driven by the central nervous system through its secretion of the gonadotropin-releasing hormone, which reaches the pituitary by way of the hypophyseal portal blood vessels (see Chapter 2). Separation of the pituitary gland from its vascular linkage to the hypothalamus results in total cessation of gonadotropin secretion and testicular atrophy. The central nervous system and the pituitary gland are kept apprised of testicular activity by signals related to each of the testicular functions: steroidogenesis and gametogenesis. Characteristic of negative feedback, signals from the testis are inhibitory. Castration results in a prompt increase in secretion of both FSH and LH. The central nervous system also receives and integrates other information from the internal and external environments and modifies GnRH secretion accordingly.

GnRH AND THE HYPOTHALAMIC PULSE GENERATOR

Gonadotropin-releasing hormone is a decapeptide produced by a diffuse network of about 2000 neurons; the neuronal perikarya are located primarily in the arcuate nuclei in the medial basal hypothalamus, and their axons terminate in the median eminence in the vicinity of the hypophyseal portal capillaries. GnRH-secreting neurons also project to other parts of the brain and may mediate some aspects of sexual behavior. GnRH is released into the hypophyseal portal circulation in discrete pulses at regular intervals, ranging from about one every hour to one every 3 hours or longer. Each pulse lasts only a few minutes and the secreted GnRH disappears rapidly with a half-life of about 4 minutes. GnRH secretion is difficult to monitor directly because hypophyseal portal blood is inaccessible and because its concentration in peripheral blood is too low to measure even with the most sensitive assays. The pulsatile nature of GnRH secretion has been inferred from results of frequent measurements of LH concentrations in peripheral blood (Figure 12). FSH concentrations tend to fluctuate much less, largely because FSH has a longer half-life than LH, 2–3 hours compared to 20–30 minutes.

Pulsatile secretion requires synchronous firing of many neurons, which therefore must be in communication with each other and with a common pulse

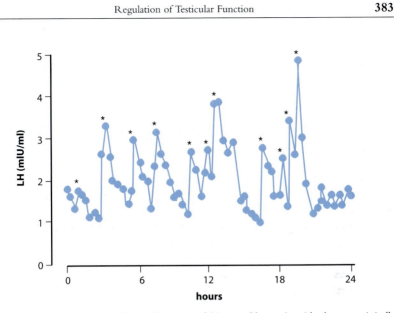

Figure 12   LH secretory pattern observed in a normal 36-year-old man. Asterisks denote statistically significant discrete pulses. (From Crowley, W. F., Jr., "Current Topics in Endocrinology and Metabolism," p. 157, copyright Marcel Decker, New York, 1985.)

generator. Because pulsatile secretion of GnRH continues even after experimental disconnection of the medial basal hypothalamus from the rest of the central nervous system, the pulse generator must be located within this small portion of the hypothalamus. Pulsatile secretion of GnRH by neurons maintained in tissue culture indicate that episodic secretion is an intrinsic property of GnRH neurons. There is good correspondence between electrical activity in the arcuate nuclei and LH concentrations in blood as determined in rhesus monkeys fitted with permanently implanted electrodes. The frequency and amplitude of secretory pulses and corresponding electrical activity can be modified experimentally (Figure 13) and are regulated physiologically by gonadal steroids and probably by other information processed within the central nervous system.

The significance of the pulsatile nature of GnRH secretion became evident in studies of reproductive function in rhesus monkeys whose arcuate nuclei had been destroyed and whose secretion of LH and FSH therefore came to a halt. When GnRH was given as a constant infusion, gonadotropin secretion was restored only for a short while. FSH and LH secretion soon decreased and stopped even though the infusion of GnRH continued. Only when GnRH was administered intermittently for a few minutes of each hour was it possible to sustain normal gonadotropin secretion in these monkeys. Similar results have been obtained in human patients and applied therapeutically. Persons who are deficient

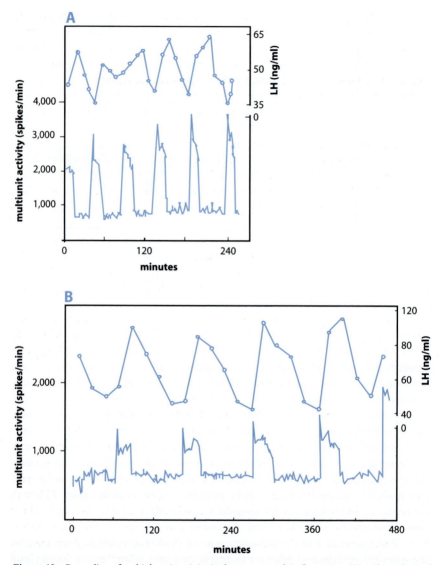

Figure 13 Recording of multiple unit activity in the arcuate nuclei of conscious (A) and anesthetized (B) monkeys fitted with permanently implanted electrodes. Simultaneous measurements of LH in peripheral blood are shown in the upper tracings. (From Wilson, R. C., Kesner, J. S., Kaufman, J. N., Uemura, T., Akema, T., and Knobil, E. *Neuroendocrinology* **39**, 256, 1984, by permission of Blackwell Publishing.)

in GnRH fail to experience pubertal development and remain sexually juvenile. Treating them with a long-acting analog of GnRH that provides constant stimulation to the pituitary is ineffective in restoring normal function. Treating GnRH deficiency with the aid of a pump that delivers GnRH under the skin in intermittent pulses every 2 hours induces pubertal development and normal reproductive function. Because treatment with a long-acting analog of GnRH desensitizes the pituitary gland and blocks gonadotropin secretion this regimen has been used successfully to arrest premature sexual development in children suffering from precocious puberty.

The cellular mechanisms that account for the complex effects of GnRH on gonadotropes are not fully understood. The GnRH receptor is a G-protein-coupled heptihelical receptor that activates phospholipase C through $G\alpha_q$ (Chapter 1). The resulting formation of inositol trisphosphate (IP_3) and diacylglycerol (DAG) results in mobilization of intracellular calcium and activation of protein kinase C. Transcription of genes for FSH β, LH β, and the common alpha subunit depends on increased cytosolic calcium and several protein kinases that have activation pathways that are not understood. Secretion of gonadotropins depends on the increase in intracellular calcium achieved by mobilizing calcium from intracellular stores and by activating membrane calcium channels. Desensitization of gonadotropes after prolonged uninterrupted exposure to GnRH appears to result from the combined effects of down-regulation of GnRH receptors, down-regulation of calcium channels associated with secretion, and a decrease in the releasable storage pool of gonadotropin.

NEGATIVE FEEDBACK REGULATORS

The hormones FSH and LH originate in the same pituitary cell whose secretory activity is stimulated by the same hypothalamic hormone. Nevertheless, secretion of FSH is controlled independently of LH secretion by negative feedback signals that relate to the separate functions of the two gonadotropins. Although castration is followed by increased secretion of both FSH and LH, only LH is restored to normal when physiological amounts of testosterone are given. Failure of testicular descent into the scrotum (*cryptorchidism*) may result in destruction of the germinal epithelium without affecting Leydig cells. With this condition blood levels of testosterone and LH are normal, but FSH is elevated. Thus testosterone, which is secreted in response to LH, acts as a feedback regulator of LH and hence of its own secretion. By this reasoning, we would expect that spermatogenesis, which is stimulated by FSH, might be associated with secretion of a substance that reflects gamete production. Indeed, FSH stimulates the Sertoli cells to synthesize and secrete a glycoprotein called *inhibin*, which acts as a feedback inhibitor of FSH.

Inhibin, which was originally purified from follicular fluid of the pig ovary, is a disulfide-linked heterodimer composed of an alpha subunit and either of two forms of a beta subunit, β_A or β_B. The physiologically important form of inhibin secreted by the human testis is the $\alpha\beta_B$ dimer called inhibin B. Its concentration in blood plasma is reflective of the number of functioning Sertoli cells and spermatogenesis. Both inhibin A and inhibin B are produced by the ovary (see Chapter 12). Little is known about the significance of alternate beta subunits or the factors that determine when each form is produced. All three subunits are encoded in separate genes, and presumably are regulated independently. They are members of the same family of growth factors that includes AMH and TGFβ. Of additional interest is the finding that dimers formed from two beta subunits produce effects that are opposite those of the $\alpha\beta$ dimer, and stimulate FSH release from gonadotropes maintained in tissue culture. These compounds are called *activins* and function in a paracrine mode in the testis and many other tissues. Although the production of the alpha subunit is largely confined to male and female gonads, beta subunits are produced in many extragonadal tissues where activins mediate a variety of functions. Activins are produced in the pituitary and appear to play a supportive role in FSH production. The pituitary and other tissues also produce an unrelated protein called *follistatin* that binds activins and blocks their actions.

The feedback relations that fit best with current understanding of the regulation of testicular function in the adult male are shown in Figure 14. Pulses of GnRH originating in the arcuate nuclei evoke secretion of both FSH and LH by the anterior pituitary. FSH and LH are positive effectors of testicular function and stimulate release of inhibin and testosterone, respectively. Testosterone has an intratesticular action that reinforces the effects of FSH. It also travels through the circulation to the hypothalamus, where it exerts its negative feedback effect primarily by slowing the frequency of GnRH pulses. Because secretion of LH is more sensitive to frequency of stimulation than is secretion of FSH, decreases in GnRH pulse frequency lower the ratio of LH to FSH in the gonadotropic output. In the castrate monkey the hypothalamic pulse generator discharges once per hour and slows to once every 2 hours after testosterone is replaced. This rate is about the same as that seen in normal men. The higher frequency in the castrate triggers more frequent bursts of gonadotropin secretion, resulting in higher blood levels of both FSH and LH. Testosterone may also decrease the amplitude of the GnRH pulses somewhat and may also exert some direct restraint on LH release from gonadotropes. In high enough concentrations, testosterone may inhibit GnRH release sufficiently to shut off secretion of both gonadotropic hormones. The negative feedback effect of inhibin appears to be exerted exclusively on gonadotropes to inhibit FSH β transcription and FSH secretion in response to

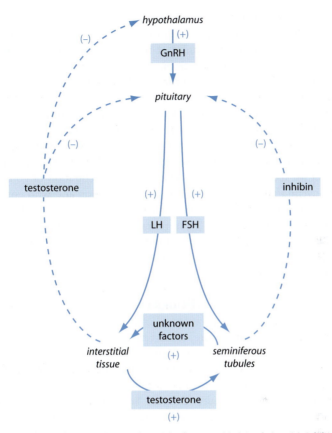

Figure 14 Negative feedback regulation of testicular function. (+), Stimulation; (−), inhibition. Direct effects of testosterone on the pituitary gland are still uncertain.

GnRH. Some evidence indicates that inhibin may also exert local effects on Leydig cells to enhance testosterone production.

PREPUBERTAL PERIOD

Testicular function is critical for development of the normal masculine phenotype early in the prenatal period. All of the elements of the control system are present in the early embryo. GnRH and gonadotropins are detectable at about the time that testosterone begins stimulating wolffian duct development. The hypothalamic GnRH pulse generator and its negative feedback control are functional in the newborn. Both the frequency and the amplitude of GnRH and LH pulses are similar to those observed in the adult. After about the sixth month of postnatal life and for the remainder of the juvenile period, the GnRH pulse generator is restrained and gonadotropin secretion is low. The amplitude and frequency of GnRH pulses decline, but do not disappear, and responsiveness of the gonadotropes to GnRH diminishes. It is evident that negative feedback regulation remains operative, however, because blood levels of gonadotropins increase after gonadectomy in prepubertal subjects and fall with gonadal hormone administration. The system is extremely sensitive to feedback inhibition during this time, but suppression of the pulse generator cannot be explained simply as a change in the set point for feedback inhibition. The plasma concentration of gonadotropins is high in juvenile subjects whose testes failed to develop and who consequently lack testosterone, but rises even higher when these subjects reach the age when puberty would normally occur. Thus restraint of the GnRH pulse generator imposed by the central nervous system diminishes at the onset of puberty.

PUBERTY

Early stages of puberty are characterized by the appearance of high-amplitude pulses of LH during sleep (Figure 15). Testosterone concentrations in plasma follow the gonadotropins, and there is a distinct daynight pattern. As puberty progresses, high-amplitude pulses are distributed throughout the day at the adult frequency of about one every 2 hours. Sensitivity of the pituitary gland to GnRH increases during puberty, possibly as a result of a self-priming effect of GnRH on gonadotropes. GnRH increases the amount of releasable FSH and LH in the gonadotropes and may also increase (up-regulate) the number of its receptors on the gonadotrope surface.

The underlying neural mechanisms for suppression of the GnRH pulse generator in the juvenile period are not understood. Increased inhibitory input from neurons that secrete neuropeptide Y or γ-aminobutyric acid (GABA) has been observed, but the factors that produce and terminate such input are not understood. Clearly, the onset of reproductive capacity is influenced by, and must be coordinated with, metabolic factors and attainment of physical size. In this regard, as we have seen (Chapter 10), puberty is intimately related to growth. Onset of puberty, especially in girls, has long been associated with adequacy of body fat stores, and it appears that adequate circulating concentrations of leptin (Chapter 9) are permissive for the onset of puberty, but available evidence indicates that leptin is not the trigger. It is likely that some confluence of genetic, developmental, and nutritional factors signals readiness for reproductive development and function.

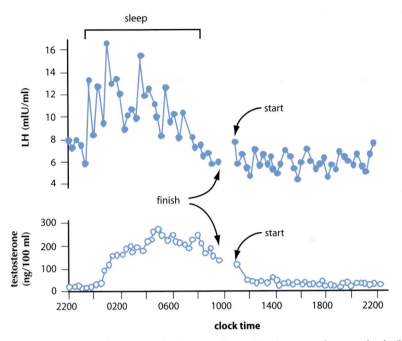

Figure 15   Plasma LH and testosterone levels measured every 20 minutes reveal nocturnal pulsatile secretion of GnRH in a pubertal 14-year-old boy. (From Boyer, R. M., Rosenfeld, R. S., Kapen, S., *et al., J. Clin. Invest.* **54**, 609, 1974, with permission.)

SUGGESTED READING

Crowley, W. F., Jr., Filicori, M., Spratt, D. I., and Santoro, N. F. (1985). The physiology of gonadotropin-releasing hormone (GnRH) in men and women. *Recent Prog. Horm. Res.* **41**, 473–526.

Crowley, W. F., Jr., Whitcomb, R. W., Jameson, J. L., Weiss, J., Finkelstein, J. S., and O'Dea, L. S. L. (1991). Neuroendocrine control of human reproduction in the male. *Recent Prog. Horm. Res.* **47**, 349–387.

George, F. W., and Wilson, J. D. (1986). Hormonal control of sexual development. *Vitamins Horm.* **43**, 145–196.

Hayes, F. J., Hall, J. E., Boepple, P. A., and Crowley, W. F., Jr. (1998). Clinical review 96: Differential control of gonadotropin secretion in the human: Endocrine role of inhibin. *J. Clin. Endocrinol. Metab.* **83**, 1835–1841.

Huhtaniemi, I. (2000). Mutations of gonadotrophin and gonadotrophin receptor genes: What do they teach us about reproductive physiology? *J. Reprod. Fertil.* **119**, 173–186.

Kierszenbaum, A. L. (1994). Mammalian spermatogenesis *in vivo* and *in vitro*: A partnership of spermatogenic and somatic cell lineages. *Endocr. Rev.* **15**, 116–134.

Mooradian, A. D., Morley, J. E., and Korenman, S. G. (1987). Biological actions of androgens. *Endocr. Rev.* **8**, 1–28.

Naor, Z., Harris, D., and Shacham, S. (1998). Mechanism of GnRH receptor signaling: Combinatorial cross-talk of Ca^{2+} and protein kinase C. *Front. Neuroendocrinol.* **19**, 1–19.

Payne, A. H., and Youngblood, G. L. (1995). Regulation of expression of steroidogenic enzymes in Leydig cells. *Bio. Reprod.* **52**, 217–225.

Plant, T. M., and Marshall, G. R. (2001). The functional significance of FSH in spermatogenesis and control of its secretion in male primates. *Endocr. Rev.* **22**, 764–786.

Rosner, W., Hryb, D. J., Khan, M. S., Nakhla, A. M., and Romas, N. A. (1999). Sex hormone-binding globulin mediates steroid hormone signal transduction at the plasma membrane. *J. Steroid Biochem. Mol. Biol.* **69**, 481–485.

Teixeira, J., Maheswaran, S., and Donahoe, P. K. (2001). Müllerian inhibiting substance: An instructive developmental hormone with diagnostic and possible therapeutic applications. *Endocr. Rev.* **22**, 657–674.

Wilson, J. D. (1988). Androgen abuse by athletes. *Endocr. Rev.* **9**, 181–199.

Wilson, J. D., Griffin, J. E., and Russell, D. W. (1993). Steroid 5α-reductase 2 deficiency. *Endocr. Rev.* **14**, 577–593.

Ying, S.-Y. (1988). Inhibins, activins, and folliculostatins: Gonadal proteins modulating the secretion of follicle-stimulating hormone. *Endocr. Rev.* **9**, 267–293.

CHAPTER 12

Hormonal Control of Reproduction in the Female: The Menstrual Cycle

OVERVIEW

The ovaries serve the dual function of producing eggs and the hormones that support reproductive functions. Unlike men, in whom large numbers of gametes are produced continuously from stem cells, women release only one gamete at a time from a limited pool of preformed gametes in a process that is repeated at regular monthly intervals. Each interval encompasses the time needed for the ovum to develop, for preparation of the reproductive tract to receive the fertilized ovum for the ovum to become fertilized, and for pregnancy to be established. If the ovary does not receive a signal that an embryo has begun to develop, the process of gamete maturation begins anew. The principal ovarian hormones are the steroids *estradiol* and *progesterone* and the peptide *inhibin*. Together these hormones orchestrate the cyclic series of events that unfold in the ovary, pituitary, and reproductive tract each month. As the ovum develops within its follicle, estradiol stimulates growth of the structures of the reproductive tract that receive the sperm, facilitate fertilization, and ultimately house the developing embryo. Estradiol, in conjunction with a variety of peptide growth factors, acts within the follicle to stimulate proliferation and secretory activity of granulosa cells and thereby enhances its own production. Progesterone is produced by the corpus luteum that develops from the follicle after the egg is shed. It prepares the uterus for successful implantation and growth of the embryo, and is absolutely required for the maintenance of pregnancy. Ovarian function is driven by the two pituitary gonadotropins, follicle-stimulating hormone (FSH) and luteinizing hormone (LH), which stimulate ovarian steroid production, growth of the follicle, ovulation, and development of the corpus luteum. Secretion of these hormones depends on stimulatory input from the hypothalamus through the gonadotropin releasing hormone (GnRH) and complex inhibitory and stimulatory input from ovarian steroid and peptide hormones.

FEMALE REPRODUCTIVE TRACT

OVARIES

The adult human ovaries are paired, flattened ellipsoid structures that measure about 5 cm in their longest dimension. They lie within the pelvic area of the abdominal cavity attached to the broad ligaments that extend from either side of the uterus by peritoneal folds called the *mesovaria*. Both the gamete-producing and hormone-producing functions of the ovary take place in the outer (cortical) portion. It is within the ovarian cortex that the precursors of the female gametes, the oocytes, are stored and develop into ova (eggs). The functional unit is the *ovarian follicle*, which initially consists of a single oocyte surrounded by a layer of *granulosa cells* enclosed within a basement membrane, the *basal lamina*, that separates the follicle from cortical stroma. When they emerge from the resting stage, follicles become ensheathed in a layer of specialized cells called the *theca folliculi*. Follicles in many stages of development are found in the cortex of the adult ovary along with structures that form when the mature ovum is released by the process of ovulation. Ovarian follicles, in which the ova develop, and the corpora lutea derived from them are also the sites of ovarian hormone production. The inner portion of the ovary, the medulla, consists chiefly of vascular elements that arise from anastomoses of the uterine and ovarian arteries. A rich supply of unmyelinated nerve fibers also enters the medulla along with blood vessels (Figure 1).

Folliculogenesis

In contrast to the testis, which produces of hundreds of millions of sperm each day, the ovary normally produces a single mature ovum about once each month. The testis must continuously renew its pool of germ cell precursors throughout reproductive life in order to sustain this rate of sperm production, whereas the ovary needs only to draw on its initial endowment of primordial oocytes to provide the approximately 400 mature ova ovulated during the four decades of a woman's reproductive life. Ovulation, the hallmark of ovarian activity, occurs episodically at 28-day intervals, but examination of the ovary at any time during childhood or the reproductive life of a mature woman reveals continuous activity, with multiple follicles at various stages in their life cycle.

Folliculogenesis begins in fetal life. Primordial germ cells multiply by mitosis and begin to differentiate into primary oocytes that enter meiosis between the eleventh and twentieth weeks after fertilization. Primary oocytes remain arrested in prophase of the first meiotic division until meiosis resumes at time of ovulation and is not completed until fertilization, which may be more than four decades later for some oocytes. Around the twentieth week of fetal life there are about 6–7 million oocytes available to form primordial follicles, but the human female is born with about only 300,000–400,000 primordial follicles in each ovary. Oocytes that fail to form into primordial follicles are lost by apoptosis, and many primordial follicles are also lost during fetal life in a process called *atresia*. The vast majority of primordial follicles remain in a resting state for many years. Each day by some seemingly random process, either because they are relieved of inhibition or are activated by still unknown factors, some follicles enter into a growth phase and begin the long

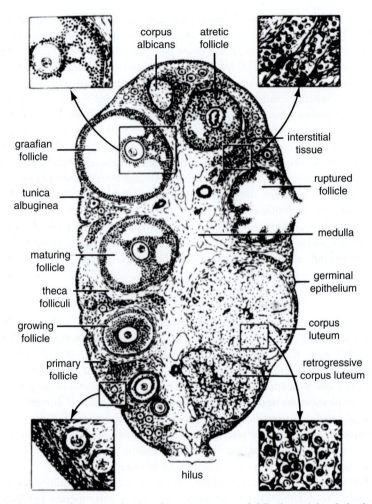

Figure 1 Mammalian ovary showing the various stages of follicular and luteal development. Obviously, events depicted occur sequentially and are not all present in any one section of a human ovary. (From Turner, C. D., and Bagnara, J. T., "General Endocrinology," 6th Ed., p. 453. W. B. Saunders, Philadelphia, 1976, with permission.)

journey toward ovulation, but the vast majority become atretic and die at various stages of development. This process begins during the fetal period and continues until menopause at around age 50, when all of the follicles are exhausted.

As each primordial follicle emerges from the resting stage its oocyte grows from a diameter of about 20 to about 100 µm and a layer of extracellular mucopolysaccharides and proteins called the *zona pellucida* forms around it

(Figure 2). Its granulosa cells change in morphology from squamous to cuboidal and begin to divide. Growth of primary follicles is accompanied by migration and differentiation of mesenchymal cells to form the theca folliculi. Its inner layer, the *theca interna*, is composed of secretory cells with an extensive smooth endoplasmic reticulum that is characteristic of steroidogenic cells. The theca externa is formed by reorganization of surrounding stromal cells. At this time a dense capillary network develops around the follicle. Oocytes complete their growth by accumulating stored nutrients and the messenger RNA and protein-synthesizing apparatus that will be activated on fertilization. As follicles continue to grow, granulosa cells increase in number and begin to form multiple layers. The innermost granulosa cells are in intimate contact with the oocyte through cellular processes that penetrate the zona pellucida and form gap junctions with its plasma membrane. Granulosa cells also form gap junctions with each other. They function as nurse cells, providing nutrients to the oocyte, which is separated from direct contact with capillaries by the basal lamina and layers of granulosa cells.

Follicular development continues, with further proliferation of granulosa cells and gradual elaboration of fluid within the follicle. Follicular fluid is derived from blood plasma and contains plasma proteins, including hormones, and various peptides and steroids secreted by the granulosa cells and the ovum. Accumulation of fluid brings about further follicular enlargement and the formation of a central fluid-filled cavity called the *antrum*. Follicular growth up to this stage is independent of pituitary hormones, but without support from follicle-stimulating hormone (see Chapters 2 and 11) further development is not possible and the follicles become atretic. Atresia is the fate of all of the follicles that enter the growth phase before puberty, and more than 99% of the 200,000 to 400,000 remaining at puberty. The physiological mechanisms that control this seemingly wasteful process are poorly understood.

In the presence of FSH antral follicles continue to develop slowly for about 2 months and grow from about 0.2 to about 2 mm in diameter. A group, or cohort, of 6–12 of these follicles enters into the final rapid growth phase about 20 days before ovulation, but in each cycle normally only one survives and ovulates. The others become atretic and die (Figure 3). The surviving follicle has been called the dominant follicle because it may contribute to the demise of other follicles in the cohort. As the dominant follicle matures, the fluid content in the antrum increases rapidly, possibly in response to increased colloid osmotic pressure created by partial hydrolysis of dissolved mucopolysaccharides. The ripe, preovulatory follicle reaches a diameter of 20 to 30 mm and bulges into the peritoneal cavity. At this time it consists of about 60 million granulosa cells arranged in multiple layers around the periphery. The ovum and its surrounding layers of granulosa cells, the *corona radiata*, are suspended by a narrow bridge of granulosa cells (the *cumulus oophorous*) in pool of more than 6 ml of follicular fluid. At ovulation a point opposite the ovum in the follicle wall erodes and the ovum with its corona of

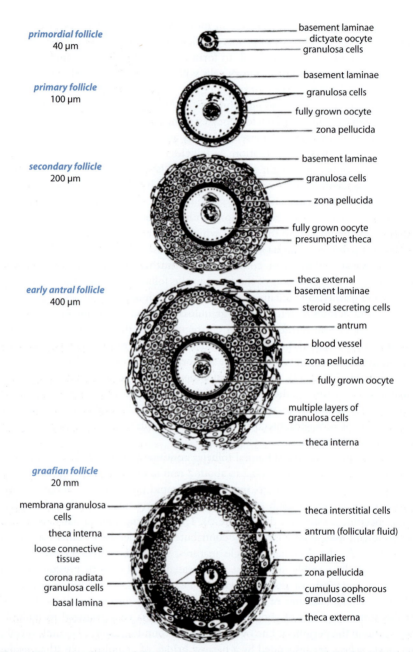

Figure 2   Stages of human follicular development. (From Erickson, G. F., "Endocrinology and Metabolism," 3rd Ed., pp. 973–1015. McGraw Hill, New York, 1995, with permission of The McGraw-Hill Companies.)

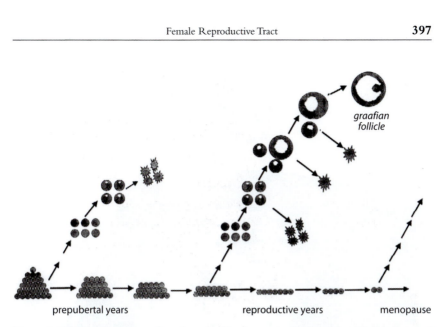

graafian
follicle

prepubertal years   reproductive years   menopause

Figure 3 Follicular development throughout the life of a woman. (Redrawn from McGee, E. A., and Hsueh, A. J. W., *Endocr. Rev.* **21**, 200–214, 2000, by permission of The Endocrine Society.)

granulosa cells is extruded into the peritoneal cavity in a bolus of follicular fluid (Figure 4).

Following ovulation there is ingrowth and differentiation of the remaining mural granulosa cells, thecal cells, and some stromal cells, which fill the cavity of the collapsed follicle to form a new endocrine structure, the *corpus luteum*. The process by which granulosa and thecal cells are converted to luteal cells is called luteinization (meaning yellowing) and is the morphological reflection of the accumulation of lipid. Luteinization also involves biochemical changes that enable the corpus luteum to become the most active of all steroid-producing tissues per unit weight. The corpus luteum consists of large polygonal cells containing smooth endoplasmic reticulum and a rich supply of fenestrated capillaries. Unless pregnancy ensues, the corpus luteum regresses after 2 weeks, leaving a scar on the surface of the ovary.

OVIDUCTS AND UTERUS

The primitive müllerian ducts that develop during early embryonic life give rise to the duct system that in primitive organisms provides the route for ova to escape to the outside (Figure 5). In mammals these tubes are adapted to provide a site for fertilization and nurture of embryos. In female embryos the müllerian ducts are not subjected to the destructive effects of the antimüllerian hormone

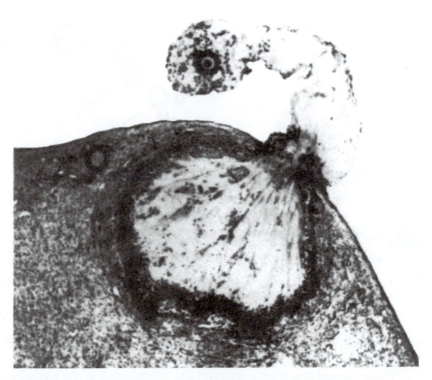

Figure 4 Ovulation in a rabbit. Follicular fluid, granulosa cells, some blood, and cellular debris continue to ooze out of the follicle even after the egg mass has been extruded. (From Hafez, E. S. E., and Blandau, R. J., "The Mammalian Oviduct." University of Chicago Press, Chicago, 1969, with permission.)

(see Chapter 11) and, instead, develop into the oviducts, uterus, and the upper portion of the vagina. Unlike the development of the sexual duct system in the male fetus, this differentiation is independent of gonadal hormones.

The paired oviducts (*fallopian tubes*) are a conduit for transfer of the ovum to the uterus (see Chapter 13). Their proximal ends are in close contact with the ovaries and have funnel-shaped openings, the *infundibula*, surrounded by fingerlike projections called *fimbriae*. The oviducts are lined with ciliated cells whose synchronous beating plays an important role in egg transport. The lining of the oviducts also contains secretory cells whose products provide nourishment for the zygote in its 3- to 4-day journey to the uterus. The walls contain layers of longitudinally and circularly oriented smooth muscle cells, which also contribute to gamete transport.

Distal portions of the müllerian ducts fuse to give rise to the uterus. In the nonpregnant woman the uterus is a small, pear-shaped structure extending about

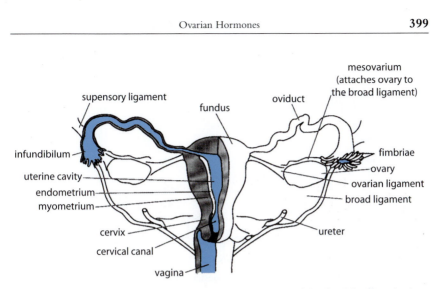

Figure 5 Uterus and associated female reproductive structures. The left side of the figure has been sectioned to show the internal structures. (From Tortora, G. J., and Anagnostakos, N. P., "Principles of Anatomy and Physiology," 3rd Ed., p. 721, copyright Harper & Row, New York, 1981.)

6 to 7 cm in its longest dimension. It is capable of enormous expansion, partly by passive stretching and partly by growth, so that at the end of pregnancy it may reach 35 cm or more in its longest dimension. Its thick walls consist mainly of smooth muscle and are called the *myometrium*. The secretory epithelial lining is called the *endometrium* and varies in thickness with changes in the hormonal environment, as discussed below. The oviducts join the uterus at the upper, rounded end. The caudal end constricts to a narrow cylinder called the *uterine cervix*, whose thick wall is composed largely of dense connective tissue rich in collagen fibers and some smooth muscle. The cervical canal is lined with mucus-producing cells and is usually filled with mucus. The cervix bulges into the upper reaches of the vagina, which forms the final link to the outside. The lower portion of the vagina, which communicates with the exterior, is formed from the embryonic urogenital sinus.

OVARIAN HORMONES

The principal hormones secreted by the ovary are estrogens (estradiol-17β and estrone) and progesterone. These hormones are steroids and are derived from cholesterol by the series of reactions depicted in Figure 6. Their biosynthesis is intricately interwoven with the events of the ovarian cycle and is discussed in the next sections. In addition, the ovary produces a large number of biologically active peptides, most of which act within the ovary as paracrine growth factors, but at

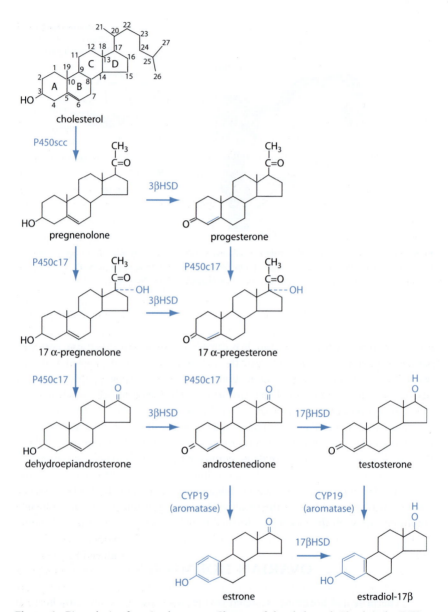

Figure 6 Biosynthesis of ovarian hormones. Cleavage of the cholesterol side chain by P450scc between carbons 21 and 22 gives rise to 21-carbon progestins. Removal of carbons 20 and 21 by the two-step reaction catalyzed by P450c17 (17α-hydroxylase/lyase) produces the 19-carbon androgen series. Aromatization of ring A catalyzed by P450cyp19 (CYP19, aromatase) eliminates carbon 19 and yields 18-carbon estrogens. 3βHSD, 3β-Hydroxysteroid dehydrogenase; 17βHSD, 17β-hydroxysteroid dehydrogenase.

least two, inhibin and relaxin, are produced in sufficient amounts to enter the blood and produce effects in distant cells.

ESTROGENS

Unlike humans, of whom it has been said, "eat when they are not hungry, drink when they are not thirsty, and make love at all seasons of the year," most vertebrate animals mate only at times of maximum fertility of the female. This period of sexual receptivity is called *estrus*, derived from the Greek word for vehement desire. Estrogens are compounds that promote estrus and were originally isolated from follicular fluid of sow ovaries. Characteristic of steroid-secreting tissues, little hormone is stored within the secretory cells. Estrogens circulate in blood loosely bound to albumin and tightly bound to the sex hormone-binding globulin (see Chapter 11), which is also called the testosterone - estrogen - binding globulin. Plasma concentrations of estrogen are considerably lower than those of other gonadal steroids and vary over an almost 20-fold range during the cycle.

The liver is the principal site of metabolic destruction of estrogens. Estradiol and estrone are completely cleared from the blood by a single passage through the liver and are inactivated by hydroxylation and conjugation with sulfate and glucuronide. About half the protein-bound estrogen in blood is conjugated with sulfate or glucuronide. Although the liver may excrete some conjugated estrogens in the bile, they are reabsorbed in the lower gut and returned to the liver in portal blood in a typical enterohepatic circulatory pattern. The kidney is the chief route of excretion of estrogenic metabolites.

PROGESTERONE

Pregnancy, or gestation, requires the presence of another ovarian steroid hormone, progesterone. In the nonpregnant woman progesterone secretion is largely confined to cells of the corpus luteum; however, because progesterone is an intermediate in the biosynthesis of all steroid hormones, small amounts may also be released from the adrenal cortex. Some progesterone is also produced by granulosa cells just before ovulation. The rate of progesterone production varies widely. Its concentration in blood ranges from virtually nil during the early preovulatory part of the ovarian cycle to as much as 2 mg/dl after the corpus luteum has formed. Progesterone circulates in blood in association with plasma proteins and has a high affinity for the corticosteroid-binding globulin (CBG). Liver is the principal site of progesterone inactivation, which is achieved by reduction of the A ring and the keto groups at carbons 3 and 20 to give *pregnanediol*, which is the chief metabolite found in urine. Considerable degradation also occurs in the uterus.

INHIBIN

As discussed in Chapter 11, inhibin is a 32-kDa disulfide-linked dimer of an alpha subunit and either of two beta subunits, β_A or β_B. It enters the circulation as either inhibin A ($\alpha\beta_A$) or inhibin B ($\alpha\beta_B$). Expression of the β_A subunit is greatest in luteal cells; the β_B subunit is a product of granulosa cells. Consequently, blood levels of inhibin B are highest during periods of preovulatory growth and expansion of granulosa cells, whereas blood levels of inhibin A are highest during peak luteal cell function. In addition to serving as a circulating hormone, inhibin probably exerts paracrine actions in the ovary, and activin formed by dimerization of two beta subunits also exerts important intraovarian paracrine actions. The activin-binding peptide follistatin, which blocks activin action, also plays an important intraovarian role. Although some activin is found in the circulation, its concentrations do not change during the ovarian cycle, and its source is primarily extraovarian.

RELAXIN

The corpus also luteum secretes a second peptide hormone called *relaxin*, which was named for its ability to relax the pubic ligament of the pregnant guinea pig. In other species, including humans, it also relaxes the myometrium and plays an important role in parturition by causing softening of the uterine cervix. Relaxin is encoded in two nonallelic genes (*H1* and *H2*) on chromosome 9. Its peptide structure, particularly the organization of its disulfide bonds, and gene organization place it in the same family as insulin and the insulin-like growth factors. Although both relaxin genes are expressed in the prostate, only the H2 gene is expressed in the ovary. A physiological role for relaxin in the nonpregnant woman has not been established.

CONTROL OF OVARIAN FUNCTION

Follicular development beyond the antral stage depends on two gonadotropic hormones secreted by the anterior pituitary gland: FSH and LH. In addition to follicular growth, gonadotropins are required for ovulation, luteinization, and steroid hormone formation by both the follicle and the corpus luteum. The relevant molecular and biochemical characteristics of these glycoprotein hormones are described in Chapters 2 and 11. Follicular growth and function also depend on paracrine effects of estrogens, androgens, possibly progesterone, as well as peptide paracrine factors, including IGF-2, activin, members of the transforming growth factor β family, and others. The sequence of rapid follicular growth, ovulation, and the subsequent formation and degeneration of the corpus luteum is repeated about every 28 days and constitutes the ovarian cycle. The part of the cycle devoted to the final rapid growth of the ovulatory follicle lasts about 14 days and is called the *follicular phase*. The remainder of the cycle is dominated by the corpus luteum and is called the *luteal phase*. It, too, lasts about 14 days. Ovulation occurs at midcycle and requires only about a day. These events are orchestrated by a complex pattern of pituitary and ovarian hormonal changes, and although ovulation is the focal point of each cycle, subsequent cohorts of follicles are being prepared for subsequent cycles in the event that fertilization does not occur.

Under normal circumstances only one follicle ovulates in each cycle and appears randomly on either the right or left ovary. Usually a cohort of 6–12 follicles will be mature enough to enter into the final preovulatory growth period that begins near the end of the luteal phase of the preceding cycle. The dominant follicle is selected early in the follicular phase that leads to its ovulation. The physiological mechanisms for selection of a single dominant follicle are not understood. Recruitment of the next cohort of follicles does not begin so long as the dominant follicle or its resultant corpus luteum is present and functional. Experimental destruction of either the dominant follicle or the corpus luteum is promptly followed by selection and development of a new ovulatory follicle from the next cohort. Clearly at the end of the luteal phase and in the first few days of the follicular phase multiple follicles have the potential to ovulate, and can be rescued by treatment with supraphysiological amounts of FSH. This accounts for the high frequency of multiple births following some therapies for infertility. Producing multiple ova in this way is used for harvesting eggs for *in vitro* fertilization technologies.

EFFECTS OF FSH AND LH ON THE DEVELOPING FOLLICLE

Estradiol Production

Synthesis of estrogens depends on both FSH and LH and on the cooperative interplay of granulosa and theca cells. Granulosa cells of the ovulatory follicle are the major and virtually only source of estradiol in the follicular phase of the ovarian cycle, and secrete estrogens in response to FSH. Granulosa cells in antral follicles are the only targets for FSH; no other ovarian cells are known to have FSH receptors. LH receptors are found only in cells of the theca interna and the stroma until about the middle of the follicular phase. Both theca and granulosa cells are required for estradiol production in the follicular phase. Neither cell type expresses the full complement of enzymes needed for synthesis of estradiol. Granulosa cells are not in contact with capillaries and hence have very limited access to cholesterol present in low-density lipoproteins (LDLs), express few LDL receptors, and have minimal levels of the P450scc enzyme that catalyzes the rate-determining conversion of cholesterol to pregnenolone (see Figure 6). In addition, granulosa cells do not express P450c17, which converts pregnenolone to 19-carbon andro-gens. Theca cells, which have a direct capillary blood supply, express high levels of LDL receptors and both P450scc and P450c17. Theca cells produce the 21-carbon pregnenolone and metabolize it to the 19-carbon androstenedione in response to LH, but because they lack aromatase, they cannot synthesize estrogens. Granulosa cells, on the other hand, express ample aromatase, and readily aromatize androgens provided by diffusion from the theca interna. This two-cell interaction is illustrated in Figure 7. Participation of two different cells, each stimulated by its own

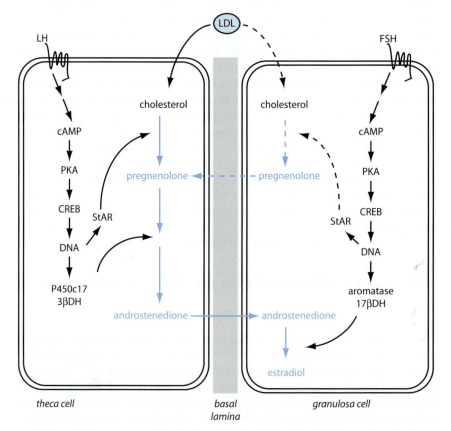

Figure 7 Theca and granulosa cell cooperation in estrogen synthesis. Theca cells produce androgens in response to luteinizing hormone (LH). Granulosa cells respond to follicle-stimulating hormone (FSH) by aromatizing androgens to estrogens. LDL, low density lipoproteins; cAMP, cyclic adenosine monophosphate; PKA, protein kinase A; DNA, deoxyribonucleic acid; CREB, cyclic AMP response element binding protein; StAR, steroid acute regulatory protein; P450c17, 17α-dehydrogenase/lyase; 3βDH, 3β-hydroxysteroid dehydrogenase; 17βDH, 17β-hydroxysteroid dehydrogenase. Dashed arrows indicate minor or questionable effect.

gonadotropin, accounts for the requirement of both pituitary hormones for adequate estrogen production and hence for follicular development.

Follicular Development

Under the influence of FSH, granulosa cells in the follicle destined to ovulate increase by more than 100-fold and the follicle expands about 10-fold in diameter, mainly because of the increase in follicular fluid. Granulosa cells in early antral follicles have few if any receptors for LH. By about the middle of the follicular phase, however, they start to express increasing amounts of LH receptor, which becomes quite abundant just prior to ovulation (Figure 8). Acquisition of LH receptors depends on FSH and enables granulosa cells to respond to both FSH and LH. Induction of LH receptors and other actions of FSH on follicular development are amplified by paracrine actions of peptide growth factors and steroid hormones. In response to FSH, granulosa cells secrete insulin-like growth factor II (IGF-II), inhibin, activin, and vascular endothelial growth factor (VEG-F), which greatly

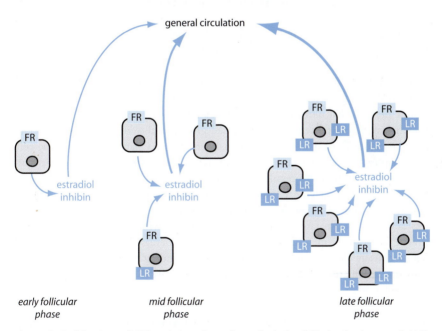

Figure 8 Proliferation and differentiation of granulosa cells during follicular development. Initially, granulosa cells are few and have receptors only for FSH on their surface. In response to continued stimulation with FSH, granulosa cells proliferate and by the midfollicular phase LH receptors begin to appear. By late in the follicular phase a large number of granulosa cells are present and they are responsive to both LH and FSH. They are now competent to secrete sufficient estradiol to trigger the ovulatory surge of gonadotropins. FR, FSH receptor; LR, LH receptor.

increases vascularization around the theca. IGF-II not only stimulates the growth and secretory capacity of granulosa cells, but also acts synergistically with LH to increase synthesis of androgens by cells of the theca interna. Inhibin, which is also produced by granulosa cells in response to FSH, also stimulates thecal cell production of androgens, and activin enhances FSH-induced expression of P450 aromatase in the granulosa cells. Thus even though theca cells lack FSH receptors, FSH nevertheless indirectly stimulates them to increase production of the androgens required by the granulosa cells for estrogen synthesis. Similarly, by stimulating thecal cells to produce androgens, LH promotes growth and development of granulosa cells through autocrine and paracrine actions of both estrogen and androgens, whose receptors are abundantly expressed in these cells. These events constitute a local positive feedback circuit that gives the follicle progressively greater capacity to produce estradiol and makes it increasingly sensitive to FSH and LH as it matures.

CELLULAR ACTIONS OF FSH AND LH

Follicle-stimulating hormone and LH each bind to specific G-protein-coupled receptors on the surface of granulosa or theca cells and activate adenylyl cyclase in the manner already described (Chapters 1 and 11). Increased concentrations of cyclic AMP activate protein kinase A, which catalyzes phosphorylation of cyclic AMP response element binding (CREB) protein and other nuclear and cytoplasmic proteins that ultimately lead to increased transcription of genes that encode growth factors and proteins critical for cell growth and steroid hormone production (see Figure 7). As described for adrenal cortical (see Chapter 4) and Leydig cells (see Chapter 11), the rate-limiting step in steroid hormone synthesis requires synthesis of the steroid acute regulatory (StAR) protein to deliver cholesterol to the intramitochondrial enzyme P450scc that converts it to pregnenolone. In addition, activation of protein kinase A results in increased expression of P450c17 in theca cells and P450 aromatase in granulosa cells. Along with increased expression of growth factors, the ripening follicle also expresses increased amounts of a protease that specifically cleaves insulin-like growth factor binding protein 4 (IGFBP-4) that is present in follicular fluid, and thereby increases availability of free IGF-II.

EFFECTS ON OVULATION

Luteinizing hormone is the physiological signal for ovulation. Its concentration in blood rises sharply and reaches a peak about 16 hours before ovulation (see below). Blood levels of FSH also increase at this time, and although large amounts of FSH can also cause ovulation, the required concentrations are not achieved during the normal reproductive cycle. The events that lead to follicular rupture and expulsion of the ovum are not fully understood, but the process is known to be initiated by increased production of cyclic AMP in theca and granulosa cells in response to LH and the consequent release of paracrine factors and enzymes.

As the follicle approaches ovulation, it accumulates follicular fluid, but despite the preovulatory swelling, intrafollicular pressure does not increase. The follicular wall becomes increasingly distensible due to activity of proteolytic enzymes that digest the collagen framework and other proteins of the intercellular matrix. One of these enzymes, plasmin, accumulates in follicular fluid in the form of its inactive precursor, plasminogen. Granulosa cells secrete plasminogen activator in response to hormonal stimulation. Because of their newly acquired receptors, granulosa cells of the preovulatory follicle respond to LH by secreting progesterone, which is thought to stimulate formation of prostaglandins by granulosa cells, which express progesterone receptors in the late follicular phase. The finding that pharmacological blockade of either prostaglandin or progesterone synthesis prevents ovulation indicates that these agents play essential roles in the ovulatory process. Prostaglandins appear to activate release of lysosomal proteases in a discrete region of the follicle wall called the *stigma*. Breakdown of the extracellular matrix of the theca and the surface epithelium of the ovary facilitates extrusion of the ovum into the abdominal cavity.

Although little or no progesterone is produced throughout most of the follicular phase, granulosa cells of the preovulatory follicle acquire the capacity for progesterone production. Stimulation by LH evokes expression of LDL receptors, P450scc, and doubtless other relevant proteins, enabling them to take up cholesterol and convert it to pregnenolone. Because the capacity to remove the side chain at carbon 17 remains limited, 21-carbon steroids are formed faster than they can be processed to estradiol, and hence are secreted as progesterone. Furthermore, as granulosa cells acquire the ability to respond to LH, they also begin to lose aromatase activity. This is reflected in the abrupt decline in estrogen production that just precedes ovulation.

EFFECTS ON CORPUS LUTEUM FORMATION

Luteinizing hormone was named for its ability to induce formation of the corpus luteum after ovulation. However, as already mentioned, luteinization may actually begin before the follicle ruptures. Granulosa cells removed from mature follicles complete their luteinization in tissue culture without further stimulation by gonadotropin. Nevertheless, luteinization within the ovary depends on LH and is accelerated by the increased concentration of LH that precedes ovulation.

Occasionally luteinization occurs in the absence of ovulation and results in the syndrome of luteinized unruptured follicles, which may be a cause of infertility in some women whose reproductive cycles seem otherwise normal.

Development of a vascular supply is critical for development of the corpus luteum and its function. Although granulosa cells of the preovulatory follicle are avascular, the corpus luteum is highly vascular, and when fully developed, each steroidogenic cell appears to be in contact with at least one capillary. After extrusion of the ovum, infolding of the collapsing follicle causes the highly vascular theca interna to interdigitate with layers of granulosa cells that line the follicular wall. Under the influence of LH, granulosa cells express high levels of VEG-F, which stimulates growth and differentiation of capillary endothelial cells. It has been estimated that vascular endothelial cells comprise fully half of the cells of the mature corpus luteum.

EFFECTS ON OOCYTE MATURATION

Granulosa cells not only provide nutrients to the ovum, but also may prevent it from completing its meiotic division until the time of ovulation. Granulosa cells are thought to secrete a substance called oocyte maturation inhibitor (OMI) into follicular fluid. LH triggers resumption of meiosis at the time of ovulation, perhaps by blocking production of this factor or interfering with its action.

EFFECTS ON CORPUS LUTEAL FUNCTION

Maintenance of steroid production by the corpus luteum depends on continued stimulation with LH. Decreased production of progesterone and premature demise of the corpus luteum are seen in women whose secretion of LH is blocked pharmacologically. In this respect, LH is said to be *luteotropic*. The corpus luteum has a finite life-span, however, and about a week after ovulation becomes progressively less sensitive to LH and finally regresses despite continued stimulation with LH. Estradiol and prostaglandin $F_{2\alpha}$, which are produced by the corpus luteum, can hasten luteolysis and may be responsible for its demise. We do not understand the mechanisms that limit the functional life-span of the human corpus luteum.

EFFECTS ON OVARIAN BLOOD FLOW

Luteinizing hormone also increases blood flow to the ovary and produces ovarian hyperemia. Luteinization is accompanied by production of VEG-F and other angiogenic factors that increase vascularization. Blood flow may be further enhanced by release of histamine or perhaps prostaglandins. Increased ovarian blood flow increases the opportunity for delivery of steroid hormones to the general circulation and for delivery to the ovary of cholesterol-laden LDL, needed to support high rates of steroidogenesis. Increased blood flow to the developing follicle may also be important for preovulatory swelling of the follicle, which depends on increased elaboration of fluid from blood plasma.

PHYSIOLOGICAL ACTIONS OF OVARIAN STEROID HORMONES

As described above, intraovarian actions of estradiol and progesterone are intimately connected to ovulation and formation of the corpus luteum. In general, extraovarian actions of these hormones ensure that the ovum reaches its potential to develop into a new individual. Ovarian steroids act on the reproductive tract to prepare it for fulfilling its role in fertilization, implantation, and development of the embryo, and they induce changes elsewhere that equip the female physically and behaviorally for conceiving, giving birth, and rearing the child. Although estrogens, perhaps in concert with progesterone, drive females of subprimate species to mate, androgens, rather than estrogens, are responsible for libido in humans of either sex. Estrogens and progesterone tend to act in concert and sometimes enhance or antagonize each other's actions. Estrogen secretion usually precedes progesterone secretion and primes the target tissues to respond to progesterone. Estrogens induce the synthesis of progesterone receptors, and without estrogen priming, progesterone has little biological effect. Conversely, progesterone down-regulates its own receptors and estrogen receptors in some tissues and thereby decreases responses to estrogens.

EFFECTS ON THE REPRODUCTIVE TRACT

At puberty estrogens promote growth and development of the oviducts, uterus, vagina, and external genitalia. Estrogens stimulate cellular proliferation in the mucosal linings as well as in the muscular coats of these structures. Even after they have matured, maintenance of size and function of internal reproductive organs requires continued stimulation by estrogen and progesterone. Prolonged deprivation after ovariectomy results in severe involution of both muscular and mucosal portions. Dramatic changes are also evident, especially in the mucosal linings of these structures, as steroid hormones wax and wane during the reproductive cycle. These effects of estrogen and progesterone are summarized in Table 1.

TABLE 1

Effects of Estrogen and Progesterone on the Reproductive Tract

Organ	Estrogen	Progesterone
Oviducts		
Lining	↑ Cilia formation and activity	↑ Secretion
Muscular wall	↑ Contractility	↓ Contractility
Uterus		
Endometrium	↑ Proliferation	↑ Differentiation and secretion
Myometrium	↑ Growth and contractility	↓ Contractility
Cervical glands	Watery secretion	Dense, viscous secretion
Vagina	↑ Epithelial proliferation	↓ Epithelial proliferation
	↑ Glycogen deposition	↑ Differentiation

MENSTRUATION

Nowhere are the effects of estrogen and progesterone more obvious than in the endometrium. Estrogens secreted by developing follicles increase the thickness of the endometrium by stimulating growth of epithelial cells, in terms of both number and height. Endometrial glands form and elongate. Endometrial growth is accompanied by increased blood flow, especially through the spiral arteries, which grow rapidly under the influence of estrogens. This stage of the uterine cycle is known as the *proliferative phase* and coincides with the follicular phase of the ovarian cycle. Progesterone secreted by the corpus luteum causes the newly proliferated endometrial lining to differentiate and become secretory. This action is consistent with its role of preparing the uterus for nurture and implantation of the newly fertilized ovum if successful mating has occurred. The so-called uterine milk secreted by the endometrium is thought to nourish the blastocyst until it can implant. This portion of the uterine cycle is called the *secretory phase* and coincides with the luteal phase of the ovarian cycle (Figure 9).

Maintaining the thickened endometrium depends on the continued presence of the ovarian steroid hormones. After the regressing corpus luteum loses its ability to produce adequate amounts of estradiol and progesterone, the outer portion of the endometrium degenerates and is sloughed into the uterine cavity. The mechanism for shedding the uterine lining is incompletely understood, although prostaglandin $F_{2\alpha}$ appears to play an important role, perhaps in producing vascular spasm and ischemia, and in stimulating release of lysosomal proteases. Loss of the proliferated endometrium is accompanied by bleeding. This monthly vaginal discharge of blood is known as *menstruation*. The typical menstrual period lasts 3 to 5 days and the total

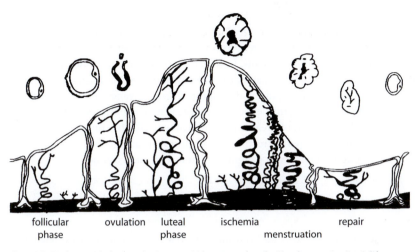

follicular   ovulation   luteal   ischemia   repair
phase   phase   menstruation

Figure 9 Endometrial changes during a typical menstrual cycle. Simultaneous events in the ovary are also indicated. The endometrium thickens during the follicular phase; uterine glands elongate, and spiral arteries grow to supply the thickened endometrium. During the early luteal phase there is further thickening of the endometrium, marked growth of the coiled arteries, and increased complexity of the uterine glands. As the corpus luteum wanes, the endometrial thickness is reduced by loss of ground substance. Increased coiling of spiral arteries causes ischemia and finally sloughing of endometrium. (From Bartelmez, G. W., *Am. J. Obstet. Gynecol.* **74**, 931, 1957, with permission.)

flow of blood seldom exceeds 50 ml. The first menstrual bleeding, called *menarche*, usually occurs at about age 13. Menstruation continues at monthly intervals until menopause, normally interrupted only by periods of pregnancy.

In the myometrium, estrogen increases expression of contractile proteins, gap junction formation, and spontaneous contractile activity. In its absence, uterine muscle is insensitive to stretch or other stimuli for contraction. Further estrogen increases the irritability of uterine smooth muscle and, in particular, increases its sensitivity to oxytocin, in part as a consequence of inducing uterine receptors for oxytocin (see Chapter 13). The latter phenomenon may be of significance during parturition. Progesterone counteracts these effects and decreases both the amplitude and frequency of spontaneous contractions. Withdrawal of progesterone prior to menstruation is accompanied by increased myometrial prostaglandin formation. Myometrial contractions in response to prostaglandins are thought to account for the discomfort that precedes menstruation.

EFFECTS ON THE MAMMARY GLANDS

Development of the breasts begins early in puberty and is due primarily to estrogen, which promotes development of the duct system and growth and pigmentation of the nipples and areolar portions of the breast. In cooperation with progesterone, estrogen may also increase the lobuloalveolar portions of the glands, but alveolar development also requires the pituitary hormone prolactin (see Chapter 13). Secretory components, however, account for only about 20% of the mass of the adult breast. The remainder is stromal tissue and fat. Estrogen also stimulates stromal proliferation and fat deposition. Responsiveness of all these tissue elements to growth-promoting effects of estrogen is of significance in neoplastic breast disease. Some forms of breast cancer remain partially or completely dependent on estrogens for growth. Treatment with estrogen antagonists may therefore have life-prolonging benefits in patients afflicted with such tumors.

OTHER EFFECTS OF OVARIAN HORMONES

Estrogen also acts on the body in ways that are not necessarily related to reproduction. As already indicated (see Chapter 10), it contributes to the pubertal growth spurt by directly stimulating growth of cartilage progenitors, closure of the epiphyseal plates, and increased growth hormone secretion. In adolescent females and males estrogens increase bone density, and in the adult they contribute to the maintenance of bone density by stimulating osteoblastic activity and inhibiting bone resorption. Estrogen can also cause selective changes in bone structure, especially widening of the pelvis, which facilitates passage of the infant through the birth canal. It promotes deposition of subcutaneous fat, particularly in the thighs and buttocks, and increases hepatic synthesis of steroid- and thyroid hormone-binding proteins. Based on epidemiological evidence, estrogens are considered to have cardiovascular protective effects. Indeed, estrogen receptors are expressed in vascular smooth muscle and cardiac myocytes, and estrogens have been found to alter plasma lipid profiles. Estradiol also acts on various cells in the central nervous system and is responsible for some behavioral patterns, especially in lower animals. More recently evidence has been brought forth that estrogens protect cognitive capacity.

Progesterone also acts on the central nervous system and, in addition to its effects on regulation of gonadotropin secretion (see below), it may produce changes in behavior or mood. Progesterone has a mild thermogenic effect and may increase basal body temperature by as much as 1°F. Because the appearance of progesterone indicates the presence of a corpus luteum, a woman can readily determine when ovulation occurred, and hence the time of maximum fertility, by monitoring her temperature daily. This simple, noninvasive procedure has been helpful for couples seeking to conceive a child or who are practicing the "rhythm method" of contraception. It is curious that more dramatic effects may result from withdrawal of progesterone than from administering it. Thus withdrawal of progesterone may trigger menstruation, lactation, parturition, and the postpartum psychic depression experienced by many women.

MECHANISM OF ACTION

Estrogens and progesterone, like other steroid hormones, readily penetrate cell membranes and bind to receptors that are members of the nuclear receptor superfamily of transcription factors (see Chapter 1). In addition, various "nongenomic" effects of both hormones have been reported, but the molecular processes involved in these actions are incompletely understood. Two separate estrogen receptors, designated ERα and ERβ, are products of different genes and are expressed in many different cells in both reproductive and nonreproductive tissues of both men and women. Their DNA-binding domains are almost identical. Although some differences are also present in their ligand-binding domains, ERα and ERβ differ mainly in the amino acid sequences of their activation function (AF) domains. Therefore they interact with different transcriptional coactivators and other nuclear regulatory proteins to control expression of different sets of genes. On binding ligand, ERα and ERβ may form homo- or heterodimers before binding to DNA in estrogen-sensitive genes. The resulting synthesis of new messenger RNA is followed, in turn, by the formation of a variety of proteins that modify cellular activity. Both receptors can also be activated by phosphorylation in the absence of ligand, and both receptors can modulate the transcription-activating properties of other nuclear regulatory proteins without directly binding to DNA. Finally, estrogen receptors have the interesting property of assuming different conformational changes, depending on the particular ligand that is bound. Because the conformation of the liganded receptor profoundly affects its ability to interact with other transcription regulators whose expression is cell type specific, some compounds produced in plants (phytoestrogens), or pharmaceutically manufactured "antiestrogens," may block the effects of estrogen in some tissues while acting as estrogen agonists in other tissues. Additionally some compounds may bind and activate ERα, but antagonize actions of ERβ. These properties have given rise to a very important category of drugs called selective estrogen receptor modulators (SERMs) which, for example, may block the undesirable proliferative or neoplastic actions of estrogens on the breast and uterus while mimicking desirable effects on maintenance of bone density in postmenopausal women. One of these compounds, tamoxifen, is routinely used in treatment of breast cancer.

Two isoforms of the progesterone receptor, PRA and PRB, are also expressed by progesterone-responsive cells. They are products of the same gene, the mRNA transcript of which contains two alternate translation start sites. PRA and PRB differ by the presence of an additional sequence of amino acids at the PRB amino terminus, which provides an additional region for interacting with nuclear regulatory proteins. Liganded PRs bind to the DNA of target genes as homodimers or heterodimers, and in different combinations may activate different subsets of genes. When expressed together in some cells, liganded PRA can repress the activity of PRB. Both PRA and PRB expression are induced by prior exposure of cells to estrogens, and can repress the activity of or expression of ERα. Studies under way in mice in which expression of specific progesterone or estrogen receptors is disrupted by genetic manipulation are determining which particular responses to ovarian steroid hormones are mediated by each receptor.

REGULATION OF THE REPRODUCTIVE CYCLE

The central event of each ovarian cycle is ovulation, which is triggered by a massive increase in blood LH concentration. This surge of LH secretion must be timed to occur when the ovum and its follicle are ready. The corpus luteum must secrete its hormones to optimize the opportunity for fertilization and establishment of pregnancy. The period after ovulation during which the ovum can be fertilized is brief and lasts less than 24 hours. If fertilization does not occur, a new follicle must be prepared. Coordination of these events requires two-way communication between the pituitary and the ovaries, and between the ovaries and the reproductive tract. Examination of the changing pattern of hormones in blood throughout the ovarian cycle provides some insight into these communications.

PATTERN OF HORMONES IN BLOOD DURING THE OVARIAN CYCLE

Figure 10 illustrates daily changes in the concentrations of major hormones in a typical cycle extending from one menstrual period to the next. The only remarkable feature of the profile of gonadotropin concentrations is the dramatic peak in LH and FSH that precedes ovulation. Except for the 2 to 3 days of the midcycle peak, LH concentrations remain at nearly constant low levels throughout the follicular and luteal phases. The concentration of FSH is also low during both phases of the cycle, but a secondary peak is evident early in the follicular phase and diminishes as ovulation approaches. Blood levels of FSH remain low throughout most of the luteal phase but begin to rise 1 or 2 days before the onset of menstruation.

The ovarian hormones follow a different pattern. Early in the follicular phase the concentration of estradiol is low. It then gradually increases at an increasing rate until it reaches its zenith about 12 hours before the peak in LH. Thereafter estradiol levels fall abruptly and reach a nadir just after the LH peak. During the luteal phase there is a secondary rise in estradiol concentration, which then falls to the early follicular level a few days before the onset of menstruation. Progesterone is barely or not at all detectable throughout most of the follicular phase and then begins to rise along with LH at the onset of the ovulatory peak. Progesterone continues to rise and reaches its maximum concentration several days after the LH peak has ended. Progesterone levels remain high for about 7 days and then gradually fall and reach almost undetectably low levels a day or two before the onset of menstruation. Inhibin B concentrations are low early in the follicular phase and then rise and fall in parallel with FSH. The apparent peak that coincides with ovulation is thought to result from the absorption into the blood of inhibin B that is already present in high concentration in the expelled follicular fluid, rather than concurrent secretion by granulosa cells. Concentrations of inhibin A reach their highest levels in the luteal phase before declining in parallel with progesterone.

REGULATION OF FSH AND LH SECRETION

At first glance, this pattern of hormone concentrations is unlike anything seen for other anterior pituitary hormones and the secretions of their target glands. Indeed, there are unique aspects, but during most of the cycle gonadotropin secretion is under negative feedback control similar to that seen for TSH (Chapter 3), ACTH (Chapter 4), and the gonadotropins in men (Chapter 11). The ovulatory burst of FSH and LH secretion is brought about by a positive feedback mechanism unlike any we have considered. Secretion of FSH and LH is also controlled by GnRH which is released in synchronized pulsatile bursts from neuronal cell bodies residing in the arcuate nuclei and the medial preoptic area of the hypothalamus (see Chapter 11).

Negative Feedback Aspects

As we have seen, FSH and LH stimulate production of ovarian hormones. In the absence of ovarian hormones, following ovariectomy or menopause, concentrations of FSH and LH in blood may increase as much as 5- to 10-fold. Treatment with low doses of estrogen lowers circulating concentrations of gonadotropins to levels seen during the follicular phase. When low doses of estradiol are given to subjects whose ovaries are intact, inhibition of gonadotropin secretion results in failure of follicular development. Progesterone alone, unlike estrogen, is ineffective in lowering high levels of FSH and LH in the blood of postmenopausal women, but it can synergize with estrogen to suppress gonadotropin secretion. These findings exemplify classical negative feedback. Inhibin may also provide some feedback inhibition of FSH secretion during the follicular phase, and may contribute the low level of FSH during the luteal phase, but its effects are probably small. In ovariectomized rhesus monkeys the normal pattern of gonadotropin concentrations can be reproduced by treating with only estradiol and progesterone. The rise in FSH concentration follows the decline in estrogen, progesterone, and inhibin secretion at the end of the luteal phase and stimulates growth of the cohort of follicles that will produce the ovum in the next cycle. Throughout the follicular and

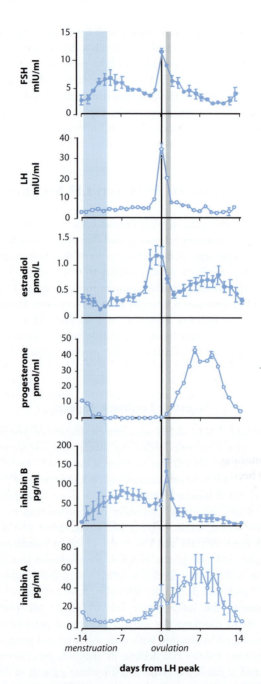

Figure 10 Mean values of LH, FSH, progesterone, estradiol, and inhibin in daily serum samples of women during ovulatory menstrual cycles. Data from various cycles are combined, using the midcycle

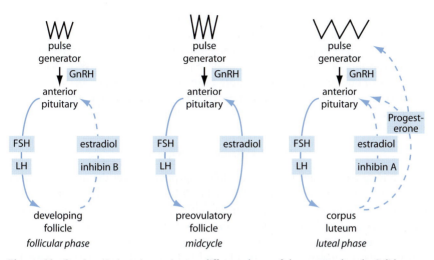

Figure 11 Ovarian–pituitary interactions at different phases of the menstrual cycle. Solid arrows indicate stimulation, dashed arrows indicate inhibition. Note that the frequency of the GnRH pulse generater slows in the luteal phase, and that the amplitude may increase at midcycle.

luteal phases of the cycle, steroid concentrations appear to be sufficient to suppress LH secretion (Figure 11).

Although the ovarian steroids and inhibin suppress FSH and LH secretion, estrogen and progesterone concentrations change during the cycle in ways that seem independent of gonadotropin concentrations. For example, estrogen rises dramatically as the follicular phase progresses, even though LH remains constant and FSH is diminishing. The mechanism for this increase in estradiol is implicit in what has already been presented and is consistent with negative feedback. Estrogen production by the maturing follicle increases without a preceding increase in gonadotropin concentration because the mass of responsive theca and granulosa cells increases, as does their sensitivity to gonadotropins. In fact, the decrease in FSH during the transition from early to late follicular phase probably results from feedback inhibition by the increasing concentration of estrogen, perhaps in conjunction with inhibin. Although luteinizing cells do not divide, progesterone and estrogen concentrations continue to increase during the early luteal phase, well after FSH and LH have returned to basal levels. Increasing steroid hormone

peak of LH as the reference point (day 0). Vertical bars indicate standard error of mean. (Redrawn from Groome, N. J., Illingworth, P. J., O'Brien, M., Pal, R., Faye, E. R., Mather, J. P., and McNeilly, A. S., *J. Clin. Endocrinol. Metab.* **81**, 1401–1405, 1996, by permission of The Endocrine Society.)

production at this time reflects completion of the luteinization process. Conversely, the gradual loss of sensitivity of luteal cells to LH accounts for the decrease in progesterone and estrogen secretion during the latter part of the luteal phase. Thus one of the unique features of the female reproductive cycle is that changes in steroid hormone production result more from changes in the number or sensitivity of competent target cells than from changes in gonadotropin concentrations (see Chapter 6).

Positive Feedback Aspects

Rising estrogen levels in the late follicular phase trigger the massive burst of LH secretion that just precedes ovulation. This LH surge can be duplicated experimentally in monkeys and women given sufficient estrogen to raise their blood levels above a critical threshold level for 2 to 3 days. This compelling evidence implicates increased estrogen secretion by the ripening follicle as the causal event that triggers the massive release of LH and FSH from the pituitary (Figure 11, midcycle). It can be considered positive feedback because LH stimulates estrogen secretion, which in turn stimulates more LH secretion in a self-generating explosive pattern.

Progesterone concentrations begin to rise about 6 hours before ovulation. This change is probably a response to the increase in LH, rather than its cause. It is significant that large doses of progesterone given experimentally block the estrogen-induced surge of LH in women, which may account for the absence of repeated LH surges during the luteal phase, when the concentrations of estrogen might be high enough to trigger the positive feedback effect (Figure 11, luteal phase). This action of progesterone, which may contribute to the decline in the LH surge, also contributes to its effectiveness as an oral contraceptive agent. In this regard, progesterone also inhibits follicular growth.

Neural Control of Gonadotropin Secretion

It is clear that secretion of gonadotropins is influenced to a large measure by ovarian steroid hormones. It is equally clear that secretion of these pituitary hormones is controlled by the central nervous system. Gonadotropin secretion ceases after the vascular connection between the anterior pituitary gland and the hypothalamus is interrupted or after the arcuate nuclei of the medial basal hypothalamus are destroyed. Less drastic environmental inputs, including rapid travel across time zones, stress, anxiety, and other emotional changes, can also affect reproductive function in women, presumably through neural input to the medial basal hypothalamus. As discussed in Chapter 11, secretion of gonadotropins requires the operation of a hypothalamic pulse generator that produces intermittent stimulation of the pituitary gland by GnRH.

Sites of Feedback Control

The ovarian steroids might produce their positive or negative feedback effects by acting at the level of the hypothalamus or the anterior pituitary gland, or both. The GnRH pulse generator drives gonadotropin secretion regardless of whether negative or positive feedback prevails. Gonadotropin secretion falls to zero after bilateral destruction of the arcuate nuclei in rhesus monkeys and cannot be increased by either ovariectomy or treatment with the same amount of estradiol that evokes a surge of FSH and LH in normal animals. When such animals are fitted with a pump that delivers a constant amount of GnRH in brief pulses every hour, the normal cyclic pattern of gonadotropin is restored and the animals ovulate each month. Identical results have been obtained in women suffering from Kallman's syndrome, in which there is a developmental deficiency in GnRH production by the hypothalamus (Figure 12). In both cases administration of GnRH in pulses of constant amplitude and frequency was sufficient to produce normal ovulatory cycles. Because both positive and negative feedback aspects of gonadotropin secretion can be produced even when hypothalamic input is "clamped" at constant frequency and amplitude, these effects of estradiol must be exerted at the level of the pituitary.

Although changes in amplitude and frequency of GnRH pulses are not necessary for the normal pattern of gonadotropin secretion during an experimental or therapeutic regimen, variations in frequency and amplitude nevertheless may occur physiologically in a way that complements and reinforces the intrinsic pattern already described. During the normal reproductive cycle, GnRH pulses are considerably less frequent in the luteal phase than in the follicular phase. Complementing its negative feedback effects, estradiol may decrease the amplitude of GnRH pulses, and progesterone slows their frequency, perhaps by stimulating hypothalamic production of endogenous opioids. It appears that an increase in amplitude of GnRH pulses precedes the LH surge, and there is good evidence that progesterone acts at the level of the hypothalamus to block the estradiol-induced LH surge. Thus feedback effects of estradiol appear to be exerted primarily, but not exclusively, on the pituitary, and those of progesterone primarily, but probably not exclusively, on the hypothalamus.

We do not yet understand the intrapituitary mechanisms responsible for the negative and positive feedback effects of estradiol. As seen with the ovary, changes in hormone secretion may be brought about by changes in the sensitivity of target cells, as well as by changes in concentration of a stimulatory hormone. Women and experimental animals are more responsive to a test dose of GnRH at midcycle than at any other time. An increase in the number of receptors for GnRH at this time has been reported, and there is evidence that the high concentrations of estradiol that precede the LH surge up-regulate GnRH receptors and stimulate LH synthesis and storage.

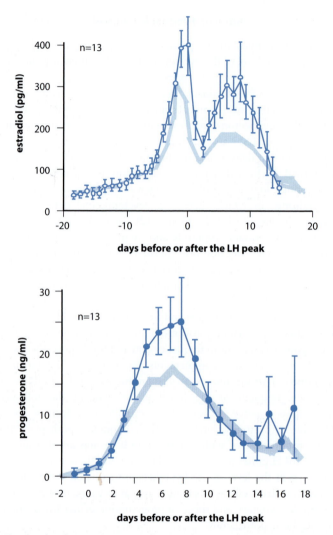

Figure 12 Results of ovulation induction employing a physiological frequency of GnRH administration to hypogonadotropic, hypogonadal women on ovarian steroid secretion. Normal values are represented by the shaded areas. (From Crowley, W. F., Jr., Filicori, M., Spratt, D. J., *et al.*, *Recent Prog. Horm. Res.* **41**, 473, 1985, with permission.)

TIMING OF REPRODUCTIVE CYCLES

Although the pacemaker for rhythmic release of GnRH resides in the hypothalamus, the timekeeper for the slower monthly rhythm of the ovarian cycle resides in the ovary. As already indicated, the corpus luteum has a built-in life-span of about 12 days and involutes despite continued stimulation with LH. A new cohort of follicles cannot arise so long as the corpus luteum remains functional. Its demise appears to relieve inhibition of follicular growth and FSH secretion, which increases sufficiently in blood to stimulate growth of the next cohort of follicles. Thus the interval between the LH surge and the emergence of the new cohort of follicles is determined by the ovary. The principal event around which the menstrual cycle revolves is ovulation, which depends on an ovulatory surge of LH. The length of the follicular phase may be somewhat variable and may be influenced by extraovarian events, but the timing of the LH surge resides in the ovary. It is only when the developing follicle signals its readiness to ovulate with increasing blood levels of estradiol that the pituitary secretes the ovulatory spike of gonadotropin. Hence throughout the cycle it is the ovary that notifies the pituitary and hypothalamus of its readiness to proceed to the next stage.

The beginning and end of cyclic ovarian activity, called menarche and menopause, occur on a longer time scale. The events associated with the onset of puberty were considered in Chapter 11. Although we still do not know what biological phenomena signal readiness for reproductive development and the end of the juvenile period, it appears that the time-keeper for this process resides in the central nervous system, which initiates sexual development and function by activating the GnRH pulse generator. Termination of cyclic ovarian activity coincides with the disappearance or exhaustion of primordial follicles, but during the final decade of a woman's reproductive life there is a paradoxical doubling of the rate of loss of follicles by atresia. Aging of the GnRH pacemaker may be a factor in this acceleration of follicular loss, based on studies in normally cycling women indicating that both the amplitude of LH pulses and the interval intervening between pulses increase with increased age.

SUGGESTED READING

Ackland, J. F., Schwartz, N. B., Mayo, K. E., and Dodson, R. E. (1992). Nonsteroidal signals originating in the gonads. *Physiol. Revs.* **72**, 731–788.

Dorrington, J. H, and Armstrong, D. T. (1979). Effects of FSH on gonadal function. *Recent Prog. Horm. Res.* **35**, 301–332.

Giudice, L. C. (1992). Insulin-like growth factors and ovarian follicular development. *Endocr. Rev.* **13**, 641–669.

Gougeon, A. (1996). Regulation of ovarian follicular development in primates: Facts and hypotheses. *Endocr. Rev.*, **17**, 121–155.

Graham, D., and Clarke, C. L. (1997). Physiological action of progesterone in target tissues. *Endocr. Rev.* **18**, 502–519.

Hayes, F. J., and Crowley, W. F., Jr. (1998). Gonadotropin pulsations across development. *Horm. Res.* **49**, 163–168.

Knobil, E., and Hotchkiss, J. (1994). The menstrual cycle and its neuroendocrine control. *In* "The Physiology of Reproduction," 2nd Ed. (E. Knobil and J. D. Neill, eds.), Vol. 2, pp. 711–749. Raven Press, Ltd., New York.

Marshall, J. C., Dalkin, A. C., Haisleder, D. J., Paul, S. J., Ortolano, G. A., and Kelch, R. P. (1991). Gonadotropin-releasing hormone pulses: Regulators of gonadotropin synthesis and ovarian cycles. *Recent Prog. Horm. Res.* **47**, 155–187.

Matzuk, M. M. (2000). Revelations of ovarian follicle biology from gene knockout mice. *Mol. Cell. Endocrinol.* **163**, 61–66.

McGee, E. A., and Hsueh, A. J. W. (2000). Initial and cyclic recruitment of ovarian follicles. *Endocr. Rev.* **21**, 200–214.

Nilsson, S., Makela, S., Treuter, E., Tujague, M., Thomsen, J., Andersson, G., Enmark. E., Pettersson, K., Warner, M., and Gustafsson, J. A. (2001). Mechanisms of estrogen action. *Physiol. Rev.* **81**, 1535–1565.

Richards, J. S., Jahnsen, T., Hedin, L., Lifka, J., Ratoosh, S., Durica, J. M., and Goldring, N. B. (1987). Ovarian follicular development: From physiology to molecular biology. *Recent Prog. Horm. Res.* **43**, 231–270.

Wise, P. M., Smith, M. J., Dubal, D. B., Wilson, M. E., Krajnak, K. M., and Rosewell, K. L. (1999). Neuroendocrine influences and repercussions of the menopause. *Endocr. Rev.* **20**, 243–248.

CHAPTER 13

Hormonal Control of Reproduction in the Female: Pregnancy and Lactation

OVERVIEW

Successful reproduction depends not only on the union of eggs and sperm but also on survival of adequate numbers of the new generation to reach reproductive age and begin the cycle again. In some species parental involvement in the reproductive process ends with fertilization of the ova; thousands or even millions of embryos may result from a single mating, with just a few surviving long enough to procreate. Higher mammals, particularly humans, have adopted the alternative strategy of producing only a few or a single fertilized ovum at a time. Prolonged parental care during the embryonic and neonatal periods substitutes for huge numbers of unattended offspring as the means for increasing the likelihood of survival. Estrogen and progesterone prepare the maternal body for successful internal fertilization and hospitable acceptance of the embryo. The conceptus then takes charge. After lodging firmly within the uterine lining and gaining access to the maternal circulation, it secretes protein and steroid hormones that ensure continued maternal acceptance, and it directs maternal functions to provide for its development. Simultaneously, the conceptus withdraws whatever nutrients it needs from the maternal circulation. At the appropriate time, the fetus signals its readiness to depart the uterus and initiates the birth process. While *in utero*, placental hormones prepare the mammary glands to produce the milk needed for nurture after birth. Finally, suckling stimulates continued milk production.

FERTILIZATION AND IMPLANTATION

GAMETE TRANSPORT

Fertilization takes place in a distal portion of the oviduct called the *ampulla*, far from the site of sperm deposition in the vagina. To reach the ovum, sperm must swim through the cervical canal, cross the entire length of the uterine cavity, and then travel up through the muscular isthmus of the oviduct. Even with the aid of contractions of the female reproductive tract, the journey is formidable. Only about one of every million sperm deposited in the vagina will reach the ampulla; once there, if they arrive first, they await the arrival of the ovum. Sperm usually remain fertile within the female reproductive tract for 1 to 2 days, but as long as 4 days is possible. Access to the upper reaches of the reproductive tract is heavily influenced by ovarian steroid hormones.

Estrogen is secreted in abundance late in the follicular phase of the ovarian cycle and prepares the reproductive tract for efficient sperm transport (Figure 1). Glycogen deposited in the vaginal mucosa under its influence provides substrate for the production of lactate, which lowers the pH of vaginal fluid. An acidic

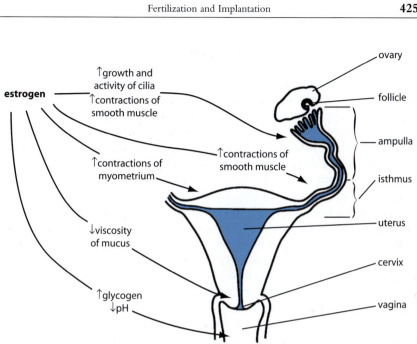

Figure 1 Actions of estrogen to promote sperm transport.

environment increases motility of sperm, which is essential for their passage through the cervical canal. In addition, the copious watery secretion produced by cells lining the cervical canal under the influence of estrogen increases access to the uterine cavity. When estrogen is absent or when its effects are opposed by progesterone, the cervical canal is filled with a viscous fluid that resists sperm penetration. Vigorous contractions of the uterus propel sperm toward the oviducts, where they may appear anywhere from 5 to 60 minutes after ejaculation. Prostaglandins present in seminal plasma and oxytocin released from the pituitary in response to intercourse may stimulate contraction of the highly responsive estrogen-dominated myometrium.

ROLE OF THE OVIDUCTS

The oviducts are uniquely adapted for facilitating transport of sperm toward the ovary and transporting the ovum in the opposite direction toward its rendezvous with sperm. It is also within the oviducts that sperm undergo a process called *capacitation*, which prepares them for a successful encounter with the ovum and its adherent mass of granulosa cells, the *cumulus oophorus*. Capacitation is an activation process that involves both the enhancement of flagellar activity and the biochemical and structural changes in the plasma membrane of the sperm head that prepare sperm to undergo the *acrosomal reaction*. The acrosome is a membranous vesicle that is positioned at the tip of the sperm head. It is filled with several hydrolytic enzymes. Contact of the sperm head with the *zona pellucida* that surrounds the ovum initiates fusion of the acrosomal membrane with the plasma membrane and the exocytotic release of enzymes that digest a path through the zona pellucida and allow a sperm to reach the ovum. The acrosome reaction and the events that produce it are highly reminiscent of the sequelae of hormone receptor interactions, and involve activation of membrane calcium channels, tyrosine phosphorylation, phospholipase C, and other intracellular signaling mechanisms. Initiation of the acrosomal reaction is facilitated by progesterone secreted by the mass of cumulus cells that surround ovum. The sperm plasma membrane appears to contain progesterone receptors that differ from the classical nuclear receptors and that trigger influx of calcium within seconds.

In response to estrogens or perhaps other local signals associated with impending ovulation, muscular activity in the distal portion of the oviduct brings the infundibulum into close contact with the surface of the ovary. At ovulation, the ovum, together with its surrounding granulosa cells, is released into the peritoneal cavity and is swept into the ostium of the oviduct by the vigorous, synchronous beating of cilia on the infundibular surface. Development of cilia in the epithelial lining and their synchronized rhythmic activity are consequences of earlier exposure to estrogens. Movement of the egg mass through the ampulla toward the site of fertilization near the ampullar–isthmic junction depends principally on currents set up in tubal fluid by the beating of cilia and to a lesser extent by contractile activity of the ampullar wall to produce a churning motion.

Propulsion of sperm through the isthmus toward the ampulla is accomplished largely by muscular contractions of the tubal wall. Circular smooth muscle of the isthmus is innervated with sympathetic fibers and has both α-adrenergic receptors, which mediate contraction, and β-adrenergic receptors, which mediate relaxation. Under the influence of estrogen, the α-adrenergic receptors predominate. Subsequently, as estrogenic effects are opposed by progesterone, the β-adrenergic receptors prevail, and isthmic smooth muscles relax. This reversal in the response to adrenergic stimulation may account for the ability of the oviduct to facilitate sperm transport through the isthmus toward the ovary and, subsequently, to promote passage of the embryo in the opposite direction toward the uterus.

After fertilization, the oviduct retains the embryo for about 3 days and nourishes it with secreted nutrients before facilitating its entry into the uterine cavity. These complex events, orchestrated by the interplay of estrogen, progesterone, and autonomic innervation, require participation of the smooth muscle of the oviduct walls as well as secretory and ciliary activity of the epithelial lining. As crucial as these mechanical actions may be, however, the oviduct does not contribute in an indispensable way to fertility of the ovum or sperm, or to their union, as evidenced by modern techniques of *in vitro* fertilization that bypass it with no ill effects.

The period of fertility is short; from the time the ovum is shed until it can no longer be fertilized is only about 6–24 hours. As soon as a sperm penetrates the ovum, the second polar body is extruded and the fertilized ovum begins to divide. By the time the fertilized egg enters the uterine cavity, it has reached the blasto-cyst stage and consists of about 100 cells. Timing of the arrival of the blastocyst in the uterine cavity is determined by the balance between antagonistic effects of estrogen and progesterone on the contractility of the oviductal wall. Under the influence of estrogen, circularly oriented smooth muscle of the isthmus is con-tracted and bars passage of the embryo to the uterus. As the corpus luteum organizes and increases its capacity to secrete progesterone, β-adrenergic receptors gain ascendancy, muscles of the isthmus relax, and the embryonic mass is allowed to pass into the uterine cavity. Ovarian steroids can thus "lock" the ovum or embryo in the oviduct or cause its delivery prematurely into the uterine cavity.

IMPLANTATION AND THE FORMATION OF THE PLACENTA

The blastocyst floats freely in the uterine cavity for about a day before it implants, normally on about the fifth day after ovulation. Experience with *in vitro* fertilization indicates that there is about a 3-day period of uterine receptivity in which implantation leads to full-term pregnancy. It should be recalled that this period of endometrial sensitivity coincides with the period of maximal proges-terone output by the corpus luteum (Figure 2). In the late luteal phase of the men-strual cycle, the outer layer of the endometrium differentiates to form the *decidua*. Decidualized stromal cells enlarge and transform from an elongated spindle shape to a rounded morphology, and accumulate glycogen. Decidualization requires high concentrations of progesterone, and may be enhanced by activity of cytokines and relaxin. Decidual cells express several proteins that may facilitate implantation, but the precise roles of these proteins either in implantation or pregnancy have not been determined definitively. One such protein is the hormone prolactin, which contin-ues to be secreted throughout pregnancy. Another is IGF-I binding protein-1.

At the time of implantation, the blastocyst consists of an inner mass of cells destined to become the fetus and an outer rim of cells called the *trophoblast*. It is the trophoblast that forms the attachment to maternal decidual tissue and gives rise to the fetal membranes (Figure 3). Cells of the trophoblast proliferate and form the multinucleated syncytial trophoblast, which has specialized functions that enable it to destroy adjacent decidual cells and allow the blastocyst to penetrate deep into the uterine endometrium. Killed decidual cells are phagocytosed by the trophoblast as the embryo penetrates the subepithelial connective tissue and even-tually becomes completely enclosed within the endometrium. Products released

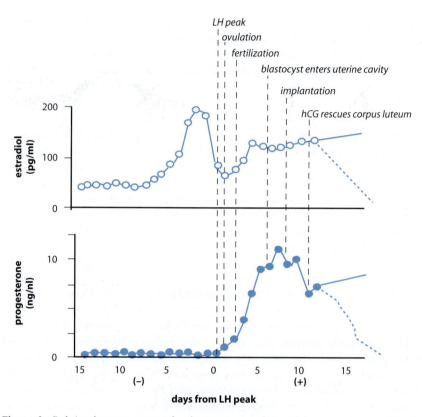

Figure 2 Relation between events of early pregnancy and steroid hormone concentrations in maternal blood. By the 10th day after the LH peak, there is sufficient hCG to maintain and increase estrogen and progesterone production, which would otherwise decrease (dotted lines) at this time. (Estradiol and progesterone concentrations are redrawn from data given in Figure 10 of Chapter 12.)

from degenerating decidual cells produce hyperemia and increased capillary permeability. Local extravasation of blood from damaged capillaries forms small pools of blood that are in direct contact with the trophoblast and provide nourishment to the embryo until the definitive placenta forms. From the time the ovum is shed until the blastocyst implants, metabolic needs are met by secretions of the oviduct and the endometrium.

The syncytial trophoblast and an inner *cytotrophoblast* layer of cells soon completely surround the inner cell mass and send out solid columns of cells that further erode the endometrium and anchor the embryo. These columns of cells differentiate into the *placental villi*. As they digest the endometrium, pools of extravasated maternal blood become more extensive and fuse into a complex

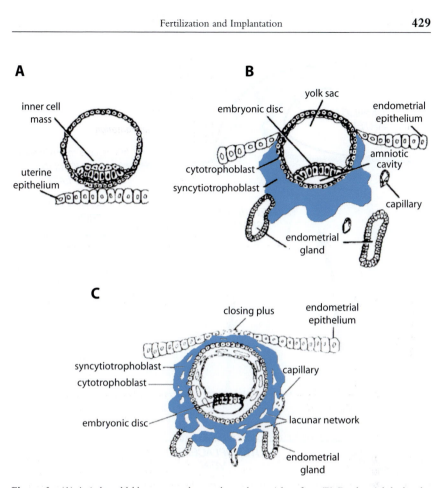

Figure 3 (A) A 6-day-old blastocyst settles on the endometrial surface. (B) By the eighth day the blastocyst has begun to penetrate the endometrium. The expanding syncytiotrophoblast (blue) invades and destroys decidualized endometrial cells. (C) By 12 days the blastocyst has completely embedded itself in the decidualized endometrium, and a clot or plug has formed to cover the site of entry. The trophoblast has continued to invade the endometrium and has eroded uterine capillaries and glands. A network of pooled extravasated blood (lacunar network) has begun to form. (Adapted from Khong, T. Y., and Pearce, J. M., "The Human Placenta: Clinical Perspectives," p. 26, Aspen Publishers, Rockville, MD, 1987.)

labyrinth that drains into venous sinuses in the endometrium. These pools expand and eventually receive an abundant supply of arterial blood. By the third week the villi are invaded by fetal blood vessels as the primitive circulatory system begins to function. As the placenta matures, trophoblastic tissue thins, reducing the barrier to diffusion between maternal and fetal blood. The syncytial trophoblast takes on specialized functions of hormone production and active bidirectional transport of

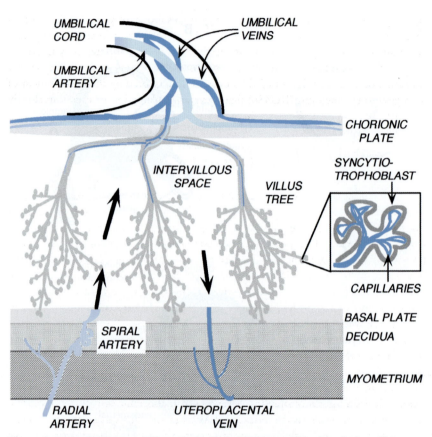

Figure 4 Placental villi are tree-like structures bathed by maternal blood in the intravillous space which is formed between the basal and chorionic plates formed from the trophoblast. Insert shows twig-like terminal villi consisting of fetal capillaries encased in a sheath of syncytiotrophoblast. Heavy black arrows indicate direction of maternal blood flow.

nutrients and metabolites (Figure 4). The overall surface area available for exchange in the mature placenta is about 10 m².

Although much uncertainty remains regarding details of implantation in humans, it is perfectly clear that progesterone secreted by the ovary at the height of luteal function is indispensable for all of these events to occur. Removal of the corpus luteum at this time or blockade of progesterone secretion or progesterone receptors prevents implantation. Progesterone is indispensable for maintenance of decidual cells, quiescence of the myometrium, and the formation of the dense, viscous cervical mucus that essentially seals off the uterine cavity from the outside. It is noteworthy that the implanting trophoblast and the fetus are genetically distinct from the mother and yet the maternal immune system does not reject the implanted embryo as a foreign body. Progesterone plays a decisive role in immunological acceptance of the embryo by promoting tolerance. It regulates accumulation of lymphocytes in the uterine cavity and suppresses lymphocyte toxicity and production of cytolytic cytokines. The importance of progesterone for implantation and retention of the blastocyst is underscored by the development of a progesterone antagonist (RU486) that prevents implantation or causes an already implanted conceptus to be shed along with the uterine lining.

THE PLACENTA

The placenta is a complex, primarily vascular organ adapted to optimize exchange of gases, nutrients, and electrolytes between maternal and fetal circulations. In humans the placenta is also a major endocrine gland capable of producing large amounts of both steroid and peptide hormones and neurohormones. The placenta is the most recently evolved of all mammalian organs, and its endocrine function is highly developed in primates. It is unique among endocrine glands in that, as far as is known, its secretory activity is autonomous and not subject to regulation by maternal or fetal signals. In experimental animals such as the rat, pregnancy is terminated if the pituitary gland is removed during the first half of gestation or if the ovaries, and consequently the corpora lutea, are removed at any time. In primates the pituitary gland and ovaries are essential only for a brief period after fertilization. After about 7 weeks, the placenta produces enough progesterone to maintain pregnancy. In addition, it also produces large amounts of estrogen, human chorionic gonadotropin (hCG), and human chorionic somatomammotropin (hCS), which is also called human placental lactogen (hPL). It can also secrete growth hormone (GH), thyroid-stimulating hormone (TSH), adrenocorticotropic hormone (ACTH), gonadotropin-releasing hormone (GnRH), corticotropin-releasing hormone (CRH), and a long list of other biologically active peptides. During pregnancy, there is the unique situation of hormones secreted by one individual, the fetus, regulating the physiology of another, the mother. By extracting needed nutrients and adding hormones to the maternal circulation, the placenta redirects some aspects of maternal function to accommodate the growing fetus.

PLACENTAL HORMONES

Human Chorionic Gonadotropin

As already discussed (see Chapter 12), the functional life of the corpus luteum in infertile cycles ends by the twelfth day after ovulation. About 2 days later the endometrium is shed, and menstruation begins. For pregnancy to develop, the endometrium must be maintained, and therefore the ovary must be notified that fertilization has occurred. The signal to the ovary in humans is hCG, a luteotropic substance secreted by the conceptus. Human chorionic gonadotropin rescues the corpus luteum (i.e., extends its life-span) and stimulates it to continue secreting progesterone and estrogen, which in turn maintain the endometrium in a state favorable for implantation and placentation (Figure 5). Continued secretion of luteal steroids and inhibin notifies the pituitary gland that pregnancy has begun and inhibits secretion of the gonadotropins, which would otherwise stimulate development of the next cohort of follicles. Pituitary gonadotropins remain virtually undetectable in maternal blood throughout pregnancy as a result of the negative feedback effects of high circulating concentrations of estrogens and progesterone. Relaxin secretion by the corpus luteum increases in early pregnancy and becomes maximum at around the end of the first trimester, and then declines somewhat, but continues throughout pregnancy. Relaxin may synergize with progesterone in early pregnancy to suppress contractile activity of uterine smooth muscle.

Human chorionic gonadotropin is a glycoprotein that is closely related to the pituitary glycoprotein hormones (see Chapter 2). Although there are wide

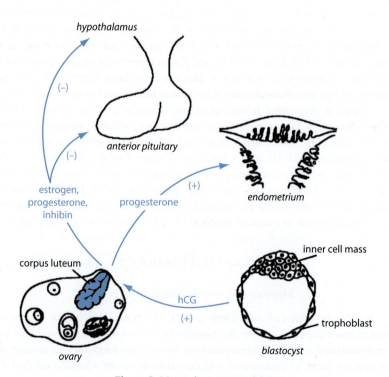

Figure 5 Maternal responses to hCG.

variations in the carbohydrate components, the peptide backbones of the glyco-protein hormones are closely related and consist of a common alpha subunit and activity-specific beta subunits. The alpha subunits of FSH, LH, TSH, and hCG have the same amino acid sequence and are encoded in the same gene. In humans seven genes or pseudogenes on chromosome 19 code for hCG-β, but only two or three of them are expressed. The beta subunit of hCG is almost identical to the beta subunit of LH, differing only by a 32-amino-acid extension at the carboxyl termi-nus of hCG. It is not surprising, therefore, that hCG and LH act through a common receptor and that hCG has LH-like bioactivity. hCG contains consider-ably more carbohydrate, particularly sialic acid residues, than do its pituitary coun-terparts, which accounts for its extraordinary stability in blood. The half-life of hCG is more than 30 hours, as compared to just a few minutes for the pituitary glycoprotein hormones. The long half-life facilitates rapid buildup of adequate concentrations of this vital signal produced by a few vulnerable cells.

Trophoblast cells of the developing placenta begin to secrete hCG early, with detectable amounts already present in blood by about the eighth day after ovula-tion, when luteal function, under the influence of LH, is still at its height. Production of hCG increases dramatically during the early weeks of pregnancy (Figure 6). Blood levels continue to rise and during the third month of pregnancy reach peak values that are perhaps 200–1000 times that of LH at the height of the ovulatory surge. Presumably because of its high concentration, hCG, which acts through the same receptor as LH, is able to prolong the functional life of the corpus luteum, whereas LH, at the prevailing concentrations in the luteal phase of an infertile cycle, cannot. High concentrations of hCG at this early stage of fetal development are also critical for male sexual differentiation, which occurs before the fetal pituitary can produce adequate amounts of LH to stimulate testosterone synthesis by the developing testis. Secretion of testosterone by the fetal testes is crucial for survival of the wolffian duct system and formation of the male internal genitalia (see Chapter 11). Human chorionic gonadotropin stimulation of the fetal adrenal gland may augment estrogen production later in pregnancy (see below). Finally, it is the appearance of hCG in large amounts in urine that is used as a diagnostic test for pregnancy. Because its biological activity is like that of LH, urine containing hCG induces ovulation when injected into estrous rabbits in the classic rabbit test. Now hCG can be measured with a simple sensitive immunological test, and pregnancy can be detected even before the next expected menstrual period.

Secretion of significant amounts of progesterone by the corpus luteum diminishes after about the eighth week of pregnancy despite continued stimulation by hCG. Measurements of progesterone in human ovarian venous blood indicate that the corpus luteum remains functional throughout most of the first trimester, and although some capacity to produce progesterone persists throughout preg-nancy, continued presence of the ovary is not required for a successful outcome. Well before the decline in luteal steroidogenesis, placental production of progesterone becomes adequate to maintain pregnancy.

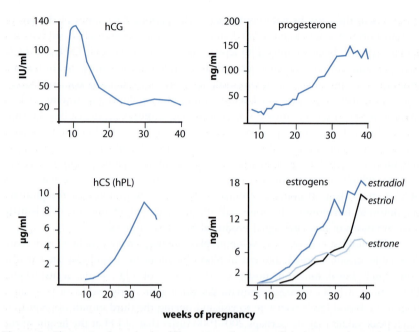

Figure 6 Changes in plasma levels of "hormones of pregnancy" during normal gestation. hCG, Human chorionic gonadotropin; hCS, human chorionic somatomammotropin; hPL, human placental lactogen. (From Freinkel, N., and Metzger, B. E., "Williams Textbook of Endocrinology," 8th Ed., p. 995, D. W. Saunders, Philadelphia, 1992, with permission.)

Human Chorionic Somatomammotropin

The other placental protein hormone that is secreted in large amounts is hCS. Like hCG, hCS is produced by the syncytial trophoblast and becomes detectable in maternal plasma early in pregnancy. Its concentration in maternal plasma increases steadily from about the third week after fertilization and reaches a plateau in the last month of pregnancy (Figure 6), when the placenta produces about 1 g of hCS each day. The concentration of hCG in maternal blood at this time is about 100 times higher than that normally seen for other protein hormones in women or men. Human chorionic somatomammotropin has a short half-life and, despite its high concentration at parturition, is undetectable in plasma after the first postpartum day.

Despite its abundance and its ability to produce a number of biological actions in the laboratory, the physiological role of hCS has not been established definitively. Human chorionic somatomammotropin has strong prolactin-like activity and can induce lactation in test animals, but lactation normally does not begin until long enough after parturition for hCS to be cleared from maternal blood. However, it is likely that hCS promotes mammary growth in preparation for lactation. It is also likely that hCS contributes to the availability of nutrients for the developing fetus by operating, like GH to mobilize maternal fat and decrease maternal glucose consumption (see Chapter 9). In this context, hCS may be responsible for the decreased glucose tolerance, the so-called gestational diabetes, experienced by many women during pregnancy. Although secretion of hCS is directed predominantly into maternal blood, appreciable concentrations are also found in fetal blood in midgestation. Receptors for hCS are present in human fetal fibroblasts and myoblasts, and these cells release IGF-II when stimulated by hCS. As already discussed (see Chapter 10), fetal growth is independent of GH, but the role of hCS in this regard is unknown.

Despite these observations, evidence from genetic studies makes it unlikely that hCS is indispensable for the successful outcome of pregnancy. Human chorionic somatomammotropin is a member of the growth hormone–prolactin family (see Chapter 2) and shares large regions of structural homology with both of these pituitary hormones. Five genes of this family are clustered on chromosome 17, including three that encode hCS and two that encode GH. Two of the hCS genes are expressed and code for identical secretory products. The third hCS gene appears to be a pseudogene that does not produce fully processed mRNA when transcribed. No adverse consequences for pregnancy, parturition, or early postnatal development were seen in three cases in which a stretch of DNA that contains both hCS genes and one hGH gene was missing from both chromosomes. No immunoassayable hCS was present in maternal plasma, but it is possible that the remaining hCS pseudogene was expressed under these circumstances or that recombination of remaining fragments of these genes produced a chimeric protein with hCS-like activity. Regardless of whether hCS is indispensable for normal gestation, important functions are often governed by redundant mechanisms, and it is likely that hCS contributes in some way to a successful outcome of pregnancy.

Progesterone

As progesterone secretion by the corpus luteum declines, the trophoblast becomes the major producer of progesterone. Placental production of progesterone increases as pregnancy progresses, so that during the final months upward of 250 mg may be secreted per day. This huge amount is more than 10 times the daily production by the corpus luteum at the height of its activity, and may be even greater in women bearing more than one fetus. Because the placenta cannot synthesize cholesterol, it imports cholesterol in the form of low-density lipoproteins (LDLs) from the maternal circulation. In late pregnancy progesterone production consumes an amount of cholesterol equivalent to about 25% of the daily turnover in a normal nonpregnant woman.

Production of progesterone by the placenta is not subject to regulation by any known extraplacental factors other than availability of substrate. As in the adrenals and gonads, the rate of conversion of cholesterol to pregnenolone by P450scc determines the rate of progesterone production. In the adrenals and gonads ACTH and LH stimulate synthesis of the steroid acute regulatory (StAR) protein, which is required for transfer of cholesterol from cytosol to the mitochondrial matrix where P450scc resides (Chapter 4). The placenta does not express StAR protein. Access of cholesterol to the interior of mitochondria is thought to be provided by a similar protein that is constitutively expressed in the trophoblast. Consequently, placental conversion of cholesterol to pregnenolone bypasses the step that is regulated in all other steroid hormone–producing tissues. Ample expression of 3β-hydroxysteroid dehydrogenase allows rapid conversion of pregnenolone to progesterone. All of the pregnenolone produced is either secreted as progesterone or exported to the fetal adrenal glands to serve as substrate for adrenal steroidogenesis (Figure 7).

Estrogens

The human placenta is virtually the only site of estrogen production after the corpus luteum declines. However, the placenta cannot synthesize estrogens from cholesterol or use progesterone or pregnenolone as substrate for estrogen synthesis. The placenta does not express P450c17, which cleaves the C20,21 side chain to produce the requisite 19-carbon androgen precursor. Reminiscent of the dependence of granulosa cells on thecal cell production of androgens in ovarian follicles (see Chapter 12), estrogen synthesis by the trophoblast depends on import of 19-carbon androgen substrates, which are secreted by the adrenal glands of the fetus and, to a lesser extent, the mother (Figure 8). The trophoblast expresses an abundance of P450 aromatase, which has activity sufficient to aromatize all of the available substrate. The cooperative interaction between the fetal adrenal glands and the placenta has given rise to the term *fetoplacental unit* as the source of estrogen production in pregnancy. The placental estrogens are estradiol, estrone, and estriol; estriol differs from estradiol by the presence of an additional hydroxyl group on carbon 16. Of these, estriol is by far the major estrogenic product. Its rate of synthesis may exceed 45 mg per day by the end of pregnancy.

Despite its high rate of production, however, concentrations of unconjugated estriol in blood are lower than those of estradiol (Figure 6) due to the high rate of metabolism and excretion of estriol. Although estriol can bind to estrogen receptors, it contributes little to overall estrogenic bioactivity, because it is only about 1% as potent as estradiol and 10% as potent as estrone in most assays. However, estriol is almost as potent as estradiol in stimulating uterine blood flow. It is possible that the fetus uses this elaborate mechanism of estriol production to ensure that uterine blood flow remains adequate for its survival.

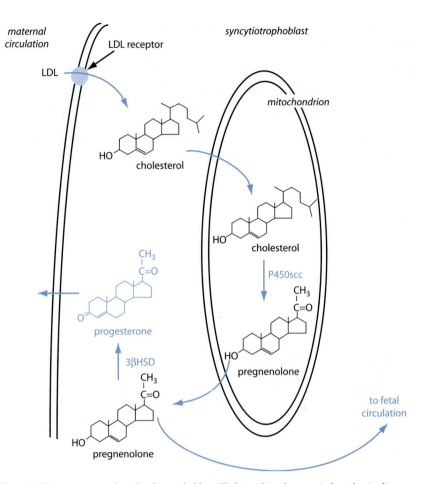

Figure 7 Progesterone synthesis by the trophoblast. Cholesterol is taken up via low-density lipoprotein (LDL) receptors and transferred to the inner mitochondrial matrix by constitutively expressed protein(s), where its C22–C27 side chain is removed by P450scc. Pregnenolone exits the mitochondria and is oxidized to progesterone by 3β-hydroxysteroid dehydrogenase (3 βHSD).

Role of the Fetal Adrenal Cortex

The fetal adrenal glands play a central role in placental steroidogenesis and hence maintenance of pregnancy, and may also have a role in provoking the onset of labor at the end of pregnancy. The adrenal glands of the fetus differ significantly in both morphology and function from the adrenal glands of the adult. They are bounded by a thin outer region, called the definitive cortex, which will become the zona glomerulosa, and a huge, inner "fetal zone," which regresses and disappears

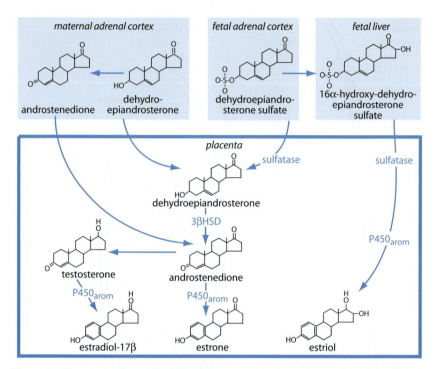

Figure 8 Biosynthesis of estrogens during pregnancy. Note that androgens formed in either the fetal or maternal adrenals are the precursors for all three estrogens, and that the placenta cannot convert progesterone to androgens. Hydroxylation of dehydroepiandrosterone sulfate on carbon 16 by the fetal liver gives rise to estriol, which is derived almost exclusively from fetal sources. Fetal androgens are secreted as sulfate esters and must be converted to free androgens by the abundant placental sulfatase before conversion to estrogens by the enzyme P450 aromatase (P450arom). 3βHSD, 3β-Hydroxysteroid dehydrogenase.

shortly after birth. The transitional zone at the interface of the two zones gives rise to the fasciculata, and reticularis of the adult (see Chapter 4). In midpregnancy the fetal adrenals are large—larger, in fact, than the kidneys—and the fetal zone constitutes 80% of its mass. Growth, differentiation, and secretory activity of the fetal adrenals are controlled by ACTH, the actions of which are augmented by a variety of fetal growth factors, including insulin-like growth factor II. The fetal pituitary is the main source of ACTH, but the placenta also secretes some ACTH. In addition, the placenta secretes CRH, which not only stimulates the fetal pituitary to secrete ACTH, but also directly stimulates steroidogenesis by the fetal adrenal glands.

The chief product of the fetal zone is the 19-carbon androgen dehydroepiandrosterone (DHEA), which is secreted as the biologically inert sulfate ester

(DHEAS). Sulfation protects against masculinization of the genitalia in female fetuses and prevents aromatization in extragonadal fetal tissues. The fetal zone produces DHEAS at an increasing rate that becomes detectable by about the eighth week of pregnancy, well before cortisol and aldosterone are produced by the definitive and transitional zones. At term, secretion of DHEAS may reach 200 mg per day. The cholesterol substrate for DHEAS production is synthesized in both the fetal liver and the fetal adrenals. It is likely that pregnenolone released into the fetal circulation by the placenta also provides substrate.

Much of the DHEAS in the fetal circulation is oxidized at carbon 16 in the fetal liver, to form 16α-DHEAS, which then is exported to the placenta. The placenta is highly efficient at extracting 19-carbon steroids from both maternal and fetal blood. It is rich in sulfatase and converts 16α-DHEAS to 16α-DHEA prior to aromatization to form estriol. Because synthesis of estriol reflects the combined activities of the fetal adrenals, the fetal liver, and the placenta, its rate of production, as reflected in maternal estriol concentrations, has been used as an indicator of fetal well-being. DHEAS that escapes 16α-hydroxylation in the fetal liver is converted to androstenedione or testosterone in the placenta after hydrolysis of the sulfate bond, and then is aromatized to form estrone and estradiol.

Role of Progesterone and Estrogens in Sustaining Pregnancy

As its name implies, progesterone is essential for maintaining all stages of pregnancy, and pharmacologic blockade of its actions at any time terminates the pregnancy. Progesterone sustains pregnancy by opposing the forces that conspire to increase uterine contractility and expel the fetus. One of these forces is physical stretch of the myometrium by the growing fetus. Stretch or tension coupled with estrogens and progesterone promotes myometrial growth and hypertrophy in parallel with growth of the conceptus. Estrogens promote expression of genes that code for contractile proteins, gap junction proteins that electrically couple myometrial cells, oxytocin receptors, receptors for prostaglandins, ion channel proteins, and no doubt many others that directly or indirectly tend to increase contractility. Throughout pregnancy, estrogen acts through a positive feedback mechanism not only to increase its own production, but also increases synthesis of progesterone, which suppresses its excitatory effects (Figure 9). Estrogens accelerate progesterone synthesis by increasing the delivery of its precursor substrate, cholesterol, to the trophoblast and by up-regulating P450scc. Estrogens increase uterine blood flow by stimulating endothelial cells to produce the potent vasodilator nitric oxide, and promote uterine formation of prostaglandin I, which is also a vasodilator. In addition they increase receptor-mediated cholesterol uptake by stimulating expression of LDL receptors. Pregnenolone and LDLs that cross the placenta and enter the fetal circulation serve as substrate for adrenal production of DHEAS. DHEAS

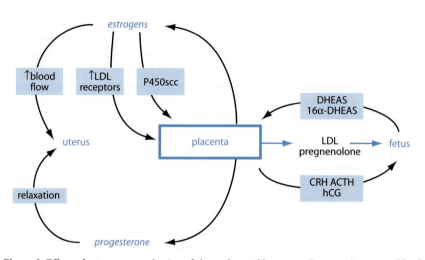

Figure 9 Effects of estrogen on production of placental steroid hormones. By increasing uterine blood flow, and inducing low-density lipoprotein (LDL) and P450 side chain cleavage (P450scc) enzyme, estrogens increase placental production of pregnenolone, which is used as substrate for androgen production in the fetal adrenals. Uptake of LDL from the maternal circulation may also transfer cholesterol to the fetal circulation. DHEAS, Dehydroepiandrosterone sulfate; 16α–DHEAS, 16α-hydroxydehydroepiandrosterone sulfate; CRH, corticotropin-releasing hormone; ACTH, adrenal corticotropic hormone; hCG, human chorionic gonadotropin.

is then converted by the placenta to estrogens in what amounts to a positive feedback system that progressively increases estrogen production in parallel with progesterone production.

PARTURITION

Pregnancy in the human lasts about 40 weeks. The process of birth, or *parturition*, is the expulsion of a viable baby from the uterus at the end of pregnancy and is the culmination of all the processes discussed in this and the previous two chapters. Study the phenomenon of parturition has revealed a surprising array of strategies that have been adopted by different species to regulate parturition. Humans and the great apes have evolved mechanisms that appear to be unique. The scarcity of experimental models that employ strategies similar to that of humans has therefore hampered efforts to study underlying mechanisms of timing and initiating parturition in humans. Consequently, our understanding of the processes that bring about this climactic event in human reproductive physiology is still incomplete.

Successful delivery of the baby can only take place after the myometrium acquires the capacity for forceful, coordinated contractions and the cervix softens and becomes distensible (called ripening), so that uterine contractions can drive the baby through the cervical canal. These changes reflect the triumph of the excitatory effects of estrogens over the suppressive effects of progesterone that prevailed hitherto. Indeed, in most animals parturition is heralded by a decline in progesterone production coincident with an increase in estrogen production. Humans and higher primates are unique in the respect that there is neither an abrupt increase in plasma concentrations of estrogens nor a fall in progesterone at the onset of parturition. It is highly likely that multiple gradual changes, gaining momentum over days or weeks, tip the precarious estrogen/progesterone balance in favor of estrogen dominance.

In theory, signals to terminate pregnancy could originate with either the mother or the fetus. Most investigators favor the idea that the fetus, which has essentially controlled events throughout the pregnancy, signals its readiness to be born. In sheep, the triggering event for parturition is an ACTH-dependent increase in cortisol production by the fetal adrenals. In this species, cortisol stimulates expression of P450c17 in the placenta and thereby shifts production of steroid hormones away from progesterone and toward estrogen. Although there is neither a stimulation of P450c17 expression in the human placenta nor a fall in progesterone, the human fetal adrenal gland may nevertheless have an essential role in orchestrating the events that lead up to parturition.

Although the ability to secrete 19-carbon androgens is acquired by the fetal zone of the adrenals early in gestation, the definitive and transitional zones mature much later. The capacity to produce significant amounts of cortisol is not acquired until about the thirtieth week of gestation. An abundant supply of cortisol in the final weeks of pregnancy is indispensable for maturation of the lungs, the GI tract, and other systems that prepares the fetus for extrauterine life. Cortisol also antagonizes the suppressive effects of progesterone on CRH production in the placenta and hence increases transcription of the CRH gene. This paradoxical effect of cortisol on placental expression of CRH is opposite to its negative feedback effects on CRH production in the hypothalamus. Instead of suppressing CRH production, fetal production of cortisol initiates a positive feedback loop (Figure 10). It may be recalled that CRH not only stimulates the fetal pituitary to secrete ACTH, but also directly stimulates steroidogenesis in the fetal adrenal cortex. Consequently, there is an increasing drive to the adrenal to increase production of cortisol and DHEAS. Accelerating secretion of DHEAS accounts for the increasingly steep rise in estrogen concentrations in maternal blood in the last weeks of pregnancy (shown in Figure 6).

Prostaglandin production is also increased in fetal membranes and the uterus in late pregnancy. Prostaglandins $F_{2\alpha}$ and E participate in or initiate events that lead to rupture of the fetal membranes, softening of the uterine cervix, and contraction of the myometrium. CRH stimulates their formation by the fetal membranes. These prostaglandins in turn stimulate placental production of CRH and establish

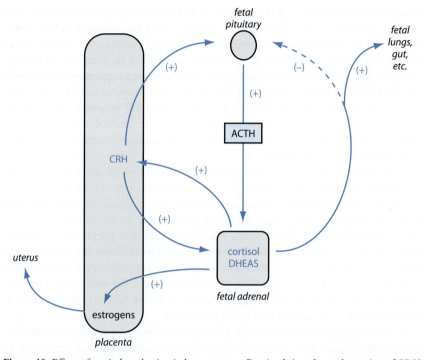

Figure 10 Effects of cortisol production in late pregnancy. By stimulating placental secretion of CRH (corticotropin-releasing hormone), cortisol initiates direct and indirect (via the fetal pituitary) positive feedback loops that enhance its own secretion and increase secretion of DHEAS (dehydroepiandrosterone sulfate). In this way cortisol-induced maturation of the fetus occurs simultaneously with increased production of estrogens, preparing the uterus for parturition. ACTH, Adrenal corticotropic-hormone.

a second positive feedback loop. We might expect cortisol to oppose prostaglandin formation in the fetus as it does in extrauterine tissues (Chapter 4). However, in fetal membranes cortisol paradoxically increases expression of the prostaglandin-synthesizing enzyme, COX 2 (see Chapter 4), and inhibits formation of the principal prostaglandin-degrading enzyme. Prostaglandins also stimulate CRH secretion by the fetal hypothalamus, increasing ACTH secretion and providing further drive for cortisol secretion and the consequent stimulation of CRH secretion.

ROLE OF CORTICOTROPIN-RELEASING HORMONE

Concentrations of CRH in maternal plasma increase exponentially as pregnancy progresses, but there is only a slight rise in ACTH and free cortisol.

Discordance between CRH plasma concentrations and pituitary and adrenal secretory activity is due largely to the presence of a CRH-binding protein (CRH-BP) that is present plasma of pregnant as well as nonpregnant women. Additionally, responsiveness of the maternal pituitary to CRH is decreased during pregnancy, possibly because of down-regulation of CRH receptors in corticotropes. Despite the somewhat blunted sensitivity to CRH, however, maternal ACTH secretion follows the normal diurnal rhythmic pattern and increases appropriately in response to stress. Until about 3 weeks before parturition, concentrations of CRH-BP in maternal plasma vastly exceed those of CRH and there is little or no free CRH. For reasons that are not understood, CRH-BP concentrations fall dramatically at the same time that placental production of CRH is increasing most rapidly and exceeds the capacity of CRH-BP. Free CRH in maternal plasma stimulates prostaglandin production in the myometrium and cervix, causing increased contractility and cervical ripening.

In addition to CRH-related positive feedback loops, a large number of genes that encode gap junction proteins, ion channels, oxytocin receptors, prostaglandin receptors, proteases that breakdown cervical collagen fibers, and a host of other proteins are activated to an increasing extent by stretch and probably paracrine and autocrine factors that arise in the placenta or decidua. There is also evidence that progesterone-inactivating enzymes that are induced in the myometrium, the placenta, and the cervix in the final weeks may lower tissue concentrations of progesterone and hence decrease its effectiveness. There appears to be no single event that precipitates parturition, but the various processes that are set in motion weeks earlier gradually build up to overwhelm progesterone dominance and unleash the excitatory forces that expel the fetus. CRH and the factors that regulate its production appear to play a crucial, but not exclusive, role (Figure 11).

ROLE OF OXYTOCIN

Oxytocin is a neurohormone secreted by nerve endings in the posterior lobe of the pituitary gland in response to neural stimuli received by cell bodies in the paraventricular and supraoptic nuclei of the hypothalamus (Chapter 2). It produces powerful synchronized contractions of the myometrium at the end of pregnancy, when uterine muscle is highly sensitive to it. Oxytocin is sometimes used clinically to induce labor. As parturition approaches, responsiveness to oxytocin increases in parallel with estrogen-induced increases in oxytocin receptors in both the endometrium and the myometrium. Oxytocin is not the physiological trigger for parturition, however, because its concentration in maternal blood normally does not increase until after labor has begun. Oxytocin is secreted in response to stretching of the uterine cervix and hastens expulsion of the fetus and the placenta as described in Chapter 1, but probably has little role in initiating parturition.

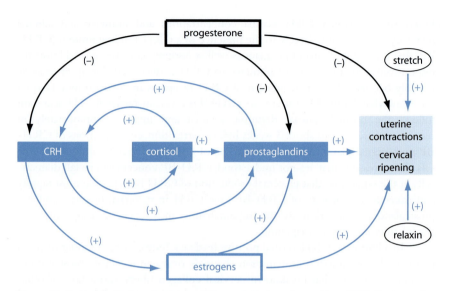

Figure 11 Positive feedback cycles that contribute to initiation of parturition. CRH, Corticotropin-releasing hormone; (+), stimulation; (−), inhibition. See text for details.

As a consequence of its action on myometrial contraction, oxytocin protects against hemorrhage after expulsion of the placenta. Just prior to delivery the uterus receives nearly 25% of the cardiac output, most of which flows through the low-resistance pathways of the maternal portion of the placenta. Intense contraction of the newly emptied uterus acts as a natural tourniquet to control loss of blood from the massive wound left when the placenta is torn away from the uterine lining.

LACTATION

The mammary glands are specialized secretory structures derived from the skin. As the name implies, they are unique to mammals. The secretory portion of the mammary glands is arranged in lobules consisting of branched tubules, the *lobuloalveolar ducts*, from which multiple evaginations, or *alveoli*, emerge in an arrangement resembling a bunch of grapes. The alveoli consist of a single layer of secretory epithelial cells surrounded by a meshwork of contractile *myoepithelial* cells (Figure 12). Many lobuloalveolar ducts converge to form a *lactiferous duct*, which carries the milk to the nipple. Each mammary gland consists of 10–15 lobules, each with its own lactiferous duct opening separately to the outside. In the inactive, nonlactating gland, alveoli are present only in rudimentary form, with the entire glandular portion consisting almost exclusively of lobuloalveolar ducts.

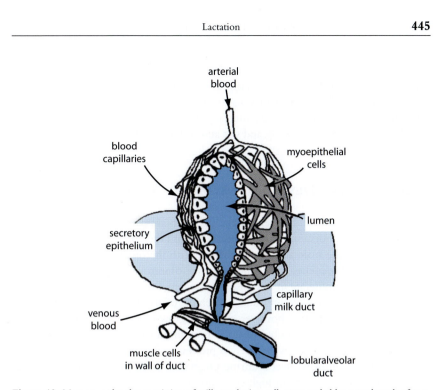

Figure 12 Mammary alveolus consisting of milk-producing cells surrounded by a meshwork of contractile myoepithelial cells. Milk-producing cells are targets for prolactin; myoepithelial cells are targets for oxytocin. (From Turner, C. W., "Harvesting your Milk Crop," p. 17. Babson Bros., Oak Brook, Illinois, 1969, by permission of Babson Bros. Co.)

The mammary glands have an abundant vascular supply and are innervated with sympathetic nerve fibers and a rich supply of sensory fibers to the nipple and areola.

Milk secreted by the mammary glands provides nourishment and immunoglobulins to offspring during the immediate postnatal period and for varying times thereafter, depending on culture and custom. Milk provides all of the basic nourishment, vitamins, minerals, fats, carbohydrates, and proteins needed by the infant until the teeth erupt. In addition, milk contains maternal immunoglobulins, which are absorbed intact by the immature intestine and provide passive immunity to common pathogens. The extraordinarily versatile cells of the mammary alveoli simultaneously synthesize large amounts of protein, fat, and lactose, and secrete these constituents by different mechanisms, along with a large volume of aqueous medium with an ionic composition that differs substantially from blood plasma. Human milk consists of about 1% protein, principally in the form casein and lactalbumin, about 4% fat, and about 7% lactose. Each liter of milk also contains about 300 mg of calcium. After lactation is established, the well-nourished woman suckling a single infant may produce about a liter of milk per day and as much as 3 liters per day if suckling twins. It should be apparent, therefore, that, in addition to hormonal regulation at the level of the mammary glands, milk production requires extramammary regulation by all those hormones responsible for compensatory adjustments in intermediary metabolism (see Chapters 5 and 9), calcium balance (see Chapter 8), and salt and water balance (see Chapter 7).

GROWTH AND DEVELOPMENT OF THE MAMMARY GLANDS

Prenatal growth and development of the mammary glands appear to be independent of sex hormones and genetic sex. Until the onset of puberty, there are no differences in the male and female breasts. With the onset of puberty, the duct system grows and branches under the influence of estrogen. Surrounding stromal and fat tissue also proliferate in response to estrogens. Progesterone, in combination with estrogen, promotes growth and branching of the lobuloalveolar tissue; for these steroids to be effective, however, prolactin, growth hormone, IGF-I, and cortisol must also be present. Lobuloalveolar growth and regression occur to some degree during each ovarian cycle. There is pronounced growth, differentiation, and proliferation of mammary alveoli during pregnancy, when estrogen, progesterone, prolactin, and hCS circulate in high concentrations.

MILK PRODUCTION

Once the secretory apparatus has developed, production of milk depends primarily on continued episodic stimulation with high concentrations of prolactin, but adrenal glucocorticoids and insulin are also important in a permissive sense that needs to be defined more precisely. All of these hormones and hCS are present in abundance during late stages of pregnancy, yet lactation does not begin until after parturition. High concentrations of estrogen and particularly progesterone in maternal blood inhibit lactation by interfering with the action of prolactin on mammary epithelium. With parturition, the precipitous fall in estrogen and progesterone levels relieves this inhibition, and prolactin receptors in alveolar epithelium may increase as much as 20-fold. Development of full secretory capacity, however, takes some time. Initially the mammary glands put out only a watery fluid called *colostrum*, which is rich in protein but poor in lactose and fat. It takes about 2 to 5 days for the mammary gland to secrete mature milk with a full complement of nutrients. It is not clear whether this delay reflects a slow acquisition of secretory capacity or a regulated sequence of events timed to coincide with the infant's capacity to utilize nutrients.

MECHANISM OF PROLACTIN ACTION

Prolactin acts on alveolar epithelial cells to stimulate expression of genes for proteins that are secreted in the milk, such as casein and lactalbumin, as well as proteins that regulate synthesis of lactose and triglycerides and the proteins that govern secretory processes. The prolactin receptor is a large peptide with a single membrane-spanning domain. It is closely related to the GH receptor and transmits its signal by activating tyrosine phosphorylation of intracellular proteins, as described for the GH receptor (see Chapter 10). Binding of prolactin causes two receptors to dimerize and activate the cytosolic enzyme, JAK-2. Some of the proteins thus phosphorylated belong to the Stat family (for signal transduction and activation of transcription), which then dimerize and migrate to the nucleus, where they activate transcription of specific genes. Prolactin may also signal through activation of a tyrosine kinase related to the src oncogene and by activating membrane ion channels. The signaling cascades set in motion the various events that accompany growth of the secretory alveoli as well as synthesis and secretion of milk. Several isoforms that result from alternative splicing of its RNA have been described, but the physiological significance of these multiple receptor isoforms is not understood.

NEUROENDOCRINE MECHANISMS

Continued lactation requires more than just the right complement of hormones. Milk must also be removed regularly by suckling. Failure to empty the mammary alveoli causes lactation to stop within about a week and the lobuloalveolar structures to involute. Involution results not only from prolactin withdrawal, but also from the presence in milk of inhibitory factors that block secretion if milk is allowed to remain in the alveolar lumens. Suckling triggers two neuroendocrine reflexes critical for the maintenance of lactation: the so-called milk let-down reflex and surges of prolactin secretion.

Milk Let-Down Reflex

Because each lactiferous duct has only a single opening to the outside and alveoli are not readily collapsible, application of negative pressure at the nipple does not cause milk to flow. The milk let-down reflex, also called the milk ejection reflex, permits the suckling infant to obtain milk. This neuroendocrine reflex involves the hormone Oxytocin, which is secreted in response to suckling. Oxytocin stimulates contraction of myoepithelial cells that surround each alveolus, creating positive pressure of about 10 to 20 mm mercury in the alveoli and the communicating duct system. Suckling merely distorts the valvelike folds of tissue in the nipple and allows the pressurized milk to be ejected into the infant's mouth. Sensory inputs from nerve endings in the nipple are transmitted to the hypothalamus by way of thoracic nerves and the spinal cord, stimulating neurons in the supraoptic and paraventricular nuclei to release oxytocin from terminals in the posterior lobe (Figure 13). These neurons can also be activated by higher brain centers, so that the mere sight of the baby, or hearing it cry, is often sufficient to produce milk let-down (Figure 14). Conversely, stressful conditions may inhibit oxytocin secretion, preventing the suckling infant from obtaining milk even though the breast is full.

Cellular Actions of Oxytocin

The oxytocin receptor is encoded in a single-copy gene that is expressed principally in uterine smooth muscle and the myoepithelial cells that surround mammary alveoli. It is a typical heptahelical receptor that is coupled through G_q to phospholipase Cβ. When activated, it stimulates formation of inositol trisphosphate (IP$_3$) and diacylglycerol (DAG) from phosphatidylinositol in the membrane. Inositol trisphosphate increases intracellular calcium by signaling its release from intracellular stores. Diacylglycerol activates protein kinase C, which

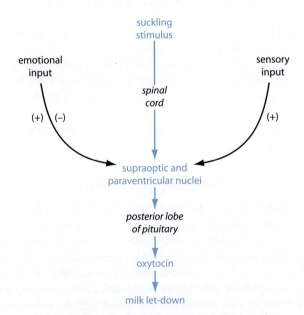

Figure 13 Control of oxytocin secretion during lactation.

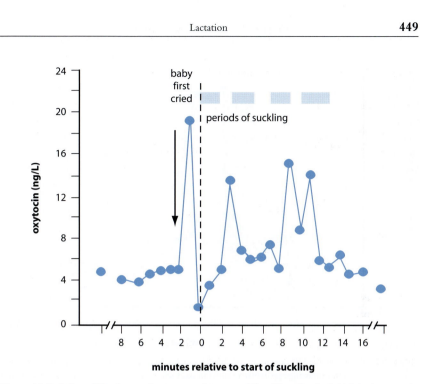

Figure 14 Relation of blood oxytocin concentrations to suckling. Note that the initial rise in oxytocin preceded the initial period of suckling. (From McNeilly, A. S., Robinson, I. C. A., Houston, M. J., *et al.*, *Br. Med. J.* **286**, 257, 1983, by permission of The British Medical Association.)

may phosphorylate membrane calcium channels and further increase intracellular calcium. Increased intracellular calcium binds to calmodulin and activates myosin light chain kinase, which initiates contraction of myoepithelial or myometrial cells.

CONTROL OF PROLACTIN SECRETION

Suckling is also an important stimulus for secretion of prolactin. During suckling, the prolactin concentration in blood may increase by 10-fold or more within just a few minutes (Figure 15). Although suckling evokes secretion of both oxytocin and prolactin, the two secretory reflexes are processed separately in the central nervous system, and the two hormones are secreted independently. Emotional signals that release oxytocin and produce milk let-down are not followed by prolactin secretion. It is unlikely that prolactin secreted during suckling can act quickly enough to increase milk production to meet current demands. Rather, such episodes of secretion are important for producing the milk needed for subsequent feedings. Milk production is thus related to frequency of suckling,

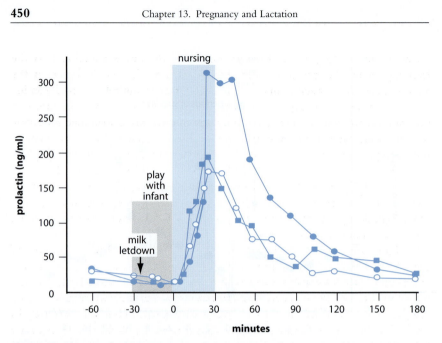

Figure 15 Plasma prolactin concentrations during nursing and anticipation of nursing. Note that although anticipation of nursing apparently resulted in oxytocin secretion, increased prolactin secretion did not occur until well after suckling began. (From Noel, G. L., Suh, H. K., and Franz, A. G., *J. Clin. Endocrinol. Metab.* **38**, 413, 1974, by permission of The Endocrine Society.)

which gives the newborn some control over its nutritional supply, and is an extension into the postnatal period of the self-serving control over maternal function that the fetus exercised *in utero*.

Increased secretion of prolactin and even milk production do not require a preceding pregnancy. Repeated stimulation of the nipples can induce lactation in some women who have never borne a child. In some cultures postmenopausal women act as wet nurses for infants whose mothers produce inadequate milk. This fact underscores the lack of involvement of the ovarian steroids in lactation once the glandular apparatus has been formed.

Prolactin is unique among the anterior pituitary hormones in the respect that its secretion is increased rather than decreased when the vascular connection between the pituitary gland and the hypothalamus is interrupted. Prolactin secretion is controlled primarily by an inhibitory hypophysiotropic hormone, dopamine. Dopamine is synthesized by sequential hydroxylation and decarboxylation of tyrosine (see Figure 21, Chapter 4). Surgical transection of the human pituitary stalk increases plasma prolactin concentrations in peripheral blood (*hyperprolactinemia*) and may lead to the onset of lactation. Stimulation of prolactin secretion by suckling results from inhibition of dopamine secretion into the hypophyseal portal circulation by dopaminergic neuronal cell bodies located in the arcuate nuclei. It has been found experimentally that abrupt relief from dopamine inhibition results in a surge of prolactin secretion. It is possible that prolactin secretion is also under positive control by way of a yet-to-be-identified prolactin-releasing factor. Experimentally, prolactin secretion is increased by neuropeptides such as thyrotropin-releasing hormone (TRH) and vasoactive inhibitory peptide. In spite of its potency as a prolactin-releasing agent, it is unlikely that TRH is a physiological regulator of prolactin secretion. Normally, TSH and prolactin are secreted independently. TSH secretion does not increase during lactation. The physiological importance of vasoactive inhibitory peptide as a prolactin-releasing hormone has not been established.

Lactotropes express estrogen receptors and increase their production of prolactin mRNA and protein in response to estrogens. Estradiol, which stimulates proliferation and hypertrophy of lactotropes, is probably responsible for the increased number of lactotropes in the pituitary and their prolactin content during pregnancy. Estradiol may therefore increase prolactin secretion by increasing its availability. In addition, although it does not act directly as a prolactin-releasing factor, estradiol decreases the sensitivity of lactotropes to dopamine. Paradoxically, however, estradiol also increases dopamine synthesis and concentration in the hypothalamus and may therefore increase dopamine secretion (Figure 16).

To date, there is no known product of prolactin action that feeds back to regulate prolactin secretion. The effects of suckling and estrogen on prolactin secretion are open loops. Experiments in animals suggest that prolactin may act as its own "short-loop" feedback inhibitor by stimulating dopaminergic neurons in the arcuate nucleus. It is not certain that such an effect is applicable to humans. If prolactin is a negative effector of its own secretion, it is not clear what mechanisms override feedback inhibition to allow prolactin to rise to high levels during pregnancy.

CELLULAR REGULATION OF PROLACTIN SECRETION

As in most other endocrine cells, secretory activity of lactotropes is enhanced by increased cytosolic concentrations of calcium and cyclic AMP (Figure 17). Dopamine, acting through G-protein-coupled receptors, inhibits prolactin secretion through several temporally distinct mechanisms. Initial inhibitory effects are detectable within seconds and result from membrane hyperpolarization, which deactivates voltage-sensitive calcium channels and lowers intracellular calcium. This effect appears to result from direct stimulation of potassium influx by G-protein-gated channels. Minutes later there is a decrease in cyclic AMP, which leads to decreased transcription of the prolactin gene. Estrogens are thought to decrease responsiveness to dopamine by uncoupling dopamine receptors from G-proteins.

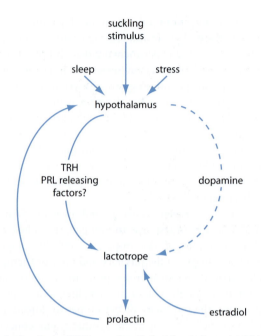

Figure 16 Control of prolactin (PRL) secretion (dashed line indicates inhibition). A physiological role for thyrotropin-releasing hormone (TRH) and other postulated releasing hormones has not been established. Estradiol stimulates secretion and may interfere with the inhibitory action of dopamine.

On a longer time scale, dopamine antagonizes the proliferative effects of estrogen by mechanisms that are not understood.

PROLACTIN IN BLOOD

Prolactin is secreted continuously at low basal rates throughout life, regardless of sex. Its concentration in blood increases during nocturnal sleep in a diurnal rhythmic pattern. Basal values are somewhat higher in women than in men and prepubertal children, presumably reflecting the effects of estrogens. Episodic increases in response to eating and stress are superimposed on this basal pattern (Figure 18). Prolactin concentrations rise steadily in maternal blood throughout pregnancy to about 20 times the nonpregnant value (Figure 19). After delivery, prolactin concentrations remain elevated, even in the absence of suckling, and slowly return to the prepregnancy range, usually within less than 2 weeks. Prolactin also increases in fetal blood as pregnancy progresses and during the final weeks reaches

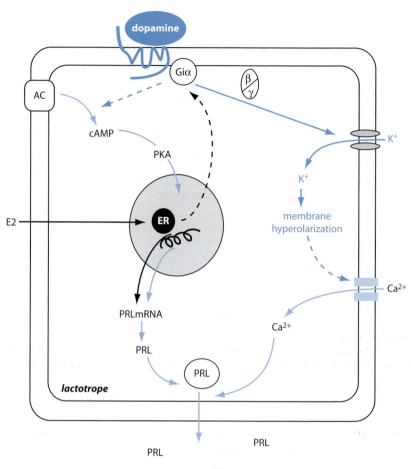

Figure 17 Cellular events in the regulation of prolactin (PRL) secretion. Steps in basal synthesis and secretion are indicated by the pale blue arrows. Effects of dopamine are shown by the dark blue arrows; effects of estradiol are shown in black (dashed lines indicate inhibition). Giα and β/γ represent the subunits of the heterotrimeric inhibitory G-protein. AC, Adenylate cyclase; cAMP, cyclic adenosine monophosphate; PKA, protein kinase A; E2, estradiol; ER, the estrogen receptor. Thyrotropin-releasing hormone (not shown) increases prolactin secretion through activation of its heptihelical receptor, which is coupled through $G\alpha_q$ to phospholipase Cβ and causes release of inositol trisphosphate and diacylglyceride, and increased intracellular calcium.

levels that are higher than those seen in maternal plasma. The fetal kidney apparently excretes prolactin into the amniotic fluid, where at midpregnancy the prolactin concentration is 5 to 10 times higher than that of either maternal or fetal blood. Although some of the prolactin in maternal blood is produced by uterine

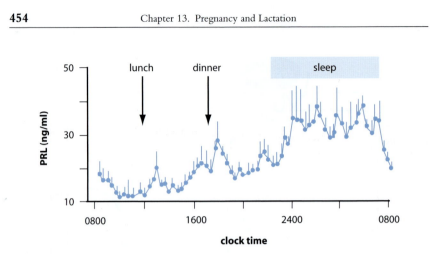

Figure 18 Around-the-clock prolactin (PRL) concentrations in eight normal women. Acute elevation of prolactin level occurs shortly after onset of sleep and begins to decrease shortly before awakening. (From Yen, S. C., and Jaffe, R. B., "Reproductive Physiology," 4th Ed., p. 268. W. B. Saunders, Philadelphia, 1999, with permission.)

cells, prolactin in fetal blood originates in the fetal pituitary and does not cross the placental barrier. The high prolactin concentration seen in the newborn decreases to the low levels of childhood within the first week after birth. The physiological importance of any of these changes in prolactin concentration in either prenatal or postnatal life is not understood. Although prolactin receptors are present in many tissues, including the gonads and reproductive tracts of both sexes, the physiological consequences of prolactin binding to these tissues in humans remain unknown. Considerable evidence has accumulated for a role of prolactin in immunity, but its actions in this regard are beyond the scope of this discussion.

LACTATION AND RESUMPTION OF OVARIAN CYCLES

Menstrual cycles resume as early as 6 to 8 weeks after delivery in women who do not nurse their babies. With breast-feeding, however, the reappearance of normal ovarian cycles may be delayed for many months. This delay, called *lactational amenorrhea*, serves as a natural, but unreliable, form of birth control. Lactational amenorrhea is related to high plasma concentrations of prolactin. Delayed resumption of fertile cycles therefore is most pronounced when breast milk is not supplemented with other foods, and consequently the frequency of suckling is high. Ovarian activity is largely limited to varying degrees of incomplete follicular development; even in those women who ovulate, luteal function is deficient. The delay in resumption of cyclicity results from decreased amplitude and frequency of

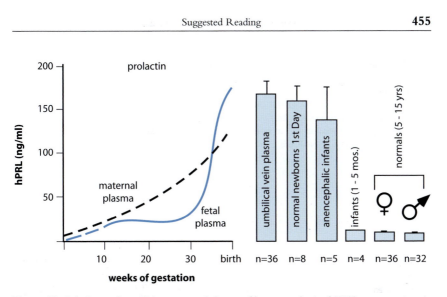

Figure 19 *Left*: Comparison of the pattern of change of human prolactin (hPRL) concentrations in fetal and maternal plasma during gestation. *Right*: Plasma levels in normal and anencephalic newborns are compared with those of normal infants and adults. The high concentrations seen in anencephalic babies presumably reflect prolactin secretion by an anterior pituitary gland uninhibited by influences of the brain. (From Aubert, M. L., Grumbach, M. M., and Kaplan, S. L., *J. Clin. Invest.* **56**, 155, 1975, with permission.)

GnRH release by the hypothalamic GnRH pulse generator (Chapter 11). Pulsatile administration of GnRH to lactating women promptly restores ovulation and normal corpus luteal function. Hyperprolactinemia, often resulting from a small prolactin-secreting pituitary tumor (microadenoma), is now recognized as a common cause of infertility and abnormal or absent menstrual cycles. Treatment with bromocriptine, a drug that activates dopamine receptors, suppresses prolactin secretion and restores normal reproductive function.

SUGGESTED READING

Ben-Jonathan, N., and Hnasko, H. (2001). Dopamine as a prolactin (PRL) inhibitor. *Endocr. Rev.* **22**, 724–763.

Challis, J. R. G., Matthews, S. G., Gibb, W., and Lye, S. J. (2000). Endocrine and paracrine regulation of birth at term and preterm. *Endocr. Rev.* **21**, 514–550.

Gimpl, G., and Fahrenholz, F. (2001). The oxytocin receptor system: Structure, function, and regulation. *Physiol. Rev.* **81**, 630–683.

Hodgen, G. D. (1986). Hormonal regulation in *in vitro* fertilization. *Vitam. Horm.* **43**, 251–282.

Jansen, R. P. S. (1984). Endocrine response in the fallopian tube. *Endocr. Rev.* **5**, 525–552.

Mesiano, S., and Jaffe, R. B. (1997). Developmental and functional biology of the primate fetal adrenal cortex. *Endocr. Rev.* **18**, 378–403.

Neill, J. D., and Nagy, G. M. (1994). Prolactin secretion and its control. *In* "The Physiology of Reproduction," 2nd Ed. (E. Knobil and J. D. Neill, eds.), Vol. 1, pp. 1833–1860. Raven Press, Ltd., New York.

Pepe, G. J., and Albrecht, E. D. (1995). Actions of placental and fetal adrenal steroid hormones in primate pregnancy. *Endocr. Rev.* **16**, 608–648.

Tucker, H. A. (1994). Lactation and its hormonal control. *In* "The Physiology of Reproduction," 2nd ed. (E. Knobil, and J. D. Neill, eds.), Vol. 2, pp. 1065–1098. Raven Press, Ltd., New York.

INDEX

A

Acetyl CoA carboxylase
  insulin effects, 190
  phosphorylative regulation, 174
Acrosome reaction, sperm, 426
ACTH, *see* Adrenocorticotropic hormone
Activins, function, 386, 402
Adipose tissue
  adipogenesis, 316
  aging effects, 316–317
  energy expenditure changes with weight loss,
    316–317
  food intake regulation, *see* Leptin
  insulin effects, 182–184, 302–303
  integrated actions of metabolic hormones,
    302–303
Adrenal glands
  adrenocortical hormones, *see specific hormones*
  anatomy
    cortex, 112–114
    medulla, 156
  blood supply, 114
  functional overview, 112
  histology, 113–114
  medullary hormones, *see* Epinephrine;
    Norepinephrine
Adrenergic receptors, types, 161–162
Adrenocorticotropic hormone (ACTH)
  adrenocortical hormone synthesis regulation,
    121–124
  fetal production, 438
  processing, 62
  pulsatile secretion, 63, 150
  receptor and signaling, 122
  secretion regulation, 148–153
  stress induction, 146–147
Adrenogenital syndrome, pathogenesis and
    treatment, 154–155
Agouti-related protein (AGRP), food intake
    regulation, 323
AGRP, *see* Agouti-related protein
Albumin
  adrenocortical hormone binding, 126
  testosterone binding, 372
  thyroid hormone binding in blood, 90–91
Aldosterone

atrial natriuretic peptide
    effects, 245–246
  deficiency, 127
  dehydration response, 251–252
  hemorrhage response, 251
  kidney effects, 128–129, 131, 235–236
  metabolism and excretion, 126–127
  mineralocorticoid receptor, 131–132
  regulation
    secretion, 132–133, 236
    synthesis, 124–126
  renin–angiotensin–aldosterone system
    regulation, 240–242
  salt intake and depletion response, 252–254
  sodium retention, 235–236
  structure, 115
  synthesis, 119–120
AMH, *see* Antimullerian hormone
Androstenedione
  metabolism and excretion, 126–127
  structure, 115
  synthesis, 120–121
Angiotensin II
  angiotensinogen synthesis and
    processing, 236
  dehydration response, 251–252
  hemorrhage response, 250
  metabolism, 236
  physiologic actions
    aldosterone synthesis regulation, 124, 132,
      236, 238
    cardiovascular effects, 239–240
    central nervous system effects, 220
    kidney, 238–239
  receptor and signaling, 124–125
  renin–angiotensin–aldosterone system
    regulation, 240–242
  salt intake and depletion response, 252–254
ANP, *see* Atrial natriuretic peptide
Antidiuretic hormone, *see* Arginine
    vasopressin
Antimullerian hormone (AMH), sexual
    differentiation, 378
Aquaporins
  antidiuretic hormone effects, 230
  kidney function, 229–230

ISBN 0-12-290421-4